Sturgeons and Paddlefish of North America

AF213462

Fish and Fisheries Series

VOLUME 27

The Kluwer Fish and Fisheries Series begins with the new millennium. The volumes in this series will cover topics ranging from the biology of individual species or groups of fishes, to broader concepts in fisheries science, conservation and management. The Series is directed to professionals and researchers in fish biology, senior undergraduate and postgraduate students, and those concerned with commercial production or harvest of fishes.

It is difficult to know if this is the best of times or the worst of times for fish and fisheries. For example, many of the historically important marine fisheries are at or very near harvest capacity, according to the best scientific data and predictions. Many of the changes in commercial harvests fit a predictable, depressing pattern. We tend to produce simplified communities, harvested by ever more efficient technology, at increasing rates of exploitation. Some would suggest that nothing can stop the apparently inevitable destruction of all commercially harvested fishes. Fish habitats seem to be increasingly degraded, and the deliberate or accidental introductions of exotic species threaten endemic native fishes. We always do things to the limit of our technology.

However, in contrast, we can point to a number of very favourable examples of current success and future promise in fish and fisheries. Our knowledge of the basic biology of fishes continues to expand. We disseminate that knowledge with ever increasing speed to libraries and personal computers around the world. Many fishes are increasingly recognized as fundamentally important subjects for basic research. Studies of the zebrafish, *Brachydanio rerio*, have produced a veritable explosion of fundamental scientific information at major research institutions around the world. Fishes as diverse as arctic charr, *Salvelinus alpinus*, and stickleback, *Gasterosteus aculeatus*, are providing insights and new understanding of the fundamental processes of natural selection and speciation. Science and technology give us a better understanding of the implications of long term climate change for fish populations. We continue to see fundamental breakthroughs in our understanding of development, genetics and evolution of fishes on almost a daily basis. Production of fishes through our increasingly sophisticated and efficient aquaculture rivals or exceeds the harvest of wild fishes in many places. Our knowledge and understanding continue to develop to the limits of our science and technology.

Science and the promise it holds for us to deal with our questions and concerns about fish and fisheries, is the basis for this Series. The future is certainly not what it used to be. This and forthcoming volumes in the Kluwer Fish and Fisheries Series will define the scientific basis for our future interactions with fishes. It is truly an exciting time.

Dr. David L. G. Noakes
Series Editor, Fish and Fisheries Series
Professor of Zoology, University of Guelph
Guelph, Canada

Sturgeons and Paddlefish of North America

Edited by

Greg T.O. LeBreton
F. William H. Beamish
and
R. Scott McKinley

KLUWER ACADEMIC PUBLISHERS
DORDRECHT / BOSTON / LONDON

A C.I.P. Catalogue record for this book is available from the Library of Congress.

ISBN 978-90-481-6729-6 (PB)
ISBN 978-1-4020-2833-5(e-book)

Published by Kluwer Academic Publishers,
P.O. Box 17, 3300 AA Dordrecht, The Netherlands.

Sold and distributed in North, Central and South America
by Kluwer Academic Publishers,
101 Philip Drive, Norwell, MA 02061, U.S.A.

In all other countries, sold and distributed
by Kluwer Academic Publishers,
P.O. Box 322, 3300 AH Dordrecht, The Netherlands.

Printed on acid-free paper

Cover design and art by Paul Vecsei.

All Rights Reserved
© 2010 Kluwer Academic Publishers
No part of this work may be reproduced, stored in a retrieval system, or transmitted
in any form or by any means, electronic, mechanical, photocopying, microfilming, recording
or otherwise, without written permission from the Publisher, with the exception
of any material supplied specifically for the purpose of being entered
and executed on a computer system, for exclusive use by the purchaser of the work.

Printed in the Netherlands.

in our understanding and appreciation of these species. It lays the groundwork for continuing scientific studies of these fishes. Most importantly, it points to the way forward for the management and conservation of these magnificent animals.

Dr. David L. G. Noakes
Series Editor, Fish and Fisheries Series
Professor of Zoology, University of Guelph
Guelph, Canada

Series Editor's Foreword

There is nothing like a vintage automobile to attract public attention anywhere in North America - except sturgeons and paddlefish. And for many of the same reasons. They are striking in appearance, with an apparently ancient body plan, as Paul Vecsei's illustrations so clearly illustrate. No matter how often we see a live sturgeon, or even the static image of one of these creatures we pause in whatever we are doing. Anyone can recognize a sturgeon or paddlefish. All sturgeons are relatively minor variations on what must be a successful morphological theme. They are the largest fishes we can encounter in freshwater. Leviathan. Monster. Huge, shadowy prehistoric figures rising slowly from the depths.

I remember my first view of a live sturgeon in the wild. My colleague Bill Beamish and I carried out a multi-year study of lake sturgeon, *Acipenser fulvescens*, in northern Ontario. On our first day in the field we set large mesh gill nets. We lifted those nets early next morning and in our first net we had a sturgeon. Its distinctly shark-like body rose from the dark water of the Groundhog River and created a mixture of awe, enthusiasm and wonder that continues to this day.

Sturgeons and paddlefish should not be living among us. Conventional wisdom says that they have all the attributes that doom them by human exploitation. They are slow-growing, long-lived, late-maturing, large sized and highly valued individually and as species. Their anadromous life cycles make them vulnerable to fishing harvest under the worst possible circumstances. Large mature adults are supremely vulnerable to capture during their spawning migration and nowadays the focus of such fishing is invariably on mature females. Females are killed at precisely the worst possible time, for the harvest of their maturing ovaries for caviar. Caviar, as Inga Saffron so aptly describes is the world's most coveted delicacy. The history of caviar is inextricably intertwined with politics, economics and intrigue. The present conservation status of sturgeons and paddlefish cannot be understood without that perspective. She admirably provides that in this volume. The history of the utilization by these fishes by native North American peoples, and the present conservation issues and multijurisdictional management are major themes in this volume.

Despite all that seems primeval and pessimistic, there is clearly reason for hope for North American sturgeons and paddlefish. Federal legislation is now in place in both the U.S.A. and Canada to identify and direct attention to the conservation and restoration of threatened species and those at risk. All sturgeons and paddlefishes of the world are probably at risk. The details might vary, but the general conclusions are the same for all species because of their common biological and ecological attributes. This volume shows the reason why there is hope, at least for North American species. Legislation and efforts to sensibly manage these species exist in North America and the social, cultural and political threats the species face elsewhere in the world are less serious here. The importance of multijurisdictional approaches to management is a key contribution of this volume. There are also contributions on the full range of scientific knowledge of North American sturgeons and paddlefish. Habitat and environmental requirements, ecomorphology, metabolism and swimming and their bioenergetics are reviewed in detail to place these fishes in their environmental context. Chapters on genetics, metabolism, respiration and embryology review those aspects of biology. Of course conservation and management are major contributions to this volume, as is the growing interest and involvement of intensive aquaculture. This volume is a landmark

Incredible levels of overfishing and habitat destruction, in combination with the late maturation and intermittent spawning of these species, resulted in all populations being greatly reduced in numbers and distribution. The story of sturgeon and paddlefish extermination is one of human gluttony and greed. To be sure it is one of the most telling ways in which humanity has failed in our stewardship of the planet.

The majority of chapters in this book focus on the various aspects of the biology, conservation, and culture of sturgeon and paddlefish and have been compiled from the latest published research material. These sections are designed to be used as a reference for readership from universities, colleges, and government agencies needing the most current of knowledge. This book also contains two chapters outlining the historical aboriginal uses and the decline of sturgeon brought on by overfishing. These chapters will appeal to anyone interested in understanding how these once prolific species have become so endangered.

Perhaps the most remarkable fact surrounding the acipenserids of North America is that, though greatly reduced, their populations have survived almost 150 years of unchecked slaughter. With modern technology and human concern sturgeon are being successfully cultured in great numbers and indeed some wild populations have begun to recover. While the history of the North American species is dark, it is not without hope. It is possible, through our advanced understanding of the biology and importance of these species, that we might prevent their complete extinction.

Many individuals have worked very hard to assist with this compilation, and each should be commended for their efforts and patience in getting this work published. The chapter written by Inga Saffron titled "The Decline of the North American Species" was graciously submitted for this work and was derived from the book "Caviar" published in 2002 by Broadway Books. The illustrations included by Paul Vecsei and photos submitted by many individuals have greatly added to the calibre of the work. It is with great appreciation that the principle editors would like to thank those involved in making this book possible. In addition, the editors would like to acknowledge the efforts of Colleen Bohan, Kevin Butterworth in collating and proof reading the final edition; Dalyce Epp for editing and production; and David Noakes for editing.

The editors are deeply indebted to Nancy Léonard (Editorial Coordinator) for her significant contributions to the review of manuscripts and the overall organizational structure of the book.

G.T.O. LeBreton, F.W.H. Beamish, and R.S. McKinley [Editors]

References

Birstein, V.J., J. Waldman and W.E. Bemis. eds. 1997. Sturgeon Biodiversity and Conservation. Developments in Environmental Biology of Fishes 17. Kluwer Academic Publishers, Dordrecht, Holland. 425pg.

Birstein, V.J. 1993. Sturgeon and paddlefishes: Threatened fishes in need of conservation. Conservation Biology 7:773-787.

Houston, J.J. 1987. Status of the Lake Sturgeon, *Acipenser fulvescens* in Canada. Canadian Field Naturalist 101(2) 171-185.

Scott, W. B. and Crossman, E. J. 1998. Freshwater fishes of Canada. Galt House Publications, Ltd. pg 77-100.

Foreword

Sturgeon and paddlefish are strange and wondrous animals. The cartilaginous skeletons, heterocercal tail, scaleless skin, and the presence of the notochord in adults distinguish members of the order Acipenseriformes from all other bony fish (Birstein 1993, Scott and Crossman 1998). Indeed, their large size alone makes them a curiosity for anyone witnessing these increasingly rare individuals. Beyond their very unique physical characteristics these fish also possess some unexpected behavioural actions. While conducting field research we have been privileged to capture lake sturgeon of incredible size, perform biopsies, and return them to their river. We have always been struck by how remarkably placid and complacent these enormous creatures were. In the lab, juveniles demonstrate an uncanny similarity to puppies, swimming in summersaults, wagging their tails and watching with beady, reflective eyes at feeding times. Indeed, when fed a diet of floating pellets, these fish will learn to feed, while "belly-up", from the water surface. This "unfishlike" behaviour often allows an attachment between researcher and subject that is unknown with other fish species.

During the last 30 years there has been an increase in scientific interest in sturgeon and paddlefish in North America. Aquaculture and conservation of these species has spawned a large pool of scientific literature. Other research into the genetics, reproduction, morphology, and embryological development of these species has increased our understanding of these fish and provided us with a wealth of information. While there have been compilations and synthesis of literature on acipenserids published in the past these have included species from around the world (Birstein 1993, Birstein et al. 1997). The North American sturgeon and paddlefish present their own unique history, biology, ecology, culture, and conservation issues and for this reason we feel that it is important that this book, dedicated solely to those species, be published. This book, through the work and assistance of many authors has compiled what is currently known about the biology North American sturgeon and paddlefish.

For 100 million years fish similar to modern day sturgeon and paddlefish lived in the waters that were to become North America. These evolved into two families, Acipenseridae and Polyodontidae. While the latter possess only one surviving species, the paddlefish, *Polyodon spathula*, nine species in Acipenseridae survive in two genera, Acipenser and Scaphirhynchus. These are currently recognized as shortnose sturgeon, *Acipenser brevirostrum*; lake sturgeon, *A. fulvescens*; green sturgeon, *A. medirostris*; Atlantic sturgeon, *A. oxyrinchus oxyrinchus*, and their sub-species Gulf sturgeon, *A. oxyrinchus desotoi*; white sturgeon, *A. transmontanus*; pallid sturgeon, Scaphirhynchus *albus*; shovelnose sturgeon, *S. platorynchus*; and the Alabama shovelnose sturgeon, S. *suttkusi*.

These fish were an integral and important part of the aboriginal culture. Prior to European influences, humans were dependant on sturgeon for survival and they revered the species. Early settlers to North America viewed the sturgeon as a nuisance fish which damaged nets used to capture other, more valuable, species. Captured individuals were killed and discarded, used as fertilizer, or as an oily fuel in steamships (Houston 1987). Only when European descendants discovered the economic value of sturgeon flesh and roe were these fish almost completely eliminated from North America during a strikingly short period of time at the turn of the 19th century.

TABLE OF CONTENTS

INTRODUCTION

THE DECLINE OF THE NORTH AMERICAN SPECIES

INGA SAFFRON
isaffron@phillynews.com

Forth upon the gitche gumee,
On the shining Big-Sea-Water,
With his fishing-line of cedar,
Of the twisted bark of cedar,
Forth to catch the sturgeon, Nahma,
Mishe-Nahma, King of Fishes,
In his birch canoe exulting
All alone went Hiawatha.

-Henry Wadsworth Longfellow, Song of Hiawatha, 1854

When the poet Henry Longfellow dispatched his noble, Native American, savage Hiawatha on his heroic quest, it made complete sense that the object of that quest should be an equally noble, Native American beast: the sturgeon. Like Melville's whale, Longfellow's sturgeon was viewed in the popular imagination of the 19th Century as a strange and fearsome creature. Back then, the sturgeon was also a common fish in North American waters. Every spring, as throngs of sturgeon and paddlefish began their commute up the rivers to spawn, ferryboat passengers and fishermen were treated to an awesome spectacle of madly swimming fish, leaping like circus performers as they jostled for room.

Yet a mere century and a half after Longfellow's poem first inspired schoolchildren with its irresistible rhythms, it is a good bet that few non-specialists realize the sturgeon once roamed the continent's waters. Numerous as it was, the sturgeon disappeared with astonishing speed in the last quarter of the 19th Century, after fishermen discovered that big money could be made selling its salted roe to European gourmands. The story of how the sturgeon populations were wiped out so quickly is an instructive one, particularly for those who are concerned about the future of these ancient creatures in the Caspian Sea.

The tale begins some 4,000 years ago in North America. Catching a quarter-ton sturgeon was no easy thing for Native American fishermen. Their reed nets easily broke apart under the force of the sturgeons' determined swimming and its sharp

G.T.O. LeBreton et al. (eds.), Sturgeons and Paddlefish of North America, 1–21.
© 2004 *Kluwer Academic Publishers. Printed in the Netherlands.*

scutes. Eventually, though, they mastered the technology of building weirs out of tree trunks and branches. This innovation enabled Native American fishermen to trap the migrating sturgeon in watery pens where they could be more easily speared and landed. For American Indians, sturgeon was a great delicacy because of its meaty flesh. The Indians dried, pounded and smoked the flesh to preserve what could not be consumed immediately. They probably did much the same thing with the eggs, and some East Coast tribes are also believed to have baked the roe in maple leaves. But Native Americans never made the culinary leap of their Eurasian cousins living near Caspian Sea and figured out to use salt to preserve the sturgeon roe - creating the delicacy we know today as caviar.

The first European settlers to arrive in North America were both amused and repulsed by the Indians' quest for sturgeon. Some tribesmen working the East Coast rivers would attempt to lasso the sturgeon using ropes made of twisted grapevines. Others preferred to attach a harpoon to the grapevine rope. As often as not, the sturgeon would bolt to the bottom after being speared, leading the fisherman who was holding the rope on a wild ride. According to Robert Beverly, an 18th Century chronicler who witnessed the fishing technique in Virginia, any fishermen who succeeded in catching a sturgeon was "counted a cockarouse, or brave fellow." Less risky techniques often worked just as well. In the western rivers of British Columbia, the native fishermen liked to paddle their canoes in close formation and probe the river bottom with long poles until they sensed a sturgeon, which they would then attempt to harpoon. The extraordinary sight was described in 1864 by Sir Arthur Birch, the colonial secretary in British Columbia, in a letter to his brother:

> "All the Indians now fishing and it is great fun to watch them spearing sturgeon which here run to the enormous size of 500 and 600 lbs. The Indians drift down with the stream perhaps 30 canoes abreast with their long poles with spear attached kept within about a foot of the bottom of the river. When they feel a fish lying they raise the spear and thrust it at the fish seldom missing."

Despite such amusement, the early European settlers were loath to imitate the Indians and consume either the flesh or the eggs of the sturgeon. The settlers willingly devoured Chesapeake oysters, worked the New England cod banks, enjoyed enormous amounts of shad, and happily adapted other native foods such as corn and potatoes, yet they remained deeply suspicious of the large, scute-armoured sturgeon well into the 19th Century. As for the roe, they completely ignored its gourmet possibilities until the last quarter of that century. It is unclear why this was the case.

Back home in England, the sturgeon was listed in the fishing laws as a royal fish, meaning that the king or other high official was entitled to claim the catch. While caviar was rarely available in Great Britain before the 19th Century, its reputation as a high-falutin' ambrosia had already seeped into literature. Shakespeare famously used caviar as a metaphor for something was over the heads of the general public. In theory, the reputation of the sturgeon and its eggs should have made it a highly desirable source of food in the New World. Indeed, in an effort to lure potential settlers, some 17th Century shipping advertisements made a point of mentioning the abundance of sturgeon in American rivers. The James River in Virginia, one of these posters boasted, had a "deep and bold channel so stored with sturgeon and other sweet fish as no man's for time hath ever possessed the like." But when the colonists actually came face to face with the storied sturgeon in the wilds of America, and saw its enormous, rooting snout and rough armoured plates, they lost any desire to make a meal of it.

Given the emphasis on the sturgeon fishing in the official advertisements, it is especially strange that the first Jamestown colonists arrived in the Virginia wilderness with neither without fishing nets nor supplies of salt – sort of like arriving at the base of Mt. Everest without fleece and boots. During the first year in America, the Jamestown settlers made no attempt to fish at all. By 1607, however, their corn was exhausted and they were facing starvation. Hunger finally got the better of the colonial gentlemen. At the urging of Captain John Smith, the colonists made their first primitive efforts to fish for sturgeon: They waded into the James River brandishing frying pans and swords in the hope of spearing a meal. As ridiculous as they must have looked, they somehow managed to catch a few sturgeon in shallow water using this method. The fish the Europeans loved to hate kept them from starving that season.

Writing in his diary, Smith recounted: "Wee had more sturgeon than could be devoured by dogge and man." To concoct an edible dish, Smith's party dried and pounded the sturgeon meat, then mixed it with roe, sorrel and other herbs. Encouraged by their modest success at fishing, the London Company, which had dispatched the group to America, requested that the colonists send a few barrels of salted roe home to be sold in the Baltic, where the local demand for caviar was growing. By this time, the colonists must have acquired salt, and they were able to assemble something that passed for caviar. The salted eggs were packed and loaded onto boats for the long journey back to Europe in 1608. But by the time the caviar barrels arrived on the London dock in 1609, the stench of decay was evident. Somewhere along the way, the Jamestown colonists had messed up the recipe.

Meanwhile, in Jamestown, conditions deteriorated. Fed up with their patrons and preoccupied with their own survival, the colonists refused to spend any more time developing a caviar business for the London Company. As soon as the famine passed, so did their taste for sturgeon. Nothing that came close to resembling caviar would be made in North America for another 200 years.

The Jamestown experience set the tone for other colonists. While settlers in North America did occasionally consume sturgeon, it was always reluctantly, when there was no other food available. One explanation is that the English colonists were not really familiar with the fish, despite its high esteem in royal law or the enthusiasm expressed in sailing advertisements. Sturgeon landings were rare in British waters, so it is likely that few ordinary Englishmen had actually encountered the fish before sailing across the Atlantic. It also appears that the European settlers were repulsed by the sturgeon's strong, oily smell, which clung to the Indians who gorged themselves on its meat. Sturgeon, for them, was a food for more primitive people. The newcomers initially felt the same way about salmon and lobster, two sea creatures that were equally beloved by Indians. As the Europeans struggled to make their way in the wild land, it seemed that eating such a grotesque bottom-feeding fish as the sturgeon would be the equivalent of sinking into barbarism. If a sturgeon did wander into their fishing nets, they usually bashed the fish on the head, tossed the carcass on the offal heap and fed the roe to their pigs. Their resistance to the meaty fish baffled the explorer Henry Hudson. "There are plenty of sturgeon which the Christians do not make use of, but the Indians eat them greedily," he wrote in a report to the East India Company in 1609. The days of the American caviar rush were still a long way off.

In time, the American colonists did find a profitable use for sturgeon flesh: feeding slaves. Sturgeon meat was cheap and filling, nobody else seemed to want it. The sugar plantations that occupied almost all the arable land in the Caribbean had large labour forces that needed to be fed. The plantation owners were always struggling to keep their costs down. Indeed, that was one of the main justifications for slavery in the first place, and they were continually on the look out for low-cost, high-protein foods. Slaves in the West Indies needed to consume large amounts of salt to endure the crushing work of chopping cane in the broiling tropical sun. Salt cod, and then salt shad, became staples of the plantation diet. Indeed, shad fishing was big business in North America at that time. Sturgeon fishermen felt there was room in the market for their product, too. In 1753, a New Jersey merchant named Edward Bradfield started a business processing Delaware River sturgeon, which he hoped to market to the Caribbean trade. The meat was packed in barrels and transported by ox team to New York and Philadelphia before being shipped to the plantations. Jonathan Richmond soon set up a rival sturgeon business in Trenton in 1770, which he advertised in the Pennsylvania Gazette:

> "CHOICE PICKLED STURGEON, cured in the Baltick manner, by JONATHAN RICHMOND, at Trenton Falls; WHO, by his experience these seven years in the business, and not trusting to hired servants or slaves, who often spoil, by not putting up the proper part of the fish, as many so curing neglect. By observing the following directions, to preserve the fish, draw the cork or bung of the keg, and pour out the pickle in a clean basin or pan, then knock up the hoops, and take out the head that is branded, and take what fish you want out, then harden on the hoops, and put the cork in as before; then pour in the liquor into the keg, if not enough to cover the fish, add a little of the vinegar to it, and take a coarse double cloth, and spread it over the head, then put the head on the cloth, and a stone or weight over the head, to keep it close from the air, and it will keep good in the warmest climate, and warrant it good."

Jonathan Richmond struggled with the same problem that had vexed the ancient Greeks, the medieval Venetians and the Jamestown colonists: how to keep sturgeon from going bad on its way to market. Because the 18th Century sturgeon business was directed at the slave market, the early entrepreneurs didn't even bother to export the delicate roe. Perhaps if there had been a bigger domestic market for sturgeon, as there was for cod and shad, the fledgling sturgeon merchants might have considered such a boutique product. But almost 200 years after the Jamestown colonists turned up their noses at sturgeon, virtually no middle class people in America ate the fish. Sturgeon remained a food for the great unwashed - for Indians, servants and slaves.

Despite the distaste for sturgeon, the fish was an object of curiosity. On the Delaware River, 19th Century shad fishermen reported seeing thousands of Atlantic and Shortnose sturgeon heading toward Philadelphia and the spawning grounds beyond. They were compelled to pull in their nets to prevent the crazed sturgeon from tearing them apart. As early as the 1650s, William Penn had fretted that the huge beasts would capsize a fisherman's boat. We don't know how often that occurred. But one day in the early 1800s, while the steamboat Sally was heading north from Philadelphia, witnesses standing near a porthole saw, "a large sturgeon in jumping made such a leap that it passed clear through one of these windows and landed in the vessel, where it was killed."

In the 1840s, a small article appeared in a Philadelphia newspaper reporting that a Russian immigrant had settled on Ridley Creek, a Delaware tributary a few miles below Philadelphia, and was offering fishermen a dollar for each live sturgeon they brought him. What is notable is that he kept the sturgeon in a river pen, much as the Russian sturgeon fishermen do today, so the roe could be taken at its peak from a live fish. This Russian, whose name is not recorded, sold the meat in New York and made oil from the heads. Unlike his predecessors in the sturgeon business, he did not throw away the roe. It was salted, packed in tin cans, and exported to France, Germany and Russia. The anonymous Russian had started America's first successful caviar business.

*

A Russian may have figured out how to prepare good caviar from American sturgeon, but Germans made an industry out of it.

Bendix Blohm, the first of these German caviar dealers, could not have been more ill prepared for the role. Born in Holstein, the region north of Hamburg, he immigrated to America in 1852 intending to start a business exporting pickled sturgeon back to Germany, where it was prized. He set out without knowing the first thing about making caviar.

Blohm, who was already 41, headed for New York's Hudson River. He had no doubt heard that American fishermen thought so little of the sturgeon that they tossed them on the trash heap. In reality, Blohm's information was already out of date. The efforts of Jonathan Richmond had begun to pay off. By the 1850s, the influx of poor immigrants from Europe had created a potential new market for cheap sturgeon meat. Immigrants were accustomed to eating a wider range of foods than the squeamish Americans of English descent eat. Smoked halibut had been the immigrants' favourite fish, but when halibut prices shot up, they had no qualms about switching to sturgeon. Some maintained that sturgeon even tasted better than halibut. Hoping to cash in on the new demand, a New York smokehouse began in 1857 to process sturgeon, which street peddlers sold door-to-door in immigrant neighbourhoods. Hudson River fishermen nicknamed the cheap sturgeon meat "Albany Beef."

The development of a profitable domestic market for sturgeon meat finally gave American fishermen an incentive to go after the fish. They also began to look for ways to maximize their profits. Boiling down the head and other waste for oil was one way. Sturgeon yielded an oil that was less smoky than sperm oil from whales and lasted longer. Isinglass gelatine could also be made from the sturgeon's swim bladder. Before the discovery of isinglass in the early 19th Century, gelatine had been derived from boiling beef bones, a time-consuming and expensive process. The availability of cheap sturgeon gelatine coincided with the new middle-class fad for jellied concoctions, made with fruit or meat.

Fishermen now began to think about doing something with the roe besides using it for pig feed. In 1849, a Boston entrepreneur named N.K. Lombard, attempted to set up a caviar business in Woolwich on Maine's Kennebec River. He bought about 160 tons of sturgeon – roughly 1,000 fish - and extracted the meat, oil and roe. The meat and oil sold well, but the caviar was a disaster. Lombard had apparently undersalted it and the entire batch spoiled. The following year Lombard did not spare the salt. Another disaster. He abandoned the business in 1851.

When Bendix Blohm began fishing on the Hudson the following year, he was just as clueless as Lombard was about preparing the roe. The sturgeon that Blohm managed to catch were sent directly to New York City to be turned into Albany beef. Fishing on the Hudson proved much less profitable than Blohm expected. The strong tides limited the amount of fishing he could do. From the city of Troy, south to Manhattan, the

Hudson is really an extension of the sea, an estuary with powerful tides that wash up from New York harbour. The mildly salty estuarine waters are ideal for sturgeon but terrible for fishermen. There are only a few hours during the day when the water slackens enough to let out a seine net. Blohm fished this way for a decade, catching just enough sturgeon to survive, but never producing any caviar. When the country went into an economic slump following the Civil War, Blohm decided to take his boat and nets to Penns Grove, New Jersey and the calmer waters of the Delaware.

Penns Grove is an old-fashioned small town located on the western shore of New Jersey where the state bulges into the Delaware. It has straight shady streets and sprawling Victorian houses. Because the Delaware coast is so marshy, most towns in the area are located a few miles inland on the higher ground. But Penns Grove developed in one of the few places where the fast land extends right to the Delaware, making it a convenient port. Steamers from Philadelphia docked right at the foot of West Main Street, depositing families eager to spend a few weeks away from the city.

Whatever its other attractions, Penns Grove's main business was fishing. When the shad and herring started running in March, the men would cast off in their dories and not come home until the fish was piled up to the sails. The local people still viewed sturgeon with the same disdain as most American fishermen, despite the growing market for Albany beef among the inhabitants of New York's tenements. Blohm was delighted to with the situation. He was the only fisherman working the rich sturgeon grounds south of Penns Grove. The fish coursed up the Delaware in huge numbers, far more than in the Hudson, and in the calm waters, the advancing horde could be virtually skimmed off the bottom with seine nets.

Blohm still knew nothing about making caviar. But determined to expand his business, he invited a couple of Germans from New York to come down and teach him the process. Blohm rented an old sawmill where they could work, and by the spring of 1870, they were shipping caviar by the barrel back to Hamburg.

With his new foolproof recipe and his German connections, Blohm began to turn caviar into a commodity. Men all over America were eager to try new enterprises now that the ravages of the Civil War were finally over. Blohm hired six Penns Grove fishermen to fish exclusively for sturgeon, paying them a generous $25 to $35 a month. On the opposite bank of the Delaware, the bustle of the times could be gauged by the plumes of black smoked wafted over Philadelphia from its many factories. The city took pride in being the workshop of the world. Philadelphia was now the second-biggest city in America. Men who had been ordinary soldiers in the Civil War were making their fortunes by turning out locomotives, men's hats, and carpets, and they hungered for emblems of their newfound status. In the past, socially prominent Philadelphians had disdained extravagant shows of wealth, but new entrepreneurs had different tastes. The Caldwell jewellery store found it hard to keep up with orders for diamonds. An exotic food like caviar appealed to this ambitious business set.

As his caviar business grew, Blohm dug a pond near the river so he could hold live sturgeon until the roe was ripe and could be packed for its trip across the ocean, just as his Russian predecessor had done at Ridley Creek. Transatlantic steamships now ran regularly from Philadelphia to the continent. The idea that a wealthy family in Hamburg could enjoy sturgeon roe packed in New Jersey didn't seem as fantastical as it had just a few years earlier. Blohm could also send the sturgeon meat up to New York on the Pennsylvania Railroad train. The telegraph lines that had been installed in the 1850s made it possible to wire orders in the morning and have the goods on the train the same day. Instead of pickling the sturgeon meat as he had sometimes done in the past, Blohm decided there was a stronger market for smoked fish. Blohm's caviar business, which had seemed like an eccentric gamble before the war, seemed to be part of America's growing optimism as the centennial celebrations of 1876 approached. Increasingly Philadelphia's new middle class found the time to ride the ferry across the Delaware to spend a summer day in Penns Grove. Thomas Eakins, one of the greatest painters of the era, and a Philadelphia native, scoured many Jersey fishing towns for subjects. He painted a series of canvases depicting men working the shallow-bottomed shad boats and hauling in large seine nets. The equipment used to catch shad was virtually identical to that used for sturgeon. Penns Grove's shad men couldn't help but notice Bendix Blohm's success. Pretty soon, they stopped throwing away the sturgeon trapped in their nets. The American Caviar Rush had begun.

*

Like the Gold Rush out in California, the Caviar Rush was short, intense and made relatively few people rich. The Caviar Rush also produced its own boom town, a jumble of crudely built wooden structures on the New Jersey side of the Delaware, at the point where it stops being a river and starts to become a brackish bay. Fishermen were soon calling the town Caviar.

For a brief time during its brief existence, Caviar was the Astrakhan of America. The little New Jersey boomtown shipped more of the world's caviar during the 1880s than any single place on earth. It had a hotel, post office, restaurant, ice house and its own rail line, which reportedly sent 15 train cars packed with caviar up to New York every day. Nearly all of it was transferred to steamers bound for Europe. Some of New Jersey's caviar even went to Russia. Caviar's advantage over Penns Grove was that it was some 20 miles closer to the sturgeons' main spawning grounds at the head of the bay. The land at Caviar spilled right down to the water, as it did in Penns Grove, so fishermen could easily pull up to the docks and unload their catches. Recalling the town's bustle a few decades later, a fisherman named George Pyle penned a nostalgic ballad to Caviar:

> "I saw great fleets of fishing skiffs.
> Come down before the gale,
> Like a great flock of sea gulls,
> So snow-white was each sail."

Today, the town of Caviar is no longer visible. The opportunistic phragmites has run amok, carpeting the rotting foundations of the processing sheds and dormitories with a high cover, which stretches in every direction, all the way to the milk-bottle tower of the Salem nuclear power plant hovering on the horizon. Not a single one of the dozens of buildings remains. The tracks of the railroad have been asphalted over. Even the town's name has been eclipsed, replaced by the generic Bayside. Only the Delaware remains as it was, a broad, implacable, greenish expanse.

The sole evidence of Caviar's bustling port are a few wooden pilings poking their wrinkled necks out of the water near the mouth of Stow Creek, and some black-and-white photographs filed in the local historical society. They show a regiment of wharves projecting from the shore, each one thronged with activity. In one image, a group of bowler-hatted men hover at the edge of a pier, waiting to greet a skiff loaded with sturgeon. Nearby, a workman stands next to a hoist as he prepares to lift a huge fish out of the boat. Behind him in the distance you can make out the shanties where fishermen lived during the season. Fishing nets are draped like hammocks to dry on the piers. Some belong to Wyatt Miller, a descendant of slaves, who made enough money repairing nets to open a restaurant in Caviar. In one of the open-air processing sheds, the sturgeon are stacked like logs. A workman in high boots rips open a fish's belly to reveal a bulging clump of eggs. In another picture, three men tentatively massage the sturgeon eggs over a sieve. A few steps behind them, women in high-necked blouses and ankle-length skirts smile in delight as the eggs are transformed into caviar.

In 1895, Boyd's Cumberland County directory listed 22 caviar and sturgeon wholesalers with offices in the town. During the fishing season, Caviar's population swelled from a few dozen to 400 people. Fishermen bedded down in dormitories, ready to jump into their boats whenever the tides commanded. These were usually 26-foot skiffs equipped with both sails and oars, but some fishermen had the misfortune to work from houseboats. These were foul-smelling barges with a single cabin where fishermen slept, ate, gutted fish, and made caviar.

*

In 1865, when the Civil War shuddered to an end, caviar the food and Caviar the town were still unknown.

While sturgeon fishing had become an important industry on the Hudson by the late 1850s, the Caviar Rush did not hit the Delaware River until two decades later, when returning Civil War soldiers and newly freed slaves were searching for ways to make a living. One of the men looking for work was Charles A. Dolbow, a Penns Grove fishermen descended from Swedes who had settled on the Jersey side of the Delaware during the time of William Penn. The Dolbows made their livings as farmers and fishermen.

After the war ended, Charles Dolbow was lucky enough to find a job with Bendix Blohm and his new caviar enterprise, as captain on a sturgeon boat. He would have much preferred to work for himself. But like many of the town's fishermen, he lacked money and connections, not to mention experience at making caviar. If he hoped to start his own operation, Dolbow would need to save enough money to buy his own set of boats and nets.

His luck began to change in 1873, when two more Germans from Hamburg arrived on the Delaware coast. They were the sons of Dieckmann & Hansen, the big German dealer that produced caviar from European sturgeon caught in the Elbe River. Peter Hansen and John Dieckmann had left Hamburg with instruction to sign up American caviar suppliers.

But they quickly discovered that, other than Bendix Blohm, none of the American fishermen knew the first thing about making the delicacy in 1873. The Germans realized they would have to start from scratch if they wanted American caviar suppliers. They spent the year travelling along the East Coast, handing out nets and sieves to fishermen and teaching them the fine points of salting sturgeon roe. Charles Dolbow eagerly signed on with Dieckmann & Hansen, and in exchange the Hamburg firm lent him $500 to purchase the equipment he needed to start his own business. In 1876, Dolbow was able to leave his job with Blohm and establish his own fishing camp in Caviar. He shipped 1,300 sturgeon to New York his first year, and invested some of the profits in a schooner-full of ice from Boston. The ice, which arrived the following spring, would fill a refrigerated storehouse for caviar and sturgeon meat. In a matter of a few years, Charles Dolbow and his son Harry would become Dieckmann & Hansen's most important supplier in America. The two Penns Grove fishermen were so successful that by the 1880s, there was a good chance that a dish of caviar appearing on a German table had originated in New Jersey and was packed by a Dolbow.

Other German caviar merchants followed Dieckmann & Hansen to America. P. Feddersen, a Berlin dealer, found his way to the tiny island of Solomons on the Chesapeake in southern Maryland, where he contracted with a prominent oyster fishermen named Joseph C. Lore. Feddersen provided Lore with barrels, strict instructions about salting and a good supply of Germany's renowned Luneburg salt. Along the Atlantic seaboard, sleepy fishing ports went to work to satisfy Europe's caviar hunger: Chester in Pennsylvania, Port Penn in Delaware, and Savannah in Georgia. A decade after Dieckmann & Hansen arrived in America, there were more than 900 watermen trawling for sturgeon along the Atlantic coast.

Most of the American caviar was sent to Europe, but sometimes second-rate batches were sold domestically. A few New York bars began handing out free caviar sandwiches, in the hope that the salty snack would increase drink orders. The Denver and Rio Grande railroad offered caviar in its dining car, for the same price as a plate of olives and celery, two other novel delicacies. Most of the caviar was processed with a salt ratio of nine to ten percent, double today's standard, in an effort to prevent spoilage. The Dolbows also offered a better grade with five-percent salt, but only for the European trade.

The Delaware was where America's caviar industry began, and it was Delaware fishermen who dominated fishing up and down the East Coast. Because the sturgeon started spawning earlier in the warm, southern rivers, the northerners shipped their boats by rail to Georgia around February to meet the first runs of sturgeon. Working their way north river by river, the Delaware fishermen followed the sturgeon until they reached the town of Caviar. They started in the Savannah River, put their skiffs on a flatbed train for Virginia, moved on to the James River, hopscotching up the coast from the Potomac River, to the Patuxent, Chesapeake and Delaware, where the season lasted until the end of June.

Chasing the sturgeon north, a fisherman could earn good money, more than if he stayed home to wait for the shad run. It was tough work, though. Sturgeon nets weighed 200 pounds even when empty, and quite a bit more after they were pulled up wet and loaded with fish. Fishermen spent nearly the whole day on the river when the sturgeon were running. A single ripe female brought \$2 at the wharf. That was a hefty sum when a workingman's weekly salary might be \$10. By 1897, the price for a female sturgeon swollen with eggs had jumped to \$30.

Not every sturgeon was of equal value. American fishermen learned to rank sturgeon in the 19th Century much as Russians do today, saving their highest regard for egg-laden "cows." Most fishermen had no qualms about salting the roe from "runners," female fish caught before their eggs were hard and ripe, even though the caviar was second-rate. It could be sold as sandwich caviar. As for the male "bucks" and the female "slunkers," which had already given up their burden of eggs, they were good only for meat. The juvenile "mammoses" were nearly worthless. After fishermen had paid their expenses, they could take home \$500, enough to spend the rest of year sitting around, telling stories and mending nets.

For ambitious fishermen like Charles Dolbow, who processed the fish himself, the profits were even higher. He arranged to sell everything separately, the caviar, meat, isinglass, and oil. Caviar naturally commanded the most money. German dealers were paying $9 for a 135-keg of caviar in 1885, but fifteen years later the cost shot up to $100 a keg. Fishermen had never seen such prices for any other fish. In 1908, a professor at the University of Pennsylvania named Walter Tower, published a short history of the sturgeon fishery. "Nowhere else in the whole annals of commercial fisheries," he concluded, "is there a parallel to this case of sturgeon, rising as it did in less than a quarter of a century from a fish despised and ruthlessly destroyed...to the highest rank of commercial value." Most fishermen expected the prices to keep rising.

With all that money being made off the sturgeon, fishermen failed to notice that they were catching fewer of them. In the 1870s, fishermen caught an average of 65 sturgeon each time they hauled their heavy nets out of the water. A decade later, the average had fallen to 30. At first it didn't matter if there were fewer sturgeon struggling in the gill net because the price for a cow was doubling every year. The Delaware, always the most fruitful of the east coast rivers, saw its total catch drop by half between 1890 and 1897, from five million pounds down to two and a half million, yet fishermen's incomes remained stable, or even increased.

By then Dieckmann & Hansen was firmly established in America. Ferdinand Hansen, a grandnephew of the company's founders, arrived in 1886 to run the firm's New York office and oversee operations in Penns Grove and Caviar. Although Hansen was just 17, he had enough experience in the caviar trade to recognize what the falling catches meant. It had taken barely 30 years to pick the Elbe River in Germany clean of sturgeon. In 1899, exactly 29 years after Bendix Blohm made his first batch of caviar in Penns Grove, Delaware fishermen averaged only eight sturgeon in a net. Dealers scrambled to find enough American caviar to satisfy Europe's appetite. Some fishermen made crude attempts to pass off other fish roes as sturgeon caviar. The Berlin wholesaler P. Feddersen wrote to his Chesapeake supplier Joseph Lore in June 1899 to vent his frustration, and to beg for the real thing:

> The season, "is pretty well done.... There is a big lot (of) Chesapeake caviar in the market. Where does it come from? One man wrote me he had 25 kegs and another party had 150 kegs in N.Y. I went there, but I like to buy caviar made out (of) sturgeon roe, but not dirt - shad roe - put up. I tell you there is some bad stuff in the market."

The season of 1900 marked the 30-year anniversary of the American caviar industry. After three decades of relentless fishing along the East Coast, vast numbers of spawning sturgeon had been taken out of the reproductive pool. The fish that should have been spawning for the first time in 1900 had never been born. Their eggs had long ago been eaten. That year, sturgeon stocks crashed in every river along the East Coast. In the James, in the Chesapeake, in the Delaware, fishermen pulled up empty nets. Some years later, Charles Dolbow's son Harry, who had taken over the business and changed his family name to 'Dalbow,' would recall bitterly that, "caviar never got a decent, respectable price until the failure of 1900." European dealers closed their American operations. Others moved west, to the Great Lakes and the Pacific coast. Dieckmann & Hansen did not give up their American operation entirely, but the firm's attentions were now focused on the rich Russian fishing grounds. In 1895, they started making caviar on Siberia's Amur River, and then moved to Astrakhan in 1902. The supplies from Russia more than made up for the decline in American production. Harry Dalbow continued to fish the Delaware, and Ferdinand Hansen continued to buy from him. But everyone sensed it was the end of something. The start of the 20th Century marked a watershed for the Volga as well as the Delaware: Astrakhan produced 29,800, tons of sturgeon in 1900, a record that would never again be surpassed.

*

Harry Dalbow wasn't ready to believe that caviar was over. In 1891, he had formed a partnership with one of his relatives, Joseph "Yaller" Dolbow. They started with a pair of sailboats moored at Caviar, but were confident enough in the future to invest in the new gasoline-power engines. Eventually they built a fleet of twenty boats.

Many fishermen came to believe that the sturgeon fishery needed regulation to survive, and made attempts to police the bloody free-for-all. In 1904, fishermen from Caviar, Chester and Port Penn gathered in the Davis Hotel in Philadelphia to form the Sturgeon Fisherman's Protective Society. The first item on the agenda was a proposal to throw back any sturgeon measuring less than four feet. Such young fish could not possibly have ripe eggs. But some fishermen contended the effort was pointless; most of the fish would be dead or injured before they could be separated from the nets. Ferdinand Hansen pursued another tack. He appeared before the legislatures of New Jersey, Pennsylvania and Delaware and tried to convince them to fix a season for sturgeon fishing. By cutting off fishing on May 15, he suggested that a portion of the sturgeon population would be spared and given a chance to spawn. His plan was rejected, and years later, he complained that the lawmakers sided with the fishermen, who "preferred to kill the goose that laid the golden eggs." In a way, the crash of 1900 acted as a brake on fishing, thinning the ranks at Caviar. Even Yaller Dolbow gave up fishing in 1901. But Harry had been raised on sturgeon, and like the fish he stubbornly returned to the river he had always known.

Although Harry earned $2,000 from sturgeon fishing in some years, the money was never enough to make him rich. After the fishing ended, he would often pitch in to work on Salem County vegetable farms, the gardens of the Garden State. South Jersey's vegetable industry was in its heyday. Growers sent truckloads of tomatoes, beans and asparagus to the area's booming canning factories like Campbell Soup in Camden. In the fall of 1906, Harry was helping out on a neighbour's farm on the assembly line where the tomatoes were being packed into tin cans and sealed using a new vacuum process. As Harry tamped down the lids he had a small epiphany. Why couldn't caviar be packed in small, airtight containers and sealed shut? He did not have to go far to test his idea. South Jersey was a hotbed of innovation in the new science of food preservation. While the French chef Nicolas Appert had invented the first reliable canning and bottling methods in 1809, South Jersey farmers built on his discoveries and, in 1847, they produced the first canned tomatoes for commercial sale in Jamesburg. The acceptance of canned foods expanded during the Civil War, when Union troops were fed on canned pork and beans, canned sardines and canned succotash. The American-Can Company became one of the largest employers in South Jersey. Harry decided to ask the company to design a machine that could pack caviar just like tomatoes.

The marriage of canning and caviar could not have happened a more propitious time. America's ancient sturgeon populations, so abundant at the nation's birth 200 years earlier, were vanishing in every river. The Atlantic Sturgeon had been reduced to a few scattered populations of fish by 1900. Catches of White Sturgeon in the Sacramento River peaked in 1885, and California banned sturgeon fishing altogether in 1901. Although the White Sturgeon in the Columbia River was not fished commercially until 1888, the stocks there were exhausted there in 12 years flat. Lake

Erie, which had yielded 5 million pounds of sturgeon in 1885, sent a mere 200,000 pounds to market ten years later. Fishermen resorted to making caviar from the roe of the Mississippi Paddlefish. Dry and grainy as it was, they passed it off as Russian caviar.

The Native Americans who had the sturgeon to themselves for so long found their survival threatened by insatiable demand for caviar. "A good many whitemen are yet fishing now near Harrison River, openly stealing our fish, our only food. They want to see us starving, we cannot bear it," four Chilliwack chiefs from British Columbia complained in an 1894 petition to the Canadian government. Unless something was done soon, they warned, Indian fishermen would surely resort to violence to destroy the commercial hook lines draped across the rivers. By 1902, the sturgeon on the Harrison River were hardly worth the commercial fishermen's time. What remained was left to the Indians. But the prospects for these original Americans were almost as bad as the sturgeon's.

Greedy fishermen were not alone in killing off the sturgeon. Pollution was taking its toll. The Delaware River, which had been clean enough in the early 19th Century for ships to fill their water casks in mid-channel, was now a cesspool, so slick with oil from the Philadelphia refineries that it was known to combust into flame. The building boom that followed the surge of immigrants had also damaged the river. Silt from construction, new roads and farms poured in the Delaware and the Chesapeake, forming a cushion of silt over the hard rocky surfaces where sturgeon liked to spawn. Out west, the run-off from sawmills around the Great Lakes reduced the oxygen levels, suffocating both the rivers and the sturgeon.

As fishermen scoured America's rivers and lakes for the remaining sturgeon, it was becoming increasingly difficult to fill the standard 135-pound keg. Those wooden barrels had hardly changed from the ones used in Catherine the Great's time, and looked old and crude in an age of gasoline-powered engines and airplanes. Harry Dalbow's canning operation offered a sleek and modern successor: the little glass jar. With American-Can's help, he not only mechanized the packing of caviar, he put the delicacy into individual glass jars and tins that could be sold directly to consumers. The jars were intentionally small, two and four ounces. By packing the caviar in smaller containers, Harry was able to make his limited supplies of caviar go further. He could also charge a lot more, too, when people were buying it by the ounce instead of the keg. The canning machine turned out to be a profitable form of rationing the remaining American caviar.

Dieckmann & Hansen recognized that Dalbow's idea would change the way people thought about caviar. No longer would it be ladled out of giant barrels. From then on, caviar would be sold like precious gems in the finest shops. The vacuum-sealed jars and cans also had the advantage of extending caviar's freshness. Caviar could now be pasteurized, giving it a shelf life of a year or more. The pungent smell of decomposing eggs would become mostly a memory.

In 1912, Harry sailed to Europe on the Lusitania to instruct Dieckmann & Hansen's staff in the secrets of canning. He spent several months in Astrakhan, the main port on Russia's Caspian Sea coast, teaching workers at the packing house how to use the canning equipment. Soon after he arrived home, his partner Ferdinand Hansen opened America's first retail caviar shop on the ground floor of the Waldorf-Astoria Hotel in New York. Hansen, who had previously used the trade name 'Russian Caviar,' decided that the product needed a more romantic sounding name for its label than the family names his ancestors used. He wanted something exotic, something Russian, something that evoked wealth. The obvious name was Romanoff Caviar, after the Russian royal family. With the American rivers exhausted, Hansen expected that most of the caviar in the tiny jars would soon be Russian, anyway. He was right. Unable to find enough caviar to fill a sufficient number of one-ounce jars, Harry Dalbow shut his cannery and sold his boats in 1925. Just a half-century after Americans began to make caviar, the sturgeon were virtually gone.

*

A few isolated sturgeon populations remained, however, along the Pacific coast. Recognizing the environmental damage that had been done by unchecked fishing along the East Coast, the western states gradually began to impose the kind of fishing restrictions that Ferdinand Hansen had advocated. The result was that the dominant species, the White Sturgeon, survived in more rivers and in larger numbers than its Atlantic and Midwestern counterparts. While pollution continued to take its toll on the water quality, large-scale commercial fishing fell off. Other than Indians and sports fishermen, few people had any reason to bother with North America's biggest sturgeon until the 1990s.

The collapse of the Soviet Union in 1991 changed that. For much of the 20th Century, there was simply no point to making American caviar. There were still large numbers of sturgeon in Soviet-controlled waters of the Caspian Sea, Black Sea and Amur River, and the Soviets were the world's caviar experts. The Communist government maintained total control over the market for this luxury product. Moscow determined how many fish should be caught each year, how much caviar produced, how much should be exported and at what price. Although the Soviet state was frequently unable to provide its people with adequate sausage, or design a television set that did not routinely short circuit and explode in its viewer's living room, the government managed to maintain caviar's allure as a luxury food for more than seven decades of its rule.

But when the Soviet Union collapsed, its disciplined caviar cartel soon followed. Poachers swarmed the banks of the Volga, Ural and Amur Rivers, and the caviar they made in their kitchen sinks and bathtubs was soon flooding into Europe and the U.S. New companies sprang up to distribute this bootleg caviar. With so much caviar from so many sources on the market, prices fell swiftly. For the first time in its history, caviar became cheap.

It was around this time that caviar began to seep into America's food consciousness in a big way. Caviar had always been available from gourmet shops in America, but it took a bit of effort and knowledge to locate it. That situation changed soon once the Soviet Union and its trade restrictions fell apart. Caviar stores opened in big cities across the U.S. Buying caviar became a straightforward commercial exchange. The cans and jars were stacked prominently in glass display cases. You read the price list and made your choice. A clerk would wrap your purchase in an ice pack and hand it to you in a little shopping bag, often with the store's logo printed on it. Soon you could get reasonably good Russian caviar in airport terminals, via the Internet, even in suburban supermarkets. You could also buy a lot of congealed black goo. You might hear people on the train talking about how much they were ordering for New Year's Eve or see advertisements for caviar in newspaper food sections. Middle class people began to think of caviar as an aristocratic indulgence they could afford.

North American fish purveyors also began to think of the sturgeon differently. Instead of ignoring the local varieties, they began to wonder how they could profit from them. Once again, American fishermen turned their attention to native sturgeon and paddlefish populations, and once again entrepreneurs began to make caviar from American sturgeon. But this time, much of the new caviar production occurred under the controlled conditions of sturgeon farms.

*

Chuck Edwards led me into the dark warehouse and I followed, trying not to touch anything damp or slimy. As my eyes adjusted to the moist gloom, I saw what could be a swimming pool showroom after closing. Rows of vinyl pools stretched all the way to the back of the building. There was the same steady gurgling of filters, sucking the dirty water out and pumping the oxygen in. Every once in awhile, I heard a rustling that reminded me of dry dog food escaping its bag. I made out the shadowy forms of fish inside one of the pools. They were endlessly criss-crossing its circumference. When I leaned over for a better look, a fish poked its snout out of the water and gazed at me like a puppy wanting to be petted. "Go ahead. You can touch her," Edwards said. The sturgeon's sandpaper hide grated lightly on my hand as I slid it along the fish's armoured body.

As a zoology student, Edwards dreamed of being the next Jacques Cousteau. But instead of ranging across the seas on voyages of discovery, he now covers a smaller realm, stepping from tank to tank to check on the 100,000 or so White Sturgeon that inhabit the Stolt Sea Farm in northern California. The sturgeon are grouped in pools by age and size, starting with nearly transparent hatchings that are still living off their egg sacks, and moving in stages up to the 80-pound, eight-year-old females, in whose distended bellies resides a vault of ripening eggs. Edwards estimates there are more White Sturgeon in Stolt's tanks today than there are next door in the Sacramento River, once home to behemoths weighing half a ton.

While the farm may lack the diversity and mystery of the oceans, Edwards and his staff consider themselves fortunate to be presiding over this hermetic sturgeon world, tucked among the flooded rice paddies of the San Joaquin valley. In its own way, raising sturgeon on farms is as much uncharted territory as the ocean depths. Sturgeon farming still hasn't proven itself a profitable business, despite more than a decade of trial and error, and millions of dollars in investment. Whether the farm succeeds has less to do with the fish biologists who work there, or the deep pockets of its owner, and everything to do with what happens to the sturgeon in Caspian. Stolt is betting that the future of caviar lies on the farm, although Edwards knows that the farm's success will be secured only when it is no longer worth catching sturgeon in the wild.

When I met him, Edwards was in his early 40s, with the lanky build and pensive calm of a fly-fisherman. He told me that he had moved around the aquaculture business a bit before settling down with sturgeon on the Stolt farm in Elverta, ten miles north of Sacramento. The Norwegian-owned company is one of the largest fish farmers in the world. A significant portion of the non-wild Atlantic salmon sold in the world's supermarkets is grown on Stolt farms. The company has diversified into other species like Salmon Trout, Atlantic Halibut, Turbot, and Sea Bream. These are often sold in portion-size pieces under the brand name Sterling. Raising those fish in captivity has become routine. But when Stolt first became involved with sturgeon in 1987, Edwards said, no one really was really sure that caviar was something that could be successfully produced on a farm.

To show me some of the problems the company faced, Edwards took me into another dark, warehouse building that was known around the farm as the "500" building, because it can hold 500,000 pounds of fish. The sturgeon here were bigger than in the first building, about three feet in length, and they swam around large in-ground tanks. We could see thousands of sturgeon darting in all directions in the black water. Despite their size, they were only three years old – mere grade-schoolers in sturgeon years. Edwards told me that it takes eight to ten years in captivity before the eggs of a female White Sturgeon are mature enough for caviar. This means that Stolt must feed and care for those 500,000 pounds of sturgeon, and all the rest of its fish on its farm, for almost a decade before it can realize any profit from them. In contrast, farm-raised tilapia go to market in six months, salmon in about three years. There is hardly a business in the world in which an investor will wait for ten years before he realizes any revenue, never mind a profit. One sturgeon-farming entrepreneur I interviewed compared the business to planting a forest in order to make paper. Stolt's farm had been in existence in various forms for 18 years, and it was only just anticipating its first year in the black in 2000, when I visited.

Sturgeon farming has turned out to be trickier than Stolt expected. Although the process of fertilizing sturgeon eggs was first developed in Russia in the 1860s, there were always too many fish in the sea to bother with farms. Hatcheries were built in Russia and Iran after World War II, but these were intended for restocking the sea, not for bringing up sturgeon babies. When Stolt's farm was started, no one was sure what to feed the growing sturgeon, or how many could live comfortably in a tank. They didn't know precisely how to distinguish the worthless males from the valuable females, or when the females would be ready to give up their roe. In the wild, female White Sturgeon don't spawn for the first time until they are in their late teens. Stolt has reduced the age of maturity by almost half through selective breeding and diet, but ten years is still a long time to wait for caviar.

As Edwards and I stood in the darkness of the 500 building, we could see the complex system that Stolt had introduced to keep the sturgeon happy and healthy. Because the fish live in close quarters, the water is constantly filtered to remove fish feces, uneaten food, and other waste. Computer sensors tracked the water temperature and oxygen content. If either level should vary too much from the prescribed range, the computers sound an alarm. Computers had also mechanized the feeding of the sturgeon. A basket containing pellet feed rumbled along an overhead metal track, dropping measured portions into the tanks at timed intervals. The fish would lunge toward the surface, as if they too were computer controlled. But after a few minutes of frenzied consumption, they would lose interest and drift to the bottom along with the pellets.

Most of the sturgeon on Stolt's farm are descended from 20 wild White Sturgeon that were taken from California's northern rivers in 1982, and crossed with the brood stock that the farm had created. The company tries to follow the natural rhythms of the wild sturgeon on the farm. At the same time of year that their wild cousins would be swimming upriver to spawn, a small group of Stolt's farmed sturgeon are selected as new brood stock. The females are stripped of their eggs, the males of their milt, the ichthyologic term for fish sperm. Then these two ingredients are combined with water in something that looks like a blender, shaken for a couple of minutes, and poured into trays. The fertilized eggs are continually washed by water to imitate the natural current that brings oxygen to the developing hatchlings. After a couple of days, the babies will be about a quarter-inch in size, big enough that you can make out the form of an adult sturgeon, with its pronounced snout and shark-like tail. The young sturgeon have a frisky cuteness that seems completely out of character with the adult sturgeon's sluggishness. Before a sturgeon's face becomes jowly and its hide thickens, it resembles an otter, especially as it splashes playfully around the tank. The sturgeon's front fins are parallel with their bodies, like airplane wings, enabling the fish to speed through the water like a torpedo.

From the time the babies are hatched, a good part of work on a sturgeon farm involves the sorting of fish. The sturgeon are measured and grouped into tanks according to size. When the fish are three years old, Edwards and his staff are finally able to determine the sex of the fish using a blood test. This is the end for the males. They will be culled for meat in order to reduce the number of unnecessary mouths that must be fed. Some females will also be slaughtered. "We keep only the best looking ones," Edwards said, winking at the joke. He was actually serious. Fish biologists believe that beautiful sturgeon produce beautiful eggs that can be made into beautiful caviar.

The day after we toured the farm, I found Edwards peering into a large vinyl pool that had been set up outdoors next to the 500 building. He appeared outwardly placid, but he was smoking, a sign of controlled nervousness. His staff would be taking roe from 200 adult sturgeon to make caviar, and Edwards was anxious for a good harvest. If the farm could produce four tons of caviar over the next few days, it would turn its first profit ever. Since early morning workers had been transferring mature eight-years-olds from Stolt's ponds to the outdoor holding pool. The water temperature had been reduced to near freezing. The cold would make the normally docile sturgeon even more submissive. Inside the 500 building, a crew in white lab coats had assembled in the processing room, their knives sharpened and ready to go.

The end came swiftly and efficiently - a mallet to the head, a swipe of antiseptic across the stomach. Then a worker raised his knife and made a clean lengthwise slit across the china white belly. As the skin was pulled back, another worker snipped out the two ovaries. The greyish sacks, which were about the size and colour of old canvas sneakers, were placed in a bowl and rushed next door to the kitchen to be turned into caviar.

Edwards followed the bowl into the kitchen. Three grown men hovered over the tiny eggs, picking out bits of tissue and dabs of blood. They rinsed the eggs in cold water and gently massaged the mass over a sieve to separate the beads. The eggs were still greyish, but with the addition of salt they seemed to ink up right before our eyes. When the salt content was judged to be about four percent, the American ikryanchiks – the Russian word for caviar master - spread the fresh black caviar over a plastic sheet for Edwards to inspect.

Taking a small plastic spoon, Edwards transported a cluster of eggs to his mouth and rolled them over his tongue. His lips barely moved. He stayed silent, deep in thought, while the ikryanchik stood by waiting for the verdict. Edwards handed me a spoon and I imitated him, scooping up a few eggs and letting them settle in my mouth. Most dealers say that caviar needs to sit a few weeks to be at its best. As the eggs absorb the salt, the individual globules plump up, the taste becomes stronger and more complex, like wine aging in oak. But at that moment, as waves of just-salted eggs pounded my tongue, I believed that the substance in my mouth were the most perfect caviar I had ever eaten. Edwards, less effusively, pronounced it merely good.

*

The history of the caviar farm had been so troubled that Stolt had never bothered to give its product a brand name. But in late in 1999, the company realized that the farm had crossed the line from an experimental operation to a commercial producer. Edwards was predicting the farm would turn its first profit in the spring. Everyone agreed that having a widely recognized name would help Stolt command a higher price for its caviar, just as Ferdinand Hansen had realized long ago when he started Romanoff. Stolt also thought about using a variation of the Romanov, the surname of the former Russian royal family, but concluded it was too close to that once-famous Romanoff label. Stolt considered the ridiculous Stoltinski, the Norweigian-sounding Nelson's, and the over-caffeinated Jacob's Choice. Edwards listened quietly while company officials bandied around those vaguely foreign-sounding names, then he launched his own argument: American caviar should have an American-sounding name.

Rather than obscure the origins of Stolts' caviar, he felt that the company should trumpet them. "I wanted to sell not only the best caviar, but the best American caviar," he recalled. It was also crucial that people knew the caviar came from real sturgeon, and not from its cousin, the paddlefish, which produced less succulent eggs. Since Stolt's other fish products were sold under the trade name of Sterling, Edwards suggested using the same name for the company's White Sturgeon caviar. Much to Edwards's surprise, he carried the debate. Along with the Sterling name, the words "White Sturgeon Caviar" were printed on the lids and tins.

The notion of a fine American caviar appealed to many food purveyors, particularly in California, where fresh ingredients and fine cooking had been forged into a distinct cuisine. And yet, liking the idea of American caviar and actually buying it turned out to be two different things. The farm had the misfortune to harvest its first caviar in the middle of the 1990s, when the U.S. was glutted with cheap Caspian caviar. It was hard enough for farmed caviar to beat the price of the bootlegged import, but it was impossible to beat the reputation of the wild Russian eggs, even when they had been frozen and carried to market in a suitcase. The problem, Edwards complained, is that, "people think the best caviar is Russian caviar."

While Sterling's caviar drew high praise from food writers, who marvelled at how much the White Sturgeon eggs tasted like osetra, many caviar dealers remained sceptical. "Farmed caviar grows too fast. It has a sweetwater taste. Like any farmed fish, everything tastes the same," Susanne Taylor, the former head of Dieckmann & Hansen, told me. While wild sturgeon enjoy a diet that varies constantly, farmed sturgeon eat the same processed pellets every day. Farmed sturgeon live in fresh water, not the salty brine of the sea. When I asked Armen Petrossian, of Petrossian Caviar, whether he would ever sell Sterling, he said it was unlikely. Why not? "The taste," he sputtered through a grimace. (In the fall of 2002, Petrossian cut a deal to purchase half of Stolt's caviar production. It plans to sell in its shops under the species name, Transmontanus.) I frequently heard that farmed caviar was too bland, and it was always difficult to tell whether the complaint was objective, or part of the same caprice that had elevated caviar beyond mere fish eggs in the first place. One day in the fall of 2001, Edwards telephoned me to say that Sterling had come out the favourite in a blind tasting by the Wall Street Journal, beating all the Caspian varieties. He sounded overjoyed – and a little doubtful.

Despite resistance from the old-school dealers, Sterling was starting to make inroads among American purveyors by the late 1990s. After the CITES restrictions began to turn of the spigot of bootleg caviar coming into the U.S. and Europe, the price of Caspian caviar rose significantly. Edwards started getting orders from chefs who had previously refused to take his phone calls. But despite Sterling's growing popularity, the farmed caviar still wasn't able to command the same price as the Caspian roe. Chefs and dealers consented to buy Sterling because it was cheap; the fact that it was also good was secondary. Even when Russian osetra hit $40 and $50 an ounce, Sterling was obliged to keep its retail price fixed around $30.

Until Sterling could compete equally with Caspian caviar, volume would help the bottom line. Edwards was hoping to harvest four tons in 2000, but the farm's sturgeon population was growing so fast that he could see the day when Sterling caviar might produce 10 or even 15 tons of caviar. In every Caspian country but Iran, CITES harvest quotas were being reduced. If Russia's quota kept falling, Edwards saw a chance for Sterling's farm-raised caviar to become a sizable part of the market. The shortages of Caspian caviar not only increased demand for Sterling, but it meant that the farmed caviar would someday achieve parity on price. That was the Holy Grail of caviar farming. His challenge now was to make sure the people believed that the quality of farmed caviar would always be as good as the wild kind.

While I was hanging around Edwards's office, Rod Mitchell, a major East Coast fish supplier, called in an order for his Browne Trading Company. Mitchell was a good customer and Edwards wanted to inspect the caviar himself before it was sent out. Grabbing a fist full of plastic spoons, he called over a couple of his men, and asked them if they would mind eating some caviar.

As we walked from the office to the 500 building, Edwards explained the farm was having a big problem with spoilage. Most producers can keep their caviar under refrigeration for a year or more, but Stolts' eggs barely lasted a couple of months in the refrigerators. He wondered whether this meant that farmed caviar had a shorter shelf life than the wild kind. Or was it the water? One year, he said, they made the mistake of using pond water in the tanks, instead of well water, and the eggs smelled of mud.

In a wine chateau, this kind of tasting would be organized in a cool underground cellar. At Stolt, it was took place in the supply room next to the refrigerators, amid the shelves of empty jars, sterile gloves and plastic spoons. Edwards took several 1800-gram tins out of the fridge, most of it harvested about six months earlier, and passed around the spoons. Everyone scooped out a few eggs and swirled them around on their tongues. Edwards was the first to speak: "I'd like to flush this down the toilet."

What they're looking for, Edwards explained, was the slightest evidence of imperfection. A hint of fermentation or a bitter aftertaste. Fermentation signals that the caviar is growing mould. Caspian caviar may take a beating on its way over here, but its reputation was already solid. Sterling's quality, Edwards was convinced, needed to be superior if the company hoped to knock down the resistance to farmed eggs. People who eat caviar just once a year are unlikely to notice a bit of fizz or muddy aftertaste, but the experts surely would. Edwards felt that Sterling caviar couldn't afford one bad review.

Edwards opened another tin and bent over to smell the plateau of eggs, "I smell sour," he announced. Joe Melendez, who was known to have the best taste buds on the farm, wasn't happy either. "It tastes like fresh mortar," he proclaimed after taking a sample from another tin. Melendez, a big, jovial man who worked as processor on the farm, also seemed to have the best descriptions for bad caviar: "Sour apples." "Dirty." "Cidery." "Roquefort." "Tastes like beer."

But after a couple of bad tins, the kidding around stopped. Edwards grimaced a lot and flicked his used spoons in the trash. Out of the 16 tins the group had sampled, 12 were deemed unsaleable. "Throw it out," Edwards ordered after the last one, and then abandoned the room while the others cleaned up.

It was after 5 p.m. and most of the office staff had gone home. Edwards had been at the farm since 6:30 in the morning, but he went to his desk to go over some papers. He did a search on the web, smoked a cigarette, and swivelled in wide arcs with his chair. While he was trying to decide whether to stay or go home, his manager Peter Struffenegger came rushing in, holding a caviar tin in the flat of his outstretched palm. Smoke was seeping from the bottom of the tin. Edwards started laughing. "Are you suggesting we should burn the place the down," he asked darkly.

But Peter wasn't concocting a practical joke. "Look at this," he said, pointing to the tin. There was a cigarette burning inside. "It would make a good ash tray," Edwards said. "Don't you see?" Peter demanded. "The smoke is leaking. This is why the caviar is going bad. It's the tins!"

Although the tins were the industry standard, the batch that Stolt bought were poorly made. There were small openings around the base, where the sides should have been sealed airtight to the sides of the cans. In Stolts' effort to follow the traditions of the caviar industry, it packed its wholesale caviar in the same kind of cans that had been used for the last century, since Harry Dalbow came up with his canning machine. Tins were one of the caviar business' enviable traditions. But Edwards saw that the tradition was pointless for Stolt. The company was losing too much caviar because of the faulty tins. But for sturgeon farming to succeed in the future, Stolt was going to have to abandon some of caviar's past.

Edwards had been looking at some new packing equipment. At a trade show, he had seen a machine that could pack fresh caviar in jars with a vacuum seal. Now that Stolt had proved that farmed caviar could be a good business, Edwards decided it was time to make the investment. Like Harry Dalbow, he would rely on the latest packing technology to save his caviar business.

*

Finding the right way to preserve and pack farmed caviar won't be enough save the wild sturgeon, however.

In the rivers leading the Caspian, the spring fishing continues with the same intensity as it did on the Delaware in the early 1900s, the last years when sturgeon fishing was conducted there on a commercial scale. There have been three places in history where such vigorous caviar industries have existed. On the Elbe in Germany; in America's East Coast Rivers, the Great Lakes and western rivers; and in the river systems flowing into the Caspian and Black Seas. It took 29 years to pick the Elbe clean of sturgeon. After Dieckman and Hansen arrived in South Jersey, the American caviar industry rose and fell within a span of three decades. Only the Russians have managed to keep their caviar industry going longer.

They did so thanks to strict rationing of their precious supplies. That rationing ended in 1991. Now, after a dozen years of unchecked fishing, there seems to be no consensus in the Caspian about how to control the assault. Time is running out. If history is any guide, there are about 18 years left.

References

Author's Note: I was able to locate original newspaper articles recounting the ebb and flow of South Jersey's caviar history in the archives of the Salem County (N.J.) Historical Society, the Greenwich Township (N.J.) Historical Society, and the Calvert Marine Museum (Solomons Island, MD) The corporate histories provided by Romanoff Caviar and Dieckmann & Hansen Caviar were invaluable. Research for this article was also obtained through personal interviews and site visits.
The quote from Arthur Birch, which appears elsewhere in this volume, was taken from Terry Glavin's book, A Ghost in the Water.

Allen, Brigid, ed. Food: An Oxford Anthology. New York: Oxford University Press, 1994.

Beck, Henry C. More Forgotten Towns of Southern New Jersey: New York. E. P.Dutton and Co. 1937,

Beebe, Lucius. "The Costliest Food." Holiday, September 1955.

Bemelmans, Ludwig. La Bonne Table. Boston: D. R. Godine, 1989.

Birstein, Vadim J., Waldman, John R., and Bemis, William E., eds. Sturgeon Biodiversity and Conservation, Dordecht/Boston/London: Kluwer Academic Publishers, 1997.

Birstein, Vadim J. ed. The Sturgeon Quarterly. Vol. 5, No. 1-2. June 1997.

Borodine, Dr. N., "The Ural Cossacks and their Fisheries." The Popular Science Monthly, October 1893. No. 6, P. 767.

Boss, Kit. "The Great Northwest Caviar Caper." Pacific, March 13. 1994.

Boyd's Cumberland County, N.J. Directory, 1895-96. Published by C.E. Howe Co. 236 S. 8th St. Philadelphia.

Burk, John. The History of Virginia, From its First Settlement to the Present Day. Petersburg, Va.: Dickson and Pescud, 1805.

Cobb, John N. The Sturgeon Fishery of Delaware River and Bay. Report of the Commissioner. Washington: U.S. Commission of Fish and Fisheries publication. June 1899.

Coleman, Charles Washington. "Sturgeon Fishing in the James." Cosmopolitan. 1892.

Connor, Bud. Great White Sturgeon Angling. Frank Amato Publications, 1996.

De Meulenaer, T. and Raymakers, C. Sturgeons of the Caspian Sea and The International Trade in Caviar. Cambridge: TRAFFIC International, 1996.

De Rohan-Csermak, Geza. Sturgeon Hooks of Eurasia. Chicago: Aldine Publishing, 1963.

Dieckmann & Hansen, 125 Years, 1869-1994. Hamburg, 1994.

Dodds, Richard J. Solomons Islands and Vicinity: An Illustrated History and Walking Tour. Solomons: Calvert Marine Museum, 1995.

Dorsey, Hebe. The Age of Opulence. London: Thames and Hudson, 1986.

Feddersen, P. Correspondence between P. Feddersen and M.J.C. Lore, 1898. Property of Calvert Marine Museum Archives, Solomons Island, Md.

Friedland, Susan R. Caviar: A Cookbook With 100 Recipes, A Guide to All Varieties. New York: Charles Scribner's Sons, 1986.

Gemelin, Samuel, Travels in Russia in Three Domains, Cherkass to Astrakhan, 1769 to 1770. St. Petersburg Academy of Sciences. 1773.

Georgacas, Demetrius J. Ichthyologic Terms for the Sturgeon and Etymological Terms for botargo, caviar and congeners: a linguistic, philological, and historical study. University of North Dakota Press, 1978.

Glants, Musya and Toomre, Joyce. eds. Food in Russian History and Culture. Bloomington: Indiana University Press, 1997.

Glavin, Terry. A Ghost in the Water. Vancouver: New Star Books, 1994.

Godecken, Horst. Der konigliche Kaviar. Hamburg: Heinrich Siepmann Verlag, 1969.

Goode, George Brown. The Fisheries and Fishing Industry of the United States. Washington: Government Printing Office, 1884.

Grigson, I. Fish Cookery. London: Penguin Books, 1973.

Hildebrand, Samuel F. and Schroeder, William C. Fishes of the Chesapeake Bay. U.S. Bureau of Fisheries. Bulletin 53.

Holmes, William Richard. Sketches on the Shores of the Caspian, Descriptive and Pictorial, London: R. Bentley, 1845.

Hosmer, Dorothy. "Caviar Fishermen of Romania," National Geographic, March 1940.

Ivanov, V. P. Biological Resources of the Caspian Sea. Astrakhan: KaspNIRKH, 2000.

Kraft, Herbert C. Lenni Lenape: archaeology, history, and ethnography. Newark: New Jersey Historical Society, 1986.

Kurlansky, Mark. Salt: A World History. New York: Walker and Company, 2002.

-----------. Cod: A Biography of the Fish That Changed the World. New York: Walker Publishing Company. 1997.
Lazlo, Pierre. Salt, Grain of Life. Trans. Mary Beth Mader. New York: Columbia University Press, 2001.
Mathisen, Ole A. and Bevan, Donald E. Some International Aspects of Soviet Fisheries. Columbus: Ohio State University Press, 1968.
May, Col. John. Journals, relative to a journey to the Ohio Country. Archives of the Pennsylvania Historical Society, 1789.
Molokhovets, Elena. A Gift To Young Housewives.trans Joyce Toomre. 1897. Reprint, Bloomington Indiana University Press, 1998.
Norman, Barbara. Tales of the Table: A History of Western Cuisine. Englewood Cliffs: Prentice Hall, 1972.
Popkin, Susan A. and Allen, Roger B. Gone Fishing: A History of Fishing in River, Bay and Sea. Catalogue of Exhibition, Philadelphia Maritime Museum, 1987.
Putnam, Peter. Seven Britons in Imperial Russia, 1698 –1812. Princeton: Princeton University Press, 1952.
Radcliffe, William. Fishing from Earliest Times. Chicago, ARES Publishers, 1921.
Ramade, Frederic(/accent). L'Univers du Caviar. Paris. Editions Solar, 1999.
Riddervold, Astri and Ropeid, Andreas.eds. Food Conservation, Ethnological Studies. London: Prospect Books, 1988.
Romanoff Caviar Company History. A Story of Intrigue: The 150-year history of Romanoff Caviar. With a historical overview.
Royle, J.F. On the Production of Isinglass along the Coast of India. London: Wm. H. Allen and Co. 1842.
Rundquist, Jane. "Fishing and Utilization of Sturgeon by Aboriginal Peoples of the Pacific Northwest." Private Paper prepared for Sierra Aquafarms, Inc., 1994.
Ryder, John A. The Sturgeons and Sturgeon Industries of the Eastern Coast of the United States, with an Account of Experiments bearing upon Sturgeon Culture. Pgs. 231- 328. Bulletin of the U.S. Fish Commission, 1888.
Sebold, Kimberly R. and Leach, Sara Amy. Historic Themes and Resources within the New Jersey Coastal Heritage Trail. U.S. Department of Interior publication, 1995.
Secor, D.H. and Waldman, R. Historical Abundance of Delaware Atlantic Sturgeon. Transactions of the American Fisheries Society. 1999.
Shephard, Sue. Pickled, Potted and Canned, How the Art and Science of Food Preserving Changed the World. New York: Simon and Schuster, 2001
Sim, Mary B. Commercial Canning in New Jersey: History and Early Development. Trenton: New Jersey Agricultural Society, 1951.
Skinner, John E. A Historical Review of the Fish and Wildlife Resources of the San Francisco Bay Area. California Department of Fish and Game publication, June 1962.
Spear, Lisa, Lauck, Lisa, and Pikitch, Ellen. Roe to Ruin: The Decline of Sturgeon in the Caspian Sea and the Road to Recovery. Publication of the National Resources Defense Council, Wildlife Conservation Society and SeaWeb. December 2000.
Sternin, Vulf and Dore, Ian. Caviar: The Resource Book. Moscow: Cultura, 1993.
Stockard, Charles R. "Our New Caviar Fisheries." Century Magazine, July 1908.
Stutz, Bruce. Natural Lives, Modern Times: People and Places of the Delaware River. New York. Crown Publishers Inc., 1992.
Tannahill, Reay. Food in History. New York: Stein and Day, 1973.
Taylor, William W. and Ferreri, C. Paola. Eds. Great Lakes Fisheries Policy and Management: A Binational Perspective. East Lansing: Michigan State University Press, 1999.
Thomazi, Auguste. Histoire de la Peche, des ages de la pierre a nos jours. Paris: Payot, 1947.
Toussaint-Samat, Maguelonne. A History of Food. Trans. Anthea Bell. Cambridge, Ma.: Blackwell, 1992.
Tower, Walter Sheldon. "Passing of the Sturgeon." Popular Science Monthly, October 1908
Vaisman, A. and Gorbatovsky, V. eds. Wild Animals and Plants in Commerce in Russia and CIS Countries. Moscow: TRAFFIC Europe. 1999.
Waldman, John R. and Secor, David H. "Caviar Trade in North America: An Historical Perspective." Proceedings of the Symposium on the Harvest, Trade and Conservation of North American Paddlefish and Sturgeon. Washington: TRAFFIC North America, World Wildlife Fund, 1999.
Weigley, Russell F. ed. Philadelphia: A 300-year History. New York: W.W. Norton & Company, 1982.
White, William C. "Fishy Breezes From Soviet Astrakhan." Asia, October 1931.
Wildes, Henry Emerson, The Delaware. New York: Farrar & Rinehard, 1940.
Williamson, Douglas F., Benz, George W. and Hoover, Craig M. eds. Proceedings on the Harvest, Trade and Conservation of North American Paddlefish and Sturgeon. Washington: TRAFFIC North America, World Wildlife Fund, 1999.

CHAPTER 1

NATIVE AMERICAN UTILIZATION OF STURGEON

TIM E. HOLZKAMM[1] AND LEO G. WAISBERG[2]
[1]*Tim Holzkamm Consulting, 39414 Pinewood Rd.,Ponsford, Minnesota 56575*
[2]*Seven Oaks Consulting Inc., 42 Hillhouse Rd., Winnipeg, Manitoba R2V 2V9*

1.1 Introduction

Historically, sturgeon of various species were important for Native Americans on both the Atlantic and Pacific coasts and in parts of the interior of North America. Early or first contact period accounts by European, American or Canadian merchants, colonists and soldiers record sturgeon harvests as important elements of the domestic economies of many tribal peoples. Sturgeon provided food in the form of meat, oil and roe for immediate consumption, storage or trade. Products processed from sturgeon, such as dried meat, oil, or 'pemmican-like' combinations of the two, were traded at certain locations and times and were an important component of tribal economies and trader subsistence. In portions of the northern interior, occupied by the Hudson's Bay Company and other fur trading concerns, isinglass, a by-product of the sturgeon harvest made from the dried air bladder, was for many years an important export product to the European industry. Purchase records of such companies have allowed for the quantification of sturgeon harvests prior to their decline in the early twentieth century. Throughout North America, the decline of Native American political power, coupled with the growth of non-Indian settlement, was marked by the expropriation of their sturgeon fisheries, and followed by settler or commercial overexploitation of the resource. In some cases, these changes occurred concurrently with environmental degradation.

1.2 Areas Of North America Where Native Americans Harvested Sturgeon And Paddlefish

1.2.1 ATLANTIC COAST

Atlantic sturgeon and shortnose sturgeon were formerly important food species for Native Americans residing from the Gulf of St. Lawrence to eastern Florida. The Natives of the Atlantic shore were the first to come into contact with Europeans. Thus both the Northeast and Southeast culture areas were affected by depopulation and dispossession, and had their fisheries were appropriated by Europeans at an earlier date than other Natives.

G.T.O. LeBreton et al. (eds.), Sturgeons and Paddlefish of North America, 22–39.
© 2004 *Kluwer Academic Publishers. Printed in the Netherlands.*

While no quantitative estimates of aboriginal harvests of sturgeon along the Atlantic coast have been reported in the literature, qualitative descriptions by observers have led historians to conclude that fishing was as important an activity as hunting and agriculture. Sturgeon was a major fish resource. Thomas Harriot observed during the 1580s in North Carolina that from February to May there were "plentie of Sturgeons" and herring, the two most important fish for the coastal tribes (Harriot 1972). Natives of the Atlantic coast used a variety of fishing techniques: spears, dip nets and night torch fishing, along with weirs in rivers and estuaries. Inland tribes of Virginia and the Carolinas captured quantities of sturgeon during the spring runs using clubs, nets and weirs (Feest 1978; Boyce 1978). For the more northerly Pequot, Narragansett and Massachusetts peoples, "sturgeon seem to have been particularly important," and the spring runs of spawning fish "provided opportunities for large catches with minimum effort" (Salwen 1978).

Weirs, fish traps or dams were used at many locations along the coast and on the interior rivers for sturgeon and other fish. The earliest mention of weirs was in 1562, which noted that "much skill and industry" was required for construction. Two general kinds of weir were used. Coastal weirs utilized tidal flows, while inland weirs depended upon fish movement within rivers. Inland tribes also captured sturgeon coming up the rivers to spawn using trap weirs. Rock approaches were constructed to produce a fall of water and a trap, basket or impoundment placed below the fall. Robert Beverley, noting the use of coastal and inland weirs, also described a snare technique in 1705:

> The Indian way of catching sturgeon, when they came into the narrow part of the rivers, was by a Man's clapping a noose over their tail, and by keeping fast his hold. Thus a fish finding itself entangled, would flounce, and often pull him under water, and then that man was counted a Cockarouse, or brave fellow, that would not let go, till with swimming, wading and diving, he had tired the sturgeon, and brought it ashore. These sturgeon would also often leap into their Canoes, in crossing the river, as many of them still do every Year, into the boats of the English (Swanton 1979).

In addition to traps and weirs, the use of spears was important. James Adair recorded in 1775, spear fishing for sturgeon using green cane spears in the Savannah River in Georgia (Swanton 1979). John Lawson described inland spear fishing when the sturgeon came up the rivers to spawn, along with the use of snares and nets for trapping individual fish (Swanton 1979). Harpoons were said to have been used by the Abenaki of Maine, in conjunction with the use of torches for night fishing (Snow 1978). Erhard Rostlund, who compiled a vast body of references to Native fishing in North America, noted abundant references to the use of nets as well as spears and snares for taking sturgeon in the eastern part of the continent. Rostlund (1952) particularly noted that hooks were not used for sturgeon in eastern North America.

Most of the aboriginal sturgeon fisheries on the Atlantic coast of the United States have disappeared. Tribes and autonomy declined after 1625, and many Bands were dispersed or killed. In Canada, European colonization occurred at a later date. Thus several aboriginal sturgeon fisheries, such as that of the St. John River in New Brunswick were still utilized by Maliseet Indians well into the 19th century. According to E. E. Prince, 5 years of excessive harvesting by non-Indians was sufficient to devastate the sturgeon population of that river, with yields falling from 602,500lb in 1880 to 16,264lb in 1886 (CSP 1905).

1.2.2 INTERIOR NORTH AMERICA

Lake sturgeon were especially important to Native Americans through the Northeast, Subarctic and northern Plains culture areas. Observers from the first centuries after European contact noted that sturgeon yielded extremely large harvests (Goodier 1984). Early witnesses reported a large-scale lake sturgeon fishery in the Great Lakes and Hudson Bay watersheds. Pierre Esprit Radisson in 1660 saw over 1000 sturgeon being dried on the south shore of Lake Superior, and there were still a number of sturgeon fisheries at river mouths along the shore of Lake Superior during the 19th century (Adams 1961). In 1765, Alexander Henry the Elder reported that sturgeon were so abundant at the Ontonagon River that "a month's subsistence for a regiment could be taken in a few hours" (Henry 1969). On the north-eastern Plains, his nephew, Henry the Younger, reported acquisition of 775 sturgeon, weighing from 50 to 150 pounds each, at his trading post located at confluence of the Pembina and Red Rivers between 20 April and 20 May 1808. Assuming an average weight, of 75 pounds, this would have amounted to a one month harvest of 58,125 pounds, much of which was consumed by his Indian customers. During the same year Henry noted that his post and customers also used 63,600 pounds of buffalo meat, making sturgeon nearly as important as bison (Coues 1965). On the Assiniboine River, also a tributary of the Red River, competition for sturgeon was so strong at the end of the 18th century that fur traders tried to starve their competitors by destroying weirs (HBCA B.22/a/2, fo.31). Farther north, sturgeon were also abundant at Cumberland House near the mouth of the Saskatchewan River (Tough 1996).

For the Ojibwe or Anishinaabe, inhabiting southern Manitoba and much of Ontario's Great Lakes basin, fish were important products for domestic use and for trade to outsiders. When contacted by the French in the 1630s, the Ojibwe fishing villages already dominated favourable harvest sites such as Sault Ste. Marie located between Lakes Huron and Superior. Sturgeon, along with whitefish, were important products of the Great Lakes fisheries of the Ojibwe well into the 19th century. In 1832, Lieutenant J. Allen of the U.S. Army described in detail a weir on the Ontonagon River, noting "the Indians rely so exclusively upon this fishery that they hunt but little" (Allen 1860). Elsewhere, particularly west of Lake Superior in north-western Ontario and Manitoba, sturgeon were the dominant species harvested. Jack Steinbring (1978) noted the importance of sturgeon for domestic use well into the 20th century. A similar reliance upon sturgeon existed for the Cree residing north and west of Lake Winnipeg. Studies at Cross Lake have confirmed the significance of sturgeon to the mix of fish resources available to the Cree prior to hydroelectric development of the area (Usher & Tough 1999). In the far north of Ontario, sturgeon on the Albany River, above Marten's Falls, also contributed to both commercial and domestic production of the Cree bands (Ray 1999). Further to the south, sturgeon were an important resource for the Menominee of Wisconsin and the Pottawatomi of Michigan, but for other central Algonquians south of the Great Lakes in the 17th and 18th centuries, fishing was of lesser importance than hunting and agriculture (Callendar 1978; Spindler 1978). Spears of varying sorts were used, along with pole snares and weirs.

Shovelnose sturgeon or pallid sturgeon, appear to have been little utilized by tribes located in the Upper Missouri. Maximilian, Prince of Wied, reported the capture of a sturgeon from the mouth of the Marais River in the 1830s but noted that it was "universally considered as an extraordinary rarity" (Rostlund 1952). The Hidatsa woman, Waheenee, indicated in her autobiography that sturgeon capture was opportunistic during the spring floods (Wilson 1921). Jonathan Carver (1974), observed the presence of paddlefish in the Mississippi during the 1760s, but noted that it was "not so much esteemed even by the Indians."

1.2.3 PACIFIC COAST AND PLATEAU

Historical and ethnographic accounts of the Native fishery in the Plateau, Northwest Coast and California culture areas describe anadromous white sturgeon as a major resource because of their large size and they were prized for their meat and roe. Although present along the Pacific Coast from the Aleutian Islands to Central California, white sturgeon seem to have been most abundant and of particular economic significance in the larger Columbia, Fraser, Klamath, and Sacramento River systems (Scott and Crossman 1973). Populations in the lower reaches of smaller rivers appear to have been less important to the tribes. Green sturgeon were also present, but appear to have been less significant as an economic resource.

Sturgeon were harvested using weirs, often of elaborate construction involving gateways and traps or fish corrals. Lift nets, gill nets and bag seines were also used (Rostlund 1952). The Yurok used gill nets which measured up to eighty-five feet in length three feet in width and with six inch mesh for sturgeon. A sturgeon bag net used by the Coast Yurok was described as consisting of two nets fastened together, each about seven feet long, with a bag measuring about six feet deep. The mesh was about eight inches square and the two ends of the nets were held open with vertical sticks. The ends of each net were attached to heavy lines which were dragged slowly downstream by two canoes. When the man holding the net line, in either canoe, felt a sturgeon he called out to the paddlers to bring the canoes together, and began hauling on the line to close the mouth of his end of the paired nets. This form of communal fishing was regarded by the Yurok as very efficacious, and, in the spring almost everyone participated in this type of fishing in the lower Klamath River (Kroeber and Barrett 1960).

Spears and harpoons were also important for harvesting sturgeon. In the larger coastal rivers harpoons with shafts up to fifty feet in length were used (Stewart 1977). This mode of spearing was described by Sir Arthur Birch on 7 May 1864 on the Fraser River:

> All the Indians are down now fishing and it is great fun to watch them spearing Sturgeon which here run to the enormous size of 500 & 600 lbs. The Indians drift down with the stream perhaps 30 canoes abreast with their long poles with spear attached kept within about a foot of the bottom of the River. When they feel a fish lying they raise the spear and thrust it at the fish seldom missing. The barb of the spear immediately disconnects from the pole but remains attached to a rope & you see sometimes 2 or 3 canoes being carried off at the same time down river at any pace by these huge fish (Birch 1976).

When the sturgeon were finally exhausted they were clubbed and often loaded by tipping the canoe to one side, rolling the fish over the gunwale, and bailing out the water. Sturgeon were also harvested with fishhooks, nooses or snares, or by diving and riding the fish until they were exhausted (Kroeber and Barrett 1960).

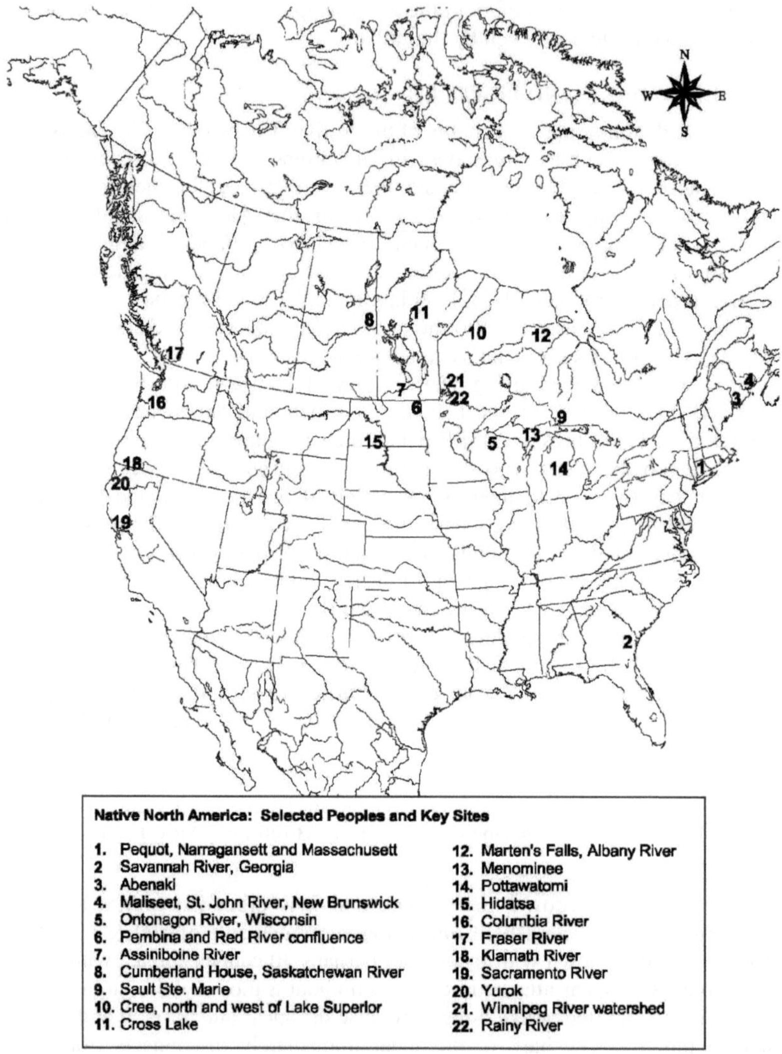

Fig. 1-1. Native North America: selected peoples and key sites

Sturgeon meat was cooked and consumed fresh or was smoked and dried for future use. Sturgeon eggs were particularly valued and were prepared by pounding and roasting. Sturgeon isinglass was also processed for use as an adhesive (Kroeber and Barrett 1960). Fur trade establishments usually provided the Indians a ready market for trading sturgeon meat and isinglass during early years (Stewart 1977; Scott and Crossman 1973).

Although Indians had fished for sturgeon in the Columbia River for centuries without degrading the resource, non-Indian commercial fishing had a substantial impact. Intensive non-Indian commercial sturgeon fishing in the Columbia River began in 1888. In that year the average size of sturgeon was about 150 pounds and over a million pounds of sturgeon were harvested. The sturgeon harvest peaked about 1892 at 5.5 million pounds and by 1895 the average size of sturgeon harvested had dropped to about 50-60 pounds. In 1899, ten years after intensive non-Indian commercial harvesting began, the total commercial sturgeon harvest was only 73,000 pounds and subsequently sturgeon were taken only as incidental by-catch to other target fish species (Larson 2002).

Case Study: Anishinaabe (Chippewa or Ojibwe) Use of Sturgeon in Rainy River During the Fur Trade Period

Recent studies have revealed the nature and extent of the reliance upon sturgeon by the Anishinaabe (also known as Ojibwa or Chippewa). Examining this reliance demonstrates the general importance of sturgeon to traditional Native economies prior to non-Indian settlement. The Winnipeg River watershed west of Lake Superior was particularly rich in sturgeon prior to Euro Canadian settlement (HBCA B.166/a/1, fo.3). During the fur trade period it was considered "the sturgeon country" and the natives were known as the "Sturgeon Indians" (Dobbs 1967). Lake of the Woods was considered "the greatest sturgeon pond in the world" (Ono, Williams and Wagner 1983). Rainy River was the major spawning ground for the Lake of the Woods sturgeon and the spawning run occurred during the months of May and June and fish concentrated near rapids, such as Long Sault and Manitou Rapids. Sturgeon were most vulnerable to traditional Anishinaabe fishing methods at this time and fish in prime condition for use as food were present in the thousands. (Holzkamm, Lytwyn, and Waisberg 1988).

Archaeological evidence of sturgeon fishing extending 2,500 years into the past has been found at sites along Rainy River. Excavations of Laurel Culture sites have unearthed large quantities of sturgeon plates, copper fish hooks, and harpoons fashioned from animal bone indicating an active fishery which provided food to support the construction of an extensive burial mound complex (Arthurs 1986).

Anishinaabe traditional fishing methods included hook and line fishing, spearing, use of weirs and nets. Hook and line fishing was only employed occasionally. Spear fishing was used primarily during the spring spawning run and was often done at night by attaching torches to canoes to attract the sturgeon to the surface. The average weight of individual sturgeon was about fifty pounds, but specimens over three hundred pounds were taken. Iron sturgeon spear tips introduced by fur traders were popular trade commodities. In 1797 John McKay, a Hudson's Bay Company trader, noted that a rival trading company sold iron spear tips to the Rainy River Anishinaabe at twenty times the usual price(HBCA B.105/a/4, fo.19; /14, fo.4d). Sturgeon were also speared from wooden platforms extending over rapids. This latter method was observed by Sir George Simpson in 1841:

> At nearly all the rapids and falls on these two rivers [Namakan and Maligne], the Indians have erected platforms, which stretch about twenty feet from the shore; and on these they fix themselves, spear in hand, for hours, as silent and motionless as possible, till some doomed fish comes within the range of their unerring weapon. If they take more fish than what they immediately require, they tether the

> supernumeraries by a string, through the mouth and gill, to the bank; and this mode of confinement, at least for a week or two, affects neither the weight nor the flavour of the prisoners (Simpson 1847).

Weirs or fish dams of stones or wooden stakes were another method of capture. One weir location was near the mouth of the Little Fork River and has been described by Elder Mary Horton of Rainy River First Nations:

In front of the spawning grounds of any fish, they built a cone shape with sticks. They caught or captured the live fish with dip nets made of zai-zub. This way of catching fish was also a lure for the fish to come in. It was built in such a way that there would be a whirl-pool or an eddy. This was made for all species of fish. Bingi-mo-don was the name of this fish trap. They built a dam in front of this trap. ... The construction was of wood or small trees. The wood was spaced just so far apart so that the large fish were trapped, but the small fish would be released. Not everyone could do this. The energetic people built these. It was a lot of work. These kinds of things were shared (Horton 1998).

Dip nets and large seines or drag-nets were also utilized by the Anishinaabe. Peter Grant, a North West Company trader, regarded Anishinaabe drag-nets as an indigenous invention, writing in 1804 that:

> They fish with nets, hooks, lines and spears, but they have a method of taking sturgeon with a kind of drag-net or seine, which I believe, is peculiar to them. The net used for this purpose is about 20 feet long by 6 feet deep, when shut double. It is dragged between two small canoes, having two men in each; while the bowmen paddle gently down the stream, the men in the sterns hold the seines by means of long cords, fixed to each end and which can be shortened or lengthened according to the depth of the water and the wishes of the seiners. Two stones are suspended from the lower ends of the seines, by which the nature of the bottom and the soundings are ascertained, a very necessary precaution to keep the whole clear of foul bottom. The course of the canoe must form an obtuse angle with the middle of the seine.
>
> These nets are mounted like the English drag nets, with small knobs of cedar fixed to the upper border instead of cork. When, by the vibrations of the cords, they perceive that fish is taken, they instantly haul up and paddle with all their might to bring the canoes together and, thereby, shut up the fish in the seine. This method of fishing is, of course, practicable only in rivers, narrow channels and small bays, where the bottom is clear (Masson 1890).

The traditional technology for harvesting sturgeon survived well into the 20th century. By 1870, when Canada included the Anishinabe traditional territory within its expanded western possessions, imported twine had generally replaced native plant materials in the manufacture of nets. Despite this replacement, specific knowledge of the aboriginal technology of fibre manufacture from native plants was still retained by Anishinaabe well into the twentieth century, when the information was given to ethnographers such as Frances Densmore of Minnesota (Densmore; 1974, 1979; Kuhm 1928).

The Rainy River sturgeon fishery produced large catches and considerable amounts were cured and stored for later use. Rainy River Anishinaabe made a product similar to bison pemmican consisting of a special blend of sturgeon oil and dried and pounded sturgeon meat packed into sturgeon skin bags. In 1826 J.D. Cameron observed this practice, noting that the sturgeon was:

Cut up in their flakes [into thin slices] and dry over a slow fire, after which they pound the dried flakes between stones until it becomes like a kind of sponge; this with the oil they gather; affords them a rich and substantial food of which they are very fond (HBCA B.105/e/6, fo.4).

Large amounts of sturgeon oil were rendered by boiling the entrails and skimming the oil. James Redsky (1972), a historian of the Lake of the Woods Anishinaabeg, recalled that sturgeon oil was considered "especially good." The oil was stored in a bag or jar made from the skin of a sturgeon. Anishinaabe had formerly manufactured such containers. The vessels were called "numay-wayan," Steinbring noted that these containers were "nearly indestructible", and "light, easily transported, and would withstand the incessant knocks of close family quarters" (Steinbring 1965a, 1965b).

Prior to the treaty with Canada in 1873 that permitted Euro Canadian settlement in their territory, large numbers of Anishinaabe from what is now northern Minnesota, north-western Ontario and eastern Manitoba gathered at Rainy River for the sturgeon fisheries. The primary tribal meeting ground was located at the pool below Chaudiere Falls, now between the town of Fort Frances, Ontario, and International Falls, Minnesota. These tribal gatherings of many bands were supported by the sturgeon fishery at this and other Rainy River meeting sites. Councils of chiefs at these sites were the centre of tribal decision-making (HBCA B.105/e/6. fol2d-3; D.4/109, fo.10-10d). For example, in June 1841, Governor Simpson of the Hudson's Bay Company found an assembly of over 500 at Fort Frances. During the council, the Chiefs, each attended by tribal soldiers, negotiated with Simpson on a matter of trade policy. Governor Simpson acquiesced: "seeing that there was no alternative I made a merit of necessity ... mustering as they do in large bodies ... during the summer ... I consider it good policy to avoid any difficulty or dispute with them" (HBCA D.3/2, fo.30d-33d).

By 1868 the usual number congregating at locations on the Rainy River was 600 to 1,500 (NAC MG11, C.O.42, vol.698 pp.126-148). Such large gatherings, fed by sturgeon, facilitated the renewal of friendships and social ties, the discussion of military and political affairs, and the holding of religious ceremonials. Rites of the Midewiwin, or Grand Medicine Society, were often held during the spring spawning run. Peter Jacobs (1857), a missionary, reported three hundred Anishinaabe engaged in the sturgeon fishery at Manitou Rapids in 1852. Sturgeon are frequently referred to in Anishinaabe myth (Vecsey 1983; Overholt and Callicott 1982). A giant sturgeon was often associated with the spiritual power controlling fishes and fisheries (Hultkrantz 1983). Anishinaabe leaders refused to let the Reverend Jacobs establish a mission at Manitou Rapids because they feared it would ruin the sturgeon fishery (HBCA D.5/25, fo.347). Elder Ron Geyshick of Lac La Croix First Nation noted the spiritual importance of sturgeon:

> Sturgeon are people, very much like us. You must be careful which ones you eat, and make offerings and feasts when you catch them. They're one of our Indian foods. When the sturgeon run is the time we have our spring feasts and ceremonies, the time when our boys and girls do their fasting. Everything must be done the right way (Geyshick 1989).

Fur traders often recorded the Rainy River Anishinaabe as difficult trading partners. Anishinaabe were often content to live upon sturgeon and other native foods rather than engage in trade. When they did engage in trade Anishinaabe sold wild rice, maple sugar, berries, venison, birch bark canoes and other items of use to fur traders, in addition to sturgeon and furs (Waisberg 1984). One Canadian official writing after 1857, noting that Sir John Richardson had earlier concluded that the Anishinaabe were "independent of the Hudson's Bay Company, from the fact that they have abundance of sturgeon and great quantities of wild rice," concurred; he described them as "formidable if inclined to be troublesome ... independent; sometimes even a little saucy" (Dawson 1968; NAC MG11 C.O.42 vol.698, pp.126-148).

Sturgeon supplied by Anishinaabe was critical for the fur trade. A shortage of sturgeon often meant extra expenses for other provisions, or even hunger at some of the posts. Control of the sturgeon supply at such times meant Anishinaabe could dictate some of the terms of trade. In 1829 competition with American Fur Company traders gave Anishinaabe an additional market to the Hudson's Bay Company, which prompted one trader to write: "The Indians below [at Manitou Rapids] have a little sturgeon but want liquor, which we have none to give, of course Chatelin [the HBC sturgeon trader] will not get much from them for ammunition and tobacco" (HBCA B.105/a/14, fo.5).

A reliable supply of sturgeon was essential to trading operations in the region. William Sinclair, in charge of the Fort Frances post, noted in 1844:

The dried sturgeon procured during the summer has been very beneficial to the support of the establishment. I really do not know what we would have done without it (HBCA D.5/2, fo.561).

Non-Indians also fished for sturgeon, but they usually found it more practical to obtain their supply from the Anishinaabe, who resided at the best fishing spots on Rainy River, such as Manitou or Long Sault Rapids, downstream from Fort Frances.

In addition to meat and oil, sturgeon also yielded isinglass, a gelatinous substance that was produced from the inner membrane of the air bladder. Anishinaabe valued isinglass as a binding agent in the preparation of paint. James Isham, a Hudson's Bay Company trader at York Factory in 1743, noted:

The Glue the Natives saves out of the Sturgeon is very strong and good, they use it in mixing with their paint, which fixes the Colours' so they never rub out (Isham 1949).

Isinglass was also a product for which there existed a market in Europe. It yielded superior glue and was also a fining agent for beers and wines. Efforts to procure isinglass from the Indians were almost as old as the establishment of the Hudson's Bay Company posts on James Bay. In 1694, James Knight at Albany Fort was informed by the company's governing committee in London that:

We should be glad you could procure us some Isinglass being only the sound of the sturgeon dryed, wee are Informed great quantitys may be had, that Comodity is also very currant here (Rich 1957).

Several years later they wrote again, advising "Isinglass is a thing much more in request here than formerly" (Rich 1957). Despite such attention the early trade in isinglass languished. In 1793 a trading post was established on the sturgeon-rich Rainy River. At this time Indians could not be persuaded to procure sufficient quantities to make the trade viable. The London Committee penned its disappointment in a 1796 letter to John McNab at Albany Fort "We are sorry the procuring of Isinglass meets so little success that so profitable an object of trade cannot be procured is much regretted" (HBCA A.6/16, fo.26).

Shortly after the union of the North West and Hudson's Bay Companies in 1821, the company's efforts to procure isinglass met with success. This was especially the case in the Lac la Pluie District which produced about fifty-six percent of the Company's annual shipment of isinglass, mostly from the Rainy River. The Hudson's Bay Company invested more manpower and goods in the district. This trade took place for the most part at the Anishinaabe fishing stations. After 1821 isinglass production in the Lac la Pluie District increased from 64 pounds in 1823 to a high of 2320 pounds in 1844 (HBCA B.239/h/1-7). Isinglass production was an activity that fitted well into the diversified traditional economy of the Anishinaabe, and its collection coincided with maximum concentration of the sturgeon and Anishinaabe populations at fishing sites.

Hudson's Bay Company financial records of the isinglass trade can give an indication of the size of the Anishinaabe sturgeon fishery. The Company's account books record the annual shipment of isinglass from each of its administrative districts. District boundaries shifted slightly during this period, but the boundaries of Lac la Pluie District usually embraced the region north of the American border to the Wabigoon River, east to the Lake Superior watershed, and west to Lake Winnipeg. Trade documents point to the Rainy River as the primary production point in the district. Isinglass was collected at the Fort Frances, Rat Portage, Fort Alexander, Shoal Lake, and Basswood Lake posts. The available account book figures indicate that Fort Frances collected between 75 and 80 percent of the isinglass produced in the district (HBCA B.105/d/60, 62, 79).

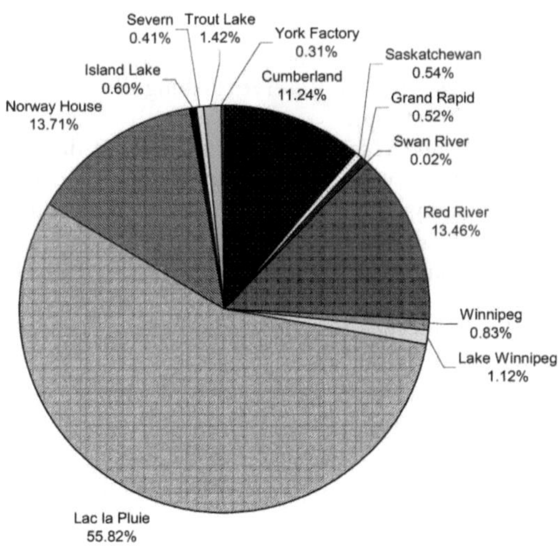

Fig. 1-2. HBC-Northern Department isinglass returns. Total percentage by district 1821-1891

Isinglass production is a fairly accurate indicator of the sturgeon harvest from the Rainy River. The amount of isinglass derived from individual sturgeon is small and preparation is labour intensive. There is general consensus that it takes about ten sturgeon to produce a pound of isinglass (Scott and Crossman 1973). Average sturgeon from Lake of the Woods when dressed for shipping weighed 25 pounds (CSP A.1897a; Russell 1898). Each pound of isinglass represents an estimated 250 pounds of dressed sturgeon. From this we can make a reasonable estimate of the amount of sturgeon represented by the isinglass harvest. Utilizing estimates derived from the isinglass returns, the average annual harvest was about 277,096 pounds of dressed sturgeon between 1823 and 1891 for the Lac la Pluie District Anishinaabe (Harkness and Dymond 1961; Holzkamm and McCarthy 1988). This is a conservative figure, as it is assumed that some sturgeon swim bladders were kept for domestic uses and not traded for isinglass (CSP 1905). Thus much larger numbers of sturgeon may have been harvested each year than is indicated by the isinglass returns. Tim Holzkamm (1987) has calculated that the average annual sturgeon catch represented by the isinglass returns would have fed a minimum of 207 people for one year, or about 1,200 for two months.

Following the union of the British North American provinces in 1867, the new federal government of Canada acquired the lands west of Lake Superior. In 1870, Canada sent a military force to Red River, and commenced negotiations for a treaty. After several years of negotiations, Canada entered into "Treaty 3" with the Anishinaabe in 1873. Simon Dawson, one of the Treaty #3 commissioners, later recalled discussions of treaty fishing rights:

> ...as an inducement to the Indians to sign the Treaty, the commissioners pointed out to them that, along with the land reserves and money payments, they would forever have the use of their fisheries. The point was strongly insisted upon and it had great weight with the Indians, who for some years previously had persistently refused to enter into any Treaty (NAC, RG10, V.3800, f.48542).

Under Treaty #3, Rainy River Chiefs included in their reserve lands the Long Sault and Manitou Rapids as "Indian Reserve Fishing Stations" (NAC, RG10, V.1922, f.2970; CLSR Plan 163 Ontario).

Treaty negotiations and reserve selections occurred during a period in which the Ojibway sturgeon fishery was not yet threatened by competing users. The first Canadian settlers on Rainy River during the mid-1880s found that "sturgeon fishing ... [was] quite an industry among the Indians" (CSP 1888). Thereafter, well-capitalized fishing companies commenced operations in western Canada with the opening of railway communications. These companies were established to catch and ship a wide range of fish products, not merely a by-product of the domestic Anishinaabe sturgeon harvest (CSP 1889, 1890, 1891).

Elsewhere in the west, such as on Lake Winnipeg, the impact of commercial fisheries on Indian subsistence was noted with alarm by officials of Canada's Department of Indian Affairs. Responding to the impact of the commercial harvest on Indians in Manitoba, Ebeneezer McColl, Manitoba Indian Inspector, stated after a personal inspection in 1888:

> The destruction of their fisheries by the white men was the burden of their speeches and the eternal nightmare of their apprehensions. They frequently pointed out to me at their councils how the buffalo, the principal source of subsistence of their kindred on the plains was destroyed by the effective weapons of destruction furnished hunters by white men, and implored me to use my influence with the Government to have their fisheries protected from being irretrievably ruined before it was too late (NAC, RG10, V.3807, f.52443).

McColl regarded fishing practices in western Canada as a "wholesale slaughter by speculators," and predicted severe impacts on Native subsistence. He noted with concern in 1888 that American commercial fishermen had begun to operate in Lake of the Woods, near the mouth of the Rainy River:

> Unless the fisheries, upon which they mainly depend now for support, are rigidly protected, not many years will elapse before they will become wholly dependent on the Government for subsistence. An extensive fishery has been established at Rat Portage, on the Lake of the Woods, and more fisheries are intended to be started on other lakes in the district (NAC, RG10, V.3807, f.52443).

In response to the American operations, Canadian fishing companies planned to commence commercial harvests on the Canadian portion of Lake of the Woods. This worried Simon Dawson, now Member of Parliament for north-western Ontario, who lobbied the Department of Indian Affairs on behalf of the Anishinaabe (NAC, RG10, V.3800, f.48542). Drawing upon his thirty years of experience in the region he recommended protection of the entire Lake of the Woods fishery to ensure Indian food supplies. Dawson warned that any decrease in the fish supply would lead to a dependence on government relief. In Parliament he raised the matter on several occasions during 1888 and 1889; on 19th May 1888, he stated to the House that it had become necessary to protect the Indian fishing right under Treaty #3: "What becomes of that stipulation if the whiteman is allowed to go wherever he likes, and to make a speculation in sweeping the fish out of the lakes and sending them to the markets of the south" (AO, Irving Papers, 31/37/11).

Limited steps were taken by the Department of Indian Affairs and the Department of Marine and Fisheries to preserve sturgeon and other fish species of the Lake of the Woods for Indian subsistence. Gill and pound nets were prohibited. Indian Agents were cross-appointed as Department of Marine and Fisheries officers with directions to protect the Anishinaabe fishery, and efforts were made to bring the issue to the attention of the American government for joint action (NAC RG10, V.3800, f. 48542; NAC, RG2, series 2, V.243; AO, Irving Papers, 31/37/110). Ultimately such efforts failed. Continued development of the fishing industry on the American portion of the lake engendered continuing complaints from the Lake of the Woods Chiefs.

Although the fisheries complained of are across the boundary in American waters, yet they destroy our fish as much as if they were carried on in Canadian waters, for the fish move backward and forward across the line. Are these fishermen allowed to destroy our fish, or have you power to stop them (CSP 1891).

Increasing political pressure from Canadian companies and settlers led to a change in government policy during 1892 which favoured commercial harvests over Anishinaabe fishing interests (NAC, RG10, V.3800, f.48542). The Canadian portion of Lake of the Woods was opened to gill and pound nets and new overseers appointed. By 1895 the Canadian industry on Lake of the Woods produced 715,950 pounds of dressed sturgeon meat, with an additional harvest of 65,800 pounds of caviar and air bladders (CSP 1897b).

This reversal of Canadian policy occurred despite continuing Indian complaints and reports from local Departmental officials. During the 1891 season on Rainy River, Indian Agent Robert Pither reported that "...where the sturgeon were always numerous in spring, very few were caught, in fact barely enough for present use" (CSP 1892). Elsewhere Pither's description of fishing conditions during 1891 was more graphic: "...On my trip up Rainy River in July, making the payments, the Indians were almost starving, as very few sturgeon went up the river to spawn and they hardly caught enough for personal use NAC, RG10, V.3800, f.48542)."

In 1892 Anishinaabe petitioned the Canadian government:

> There are also Canadians that have licences from the Government and we are of opinion that if no such licences were granted it would be easier to put a stop to the wholesale depleting of the fish in Lake of the Woods. - This one of our main resources is getting more and more scarce and we can now hardly catch enough to feed ourselves in summer. Some strong measure should be taken.... Having kept faith with the Department it is only but fair that we should expect that they would keep it towards us. We have kept our part of the Treaty, is it not hard that the Government should not keep theirs (NAC, RG10, V.3880, f.92840).

Neither this nor other complaints were heeded, as the federal government decided in favour of commercial exploitation of the fishery by licensed companies. Despite treaty promises, Canada did not intend to protect an Anishinaabe fishery oriented to a mix of subsistence and small-scale commerce (CSP 1895).

The shift on the part of Canada in favour of non-Indian large-scale commercial fisheries had a dramatic and devastating effect upon both the sturgeon resources and Anishinaabe who utilized this resource for domestic and commercial purposes. The change in federal policy marked a switch from a harvest that had the capacity for long-term sustained domestic yields under Anishinaabe management to a commercial harvesting process, directed by white entrepreneurs able to lobby the Fisheries Minister, that emphasized short-term productivity for an open-ended export market (Holzkamm and McCarthy 1988).

With essentially unlimited demand and industrial commercial extraction methods, sturgeon production continued to increase. Annual harvests averaged over one million pounds of dressed fish during the period from 1895-1899. Sturgeon production for both Indian and non-Indian fishermen dropped dramatically as the fish population declined from overharvest. Overfishing by non-Indians contributed to the near extinction of the sturgeon resource which was a mainstay of the Anishinaabe regional economy. Pollution from a pulp and paper mill at Fort Frances after 1907 adversely affected potential recovery of the species by degrading spawning grounds in the Rainy River.

Canada's Department of Indian Affairs' records show that the loss of the fisheries was a disastrous economic loss for Anishinaabe. When ethnographic research was conducted on Rainy River by Ruth Landes in 1932 and 1933, she recorded on the basis of her observations that fishing was of little importance to Anishinaabe (Landes 1969). This was a complete reversal of economic and subsistence orientation for Rainy River Anishinaabe in comparison to their orientation prior to1873 (TARR 1986).

Although one commentator has interpreted the depletion of Rainy River sturgeon as a result primarily of commercialization and market forces, this is a sterile view that overlooks Canada's social environment and political system in the late nineteenth century. Frank Tough, in an analysis of the history of commercial markets for sturgeon, explains depletion on the Rainy River as a result of "the historical role of the market in changing harvest patterns." Both Anishinaabe domestic fishers and non-Anishinaabe fishing companies were participants in a market economy, but the issue is one of allocation, permission to harvest, plus opportunity, afforded by access to capital-intensive equipment that can take a non-sustainable proportion of the resource yearly. The core issue in the depletion of sturgeon on Rainy River is not merely economic, a penetration of new market opportunities, but instead racial, the appropriation of the former Anishinaabe resource in favour of white rather than Native entrepreneurs. Tough minimizes the command aspects of Canada's political economy and the requirement for harvest approval by government. The federal government in 1892 had a choice, and exercised it to allow non-Indian large-scale commercial companies the opportunity to pillage the resource. While changing harvest patterns in regard to new markets may have provided the opportunity to pillage, they do not explain the reason for the choice by the Canadian government to allow the pillage (Tough 1999).

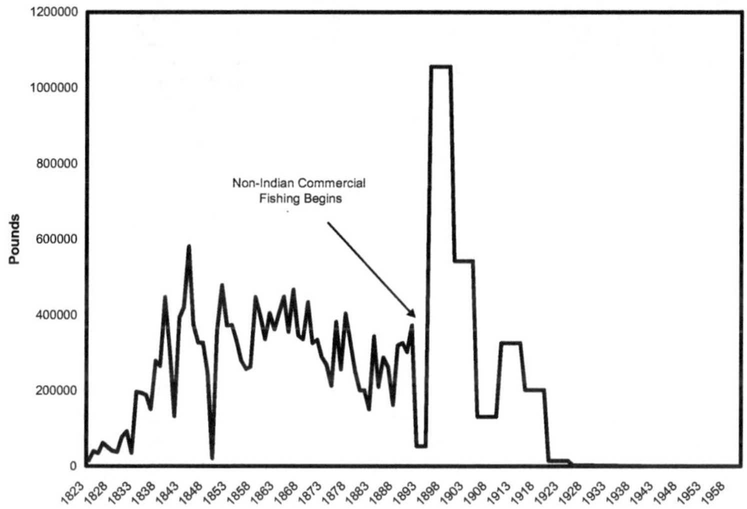

Fig. 1-3. Lake of the Woods / Rainy River Sturgeon Harvest 1823-1960

At present, efforts have been made to reverse pollution in the Rainy River and to re-establish spawning areas. Sturgeon restoration has begun with cleanup of the river and establishment of a modern sturgeon hatchery by the Rainy River First Nations at Manitou Rapids, Ontario. Renewal of the Long Sault village fishing station as a National Historic Park and interpretive heritage centre is also underway.

1.3 Conclusion

Historically, sturgeon were a major fisheries resource for Indian tribes living on the Atlantic coast, in the Great Lakes and Hudson Bay watersheds, and in the Pacific Northwest. Sturgeon were present in the Mississippi and Missouri River systems, but do not seem to have been utilized to the same extent. Sturgeon were harvested utilizing a wide variety of methods including the use of nets, weirs, spears, and snares. Hook and line fishing was also employed, but this method does not appear to have been utilized in the Atlantic coastal region.

Fig. 1-4. Rainy River First Nation members at ceremony honouring the release of sturgeon in the Rainy River

Traditionally, sturgeon were harvested as a food resource for their meat, oil, and roe. In addition, sturgeon skins were used to store preserved sturgeon and isinglass was valued as a glue. With the coming of the fur trade, sturgeon became a valued trade commodity for their meat which was an important county food and for isinglass which was exported. The advent of white settlement was marked by appropriation of the sturgeon fisheries, commercial overexploitation of the resource, and environmental degradation which quickly devastated this once critical resource.

Currently, sturgeon fisheries in many areas have been reduced to a small remnant of the original resource. Despite this reduction, sturgeon have remained an important cultural resource for Native Americans. This importance is manifested in the ongoing involvement of tribes from Wisconsin, Minnesota, Idaho, Oregon, Washington, Ontario, and elsewhere in sturgeon management programs designed to re-establish major fisheries.

1.4 References

Adams, A.T. (ed.). 1961. The explorations of Pierre Esprit Radisson, from the original manuscript in the Bodleian Library and the British Museum. Ross & Haines, Minneapolis. 342 p.

Allen, J. 1860. Report of Lieutenant Allen of the Army, of H.R. Schoolcraft's exploration of the country at and beyond the sources of the Mississippi, on a visit to the northwestern Indians in 1832. American State Papers: Class V. Military Affairs V. Gates and Seaton, Washington, DC.

Arthurs, D. 1986. Archaeological investigations at the Long Sault site (Manitou Mounds). Conservation Archaeology Report, Northwestern Region, No.7, Ontario, Ministry of Citizenship and Culture.

Birch, Sir A. 1976. A letter from New Westminster. The Beaver 42.44.

Boyce, D.W. 1978. Iroquoian tribes of the Virginia-North Carolina coastal plain. Handbook of North American Indians: Northeast. Smithsonian Institution Press, Washington, DC. 15:282-289.

Callendar, C. 1978a. Illinois. Handbook of North American Indians: Northeast. Smithsonian Institution Press, Washington, DC. 15:673-680.
Callendar, C. 1978b. Miami. Handbook of North American Indians: Northeast. Smithsonian Institution Press, Washington, DC. 15:681-689.
Canada, Sessional Papers. [CSP]
1888. Annual report of the Department of Marine and Fisheries for the year ending June 30 1887. Queen's Printer, Ottawa.
1889. Annual report of the Department of Indian Affairs for the year ended 31stDecember 1888. Queen's Printer, Ottawa.
1890. Annual report of the Department of Indian Affairs for the year ended 31st December 1889. Queen's Printer, Ottawa.
1891. Annual report of the Department of Indian Affairs for the year ended 31st December 1890. Queen's Printer, Ottawa.
1892. Annual report of the Department of Indian Affairs for the year ended 31st December 1891. No. 14. Queen's Printer, Ottawa.
1895. Annual report of the Department of Indian Affairs for the year ended 30th June 1894. No. 14. Queen's Printer, Ottawa.
1897a. Report to the Joint Commission relative to the preservation of fisheries in waters contiguous to Canada and the United States No.11D. Queen's Printer, Ottawa.
1897b. 29th Annual report of the Department of Marine and Fisheries. No. 11A. Queen's Printer, Ottawa.
1905. Special appended reports - the Canadian sturgeon and caviar industries. Canada, No.22. Queen's Printer, Ottawa.
Canada, Energy Mines and Resources, Canada Land Survey Records [CLSR] Plan 163 Ontario. Canada, National Archives [NAC]
RG10, V.1922, f.2970. J.A.N. Provencher urging the department to carry out Treaty provisions and inquiring as to the status of half-breeds.
RG10, V.3800, f.48542. Rat Portage Agency – fisheries in the Lake of the Woods.
RG10, V.3800, f.92840. Rat Portage Agency – speeches by Chiefs at annuity time concerning invasion of their fishing rights.
RG10, V.3807, f.52443. Annual report for the year 1887-1888 regarding fisheries in Manitoba and the Northwest Territories.
Carver, J. 1974. Travels through the interior parts of North-America in the years 1766, 1767, and 1768. Coles Publishing Company, Toronto. 539 p.
Coues, E. (ed.). 1965. The manuscript journals of Alexander Henry and of David Thompson. Ross & Haines, Minneapolis. 1055 p.
Dawson, S.J. 1968. Report on the exploration of the country between Lake Superior and the Red River Settlement. Greenwood Press, New York. 163 p.
Densmore, F. 1974. How Indians use wild plants for food, medicine, & crafts. Dover Publications, New York. 120 p.
Densmore, F. 1979. Chippewa customs. Minnesota Historical Society, St. Paul. 216 p.
Dobbs, A. 1967. An account of the countries adjoining to Hudson's Bay in the north-west part of America. Johnson Reprint Co., New York. 211 p.
Feest. C. F. 1978. North Carolina Algonquians. Handbook of North American Indians: Northeast. Smithsonian Institution Press, Washington, D.C. 15:271-281.
Geyshick, R. 1989. Te Bwe Win Truth. Summerhill Press, Toronto. 174 p.
Goodier, J.L. 1984. The nineteenth-century fisheries of the Hudson's Bay Company trading posts on Lake Superior: a biogeographical study. Canadian Geographer 28:341-357.
Harkness, W.J. and J.R. Dymond. 1961. The lake sturgeon: the history of its fishery and problems of conservation. Ontario Department of Lands and Forests, Fish and Wildlife Branch. 121 p.
Harriot, T. 1972. Brief and true report of the new found lands of Virginia. Ed. P Hutton. Dover Publications, New York. 91 p.
Henry, A. 1969. Travels and adventures in Canada and the Indian Territories between the years 1760 and 1776. Charles E. Tuttle Company, Rutland, VT. 388 p.
Hewes, G.W. 1998. Fishing. Handbook of North American Indians: Plateau. Smithsonian Institution Press, Washington, DC. 12:620-640.
Holzkamm, T. 1987. Sturgeon utilization by the Rainy River Ojibwa bands. Papers of the Eighteenth Algonquian Conference. Carleton University, Ottawa. 155-163 p.
Holzkamm, T. and M. McCarthy. 1988. Potential fishery for lake sturgeon (*Acipenser fulvescens*) as indicated by the returns of the Hudson's Bay Company Lac la Pluie district. Canadian Journal of Fisheries and Aquatic Sciences 45(5):921-923.

Holzkamm, T., V. Lytwyn, and L. Waisberg. 1988. Rainy River sturgeon: an Ojibway resource in the fur trade economy. Canadian Geographer 32:194-205.
Horton, M. 1998. Rainy River sturgeon and the Anishinaabeg people. Display developed by Tim Holzkamm Consulting for Koochiching County Historical Society, International Falls, Minnesota.
Hudson's Bay Company Archives [HBCA]. Provincial Archives of Manitoba.
A.6, London Office Records.
B.22/a/, Brandon House, Post Journals.
B.105/a/, Lac la Pluie, Post Journals.
B.105/d/, Lac la Pluie, Account Books.
B.105/e/, Lac la Pluie, District Reports.
B.166/a/, Portage de l'Ile, Post Journals.
B.239/h/, York Factory, District Fur Returns.
D.3, Governor Simpson's Journal.
D.4, Governor Simpson's Correspondence, Outward.
D.5, Governor Simpson's Correspondence, Inward.
Hultkrantz, A. 1983. Water sprites: The Elders of the fish in aboriginal North America. American Indian Quarterly 8(3):1-22.
Isham, J. 1949. Observations on Hudson's Bay, 1743, and notes and observations on a book entitled a voyage to Hudson's Bay in the Dobbs Galley, 1749. Champlain Society for the Hudson's Bay Record Society, Toronto. 352 p.
Jacobs, P. 1857. Journal of the Reverend Peter Jacobs. Published for the author, New York. 101 p.
Kroeber, A.L. and S.A. Barrett. 1960. Fishing among the Indians of northwestern California. University of California Anthropological Records 21:1:9-54.
Kuhm, H.W. 1928. Wisconsin Indian fishing - primitive and modern. Wisconsin Archeologist 7(2):77-79.
Landes, R. 1969. Ojibwa sociology. AMS Press, New York. 150 p.
Larson, D. 2002. Columbia River sturgeon: The historic sturgeon fisheries. Retrieved from http://www.worldstar.com/~dlarson/SturgeonoftheColumbia3.html on 30 September 2003.
Masson, L.R. (ed.). 1890. Les bourgeois de la Compagnie du Nord-Ouest: Recits des Voyages, inedits relatifs au Nord-Ouest canadien .Côt□, Quebec. 499 p.
National Archives Canada [NAC] Report of S.J. Dawson, 1870. MG11, C.O.42, vol.698, pp.126-148.
Ono, R.D., J.D. Williams and A. Wagner. 1983. Vanishing fishes of North America. Stone Wall Press, Washington, DC. 257 p.
Ontario, Archives of Ontario [AO] Sir Aemilius Irving Papers.
Overholt, T.W. and J.B. Callicott. 1982. Clothed-in-fur and other tales: an introduction to an Ojibwa world view. University Press of America, Washington, DC. 196 p.
Ray, A.J. 1999. 'Ould Betsy and her daughter:' fur tradefFisheries in Northern Ontario. Fishing places, fishing people: traditions and issues in Canadian small-scale fisheries. Ed. D. Newell and R. Ommer. University of Toronto Press Toronto. pp 80-96.
Redsky, J. 1972. Great leader of the Ojibwa: Mis-quona-queb. McClelland and Stewart, Toronto. 127 p.
Rich, E.E. (ed.). 1957. Hudson's Bay copy booke of letters, commissions, instructions outward, 1688-1696. Hudson's Bay Company Record Society, London. 357 p.
Rostlund, E. 1952. Freshwater fish and fishing in native North America. Univ. Calif. Publ. Geogr. 9: 313p.
Russell, F. 1898. Explorations in the far north. University of Iowa Press, Ames, Iowa. 290 p.
Salwen, B. 1978. Indians of southern New England and Long Island: early period. Handbook of North American Indians: Northeast. Smithsonian Institution Press, Washington, DC. 15:160-176.
Schupp, D.H. and V. Macins. 1977. Trends in percid yields from Lake of the Woods, 1888-1973. Journal of the Fisheries Research Board of Canada 34:1784-1791.
Scott, W.B. and E.J. Crossman. 1973. Freshwater fishes of Canada. Fisheries Research Board of Canada, Bulletin 184. 977 p.
Simpson, Sir G. 1847. Narrative of a journey round the world during the years 1841 and 1842. Henry Colburn, London. 362 p.
Snow, D.R. 1978. Eastern Abenaki. Handbook of North American Indians: Northeast. Smithsonian Institution Press, Washington:, DC. 15:137-147.
Spindler, L. 1978. Menominee. Handbook of North American Indians: Northeast. Smithsonian Institution Press, Washington, DC. 15:708-724.
Swanton, J.R. 1979. The Indians of the Southeastern United States. Smithsonian Institution Press, Washington, D.C. 943 p.
Steinbring, J. 1965a. The sturgeon skin jar. Manitoba Archaeological Newsletter. 2(3):12.
Steinbring, J. 1965b. Environmental knowledge among the northern Ojibwa Indians. Paper presented at a joint meeting of the Archaeological Society of Manitoba and the Natural History Society of Manitoba.

Steinbring, J. 1978. Saulteaux of Lake Winnipeg. Handbook of North American Indians: Subarctic. Smithsonian Institution Press, Washington, DC. 6:244-255.

Stewart, H. 1977. Indian fishing: early methods on the Northwest Coast. University of Washington Press, Seattle. 181 p.

Tough, F. 1996. 'As their natural resources fail': native peoples and the economic history of northern Manitoba, 1870-1930. UBC Press, Vancouver. 390 p.

Tough, F. 1999. Depletion by the market: commercialization and resource management of Manitoba's lake sturgeon (*Acipenser fulvescens*), 1885-1935. Fishing places, fishing people: traditions and issues in Canadian small-scale fisheries. Eds. Dianne Newell and Rosemary E. Ommer University of Toronto Press, Toronto. pp. 97-119.

Treaty and Aboriginal Rights Research [TARR]. 1986. The fishing rights of Treaty #3 Indians. Grand Council Treaty #3, Kenora. 69 p.

Usher, P. and F. Tough. 1999. Estimating historical sturgeon harvests on the Nelson River, Manitoba. Fishing places, Fishing people: traditions and issues in Canadian small-scale fisheries. Ed. D. Newell and R. Ommer. University of Toronto Press Toronto. pp193-216.

Vecsey, C. 1983. Traditional Ojibwa religion and its historical changes. American Philosophical Society, Philadelphia. 243 p.

Waisberg, L.G. 1984. An ethnographic and historical outline of the Rainy River Ojibwe. an historical synthesis of the Manitou Mounds site on the Rainy River. Ontario Manuscript Report Prepared for Parks Canada, Ontario Regional Office, Cornwall. W.C. Noble (ed.).

Wilson, G. 1921. Waheenee an Indian girl's story told by herself. Webb Publishing Company, St. Paul, Minnesota. 189 p.

CHAPTER 2

DISTRIBUTION, HABITAT, AND MOVEMENTS

JENNIFER A. WILSON AND R. SCOTT MCKINLEY
Centre for Aquaculture and the Environment, University of British Columbia, 4160 Marine Drive, West Vancouver, BC V7V 1N6 Canada

2.1 Introduction

Even at first glance sturgeon and paddlefish present themselves as a unique group of fishes. Species within these two families possess heavy, roughly cylindrical bodies and retain an almost completely cartilaginous skeleton (Scott and Crossman 1973, Birstein 1993), which is unusual compared to modern teleosts. These fish grow slowly, mature late, and although they spawn repeatedly, most do not exhibit annual spawning (Bemis and Kynard 1997). This combination of life history characteristics makes sturgeon and paddlefish especially vulnerable to anthropogenic influences (Boreman 1997).

In the early 1800s, sturgeon were perceived as a nuisance to the fishing community, as they were often tangled in nets, and were thought to feed on the eggs of the highly desired whitefish (Bogue 2000). By the 1860s, the economic value of sturgeon was realized and they were fished intensively for caviar and meat for international markets and more recently, they have been angled for sport (Boreman 1997, Bogue 2000). Stocks were overharvested causing a substantial decline of sturgeon populations across North America and, therefore, many sturgeon fishery-based communities began turning to paddlefish as a replacement (Boreman 1997).

Although historical populations have been drastically reduced, sturgeon and paddlefish still maintain a wide distribution across North America. Currently, paddlefish (*Polyodon spathula*) and nine species of sturgeon (shortnose sturgeon, *Acipenser brevirostrum*; lake sturgeon, *A. fulvescens*; green sturgeon, *A. medirostris*; Atlantic sturgeon, *A. oxyrinchus oxyrinchus*, and their sub-species Gulf sturgeon, *A. oxyrinchus desotoi*; white sturgeon, *A. transmontanus*; pallid sturgeon, *Scaphirhynchus albus*; shovelnose sturgeon, *S. platorynchus*; and the Alabama shovelnose sturgeon, *S. suttkusi*) are distributed throughout Canada and the United States.

The individual distribution of each of these species is directly related to the migratory strategies and habitats they have adapted. Species of the order Acipenseriformes exhibit three types of life history patterns: 1) Lake sturgeon, pallid, shovelnose, Alabama shovelnose sturgeon and paddlefish spend their entire life history in freshwater (potamodromous); 2) white and shortnose sturgeon are semi-anadromous, moving into brackish water as adults after spawning in freshwater, although some white sturgeon are land-locked by dams ; and finally, 3) green and Atlantic sturgeon

G.T.O. LeBreton et al. (eds.), Sturgeons and Paddlefish of North America, 40–72.
© 2004 *Kluwer Academic Publishers. Printed in the Netherlands.*

are anadromous, spawning in freshwater and moving into the ocean as adults (Boreman 1997).

A variety of habitats are essential for the different life stages of sturgeon and paddlefish. Although they spawn exclusively in freshwater, species of these fish can be found in fresh, brackish, and marine environments. Migrations are undertaken primarily to optimize feeding, to facilitate successful reproduction (Auer 1996a, Bemis and Kynard 1997) and to avoid unfavourable environmental conditions (Auer 1996a). The migratory behaviours of sturgeon and paddlefish are complex and species specific. The movements of these fishes have progressively been impeded by the construction of dams along their migratory routes. Hydroelectric facilities and paper mills not only block the passage of fish during migrations to spawning grounds, but also degrade essential habitats.

The destruction of fundamental spawning, feeding and overwintering habitats due to anthropogenic influences is having a negative impact on sturgeon and paddlefish numbers. Such influences include the construction of dams that block access to ideal habitats, the destruction of important habitats, especially spawning grounds, pollution and, finally, overfishing. Due to their unique life history characteristics, these fish are slow to recover from detrimental environmental impacts. Almost all species of sturgeon and paddlefish have been classified as endangered or threatened (Birstein 1993). This chapter summarizes the individual species distribution, habitat, and migratory behaviour of North American sturgeon and paddlefish and the consequent implications for management in maintaining these populations. With a thorough understanding of these topics, we can hope to make better informed management decisions to protect these species.

2.2 Distribution and Status

Sturgeon and paddlefish are widely distributed across all of North America, located throughout the majority of lake and river systems. The distribution of an individual species within a given system appears to be directly related to the migratory strategies that species exhibits. The historical distribution and abundance of sturgeon and paddlefish across this continent has been significantly reduced over time. The observed decline in these populations can be attributed mainly to anthropogenic influences that have blocked migratory routes, and destroyed essential habitats. Almost all chondrostean species are classified as endangered or threatened by various agencies in Canada and the United States (Birstein 1993).

Three organisations that have published information related to sturgeon and paddlefish status are the American Fisheries Society (AFS), the Committee on the Status of Endangered Wildlife in Canada (COSEWIC) and the International Union for Conservation of Nature (IUCN). Although definitions of the categories used to classify the status of individual species vary slightly within each organisation, the underlying concerns are the same. The AFS and COSEWIC categorize species as endangered, threatened or vulnerable. "Endangered" refers to species that are at high risk of extinction in the immediate future throughout all or a significant portion of its range (Cook and Muir 1984, Williams et al. 1989). "Threatened" refers to species that are likely to become endangered within the immediate future (Cook and Muir 1984, Williams et al. 1989). "Vulnerable", or "special concern", refers to species that exist in low numbers and may become threatened or endangered with minor habitat disturbances (Williams et al. 1989).

The IUCN categorizes species as endangered, vulnerable or lower risk. The lower risk category is further classified into conservation dependent, near threatened and least concern. "Endangered" fish are at risk of extinction in the near future, and "vulnerable" species are at risk of extinction in the medium-term future. Fish populations that are "conservation dependent" have been reduced but are currently stabilized or recovering under continuing conservation strategies (IUCN 2000). Table 2-1 summarizes the status of sturgeon and paddlefish in North America, as classified by the above agencies.

Table 2-1. Status of North American sturgeon and paddlefish. Definitions of END (Endangered), THR (Threatened), VUL (Vulnerable) and CD (Conservation Dependent) are described in the text.

Species	**Location**	**END**	**THR**	**VUL**	**CD**	**Reference**
Acipenser brevirostrum	Canada			•		Campbell 1991
	Canada & USA			•		IUCN 2000
	Canada & USA		•			Williams et al. 1989
Acipenser fulvescens	Canada & USA			•		IUCN 2000
	Canada & USA		•			Williams et al. 1989
Acipenser medirostris	Canada			•		Campbell 1991
	Canada & USA			•		IUCN 2000
Acipenser oxyrinchus oxyrinchus					•	IUCN 2000
	USA			•		Williams et al. 1989
Acipenser oxyrinchus desotoi	USA			•		IUCN 2000
	USA		•			Williams et al. 1989
Acipenser transmontanus	Canada			•		Campbell 1991
Scaphirhynchus albus	USA	•				Williams et al. 1989
Scaphirhynchus platorynchus	USA	•				Williams et al. 1989
Scaphirhynchus suttkusi	USA	•				Williams et al. 1989
Polyodon spathula	USA			•		Williams et al. 1989

Distribution maps for individual species have been approximated from those contained in the Atlas of North American Fishes (Lee et al. 1980). Base maps were constructed using ArcView GIS software (©1992-1999). Detailed river data were obtained from Environmental Systems Research Institute, Inc. (ESRI) ArcData Online (©1991-2001).

2.2.1 LAKE STURGEON (ACIPENSER FULVESCENS)

The lake sturgeon, *A. fulvescens*, has one of the most extensive geographical distributions of any North American freshwater fish (Harkness and Dymond 1961, Houston 1987, Ferguson and Duckworth 1997). It occurs throughout the Mississippi River, the Great Lakes and the Hudson Bay drainages. Figure 2-1 depicts the approximate historical distribution of lake sturgeon across North America.

In general, the lake sturgeon range extends north to the Fort George River on the east coast of James Bay, and the Seal River on the west coast of Hudson Bay; east on the St. Lawrence River to St. Roch de Aulnaires (Harkness and Dymond 1961); south on the Mississippi River and its larger tributaries to Arkansas; and west on the North Saskatchewan River in Alberta (Ferguson and Duckworth 1997).

2.2.1.1 Canada

Lake sturgeon occur almost as far upstream as Edmonton in the North Saskatchewan River, Alberta (Harkness and Dymond 1961). In Saskatchewan, it is distributed throughout the Saskatchewan, Nelson and Churchill rivers and in Cumberland Lake (Houston 1987). Lake sturgeon are found in lakes Winnipeg, Playgreen and Cross and the Assiniboine, and Red rivers of Manitoba (Houston 1987). This species is also found in lakes Manitoba and Winnipegosis, although populations in these lakes are less abundant (Scott and Crossman 1973). Quebec populations occur in most James Bay basin rivers and in larger rivers that drain into the St. Lawrence River before it reaches areas of increased salinity (Houston 1987).

The distribution of lake sturgeon distribution across Ontario is sporadic (Brousseau 1987). Brousseau (1987) suggests lake sturgeon are only found in approximately 100 bodies of water in Ontario. It occurs in the Abitibi, Mattagami and Frederick House river systems (Payne 1987) as well as the Groundhog and Missinaibi rivers, and several of their tributaries (Nowak and Jessop 1987).

2.2.1.2 United States

Distribution of *A. fulvescens* within the United States is generally limited to the Mississippi drainage and its major tributaries, although observations have been made of this species occurring in a few rivers of the Great Lakes drainage (Harkness and Dymond 1961). Lake sturgeon occur from Lake Champlain south through the Mississippi and Tennessee rivers (Scott and Crossman 1973). It is also found from lakes Manitoba and Winnipeg, south through North and South Dakota, Nebraska, Kansas, Missouri and finally Arkansas (Scott and Crossman 1973).

In New York State, lake sturgeon were reported in lakes Erie and Ontario and in the St. Lawrence and Niagara rivers between 1991 and 1994 (Carlson 1995). Sturgeon have occasionally been found in the Oswego River of New York at Varick Dam, the Cayuga and Seneca Canal and Cayuga Lake (Carlson 1995). This species has also been reported rarely in the Finger Lakes, Lake Champlain, and the Seneca, St. Regis, Genesee and Oswegatchie Rivers (Smith 1985a, Carlson 1995).

Becker (1983) reported lake sturgeon in the Wisconsin, St. Croix, Chippewa (including major tributaries), and Mississippi rivers of the Mississippi River drainage basin. In the Lake Michigan drainage it is found in Lake Michigan, the Wolf River upstream to Shawano, the Menominee River upstream to the White Rapids Dam, and the Fox River upstream to Lake Puckaway (Becker 1983). Lake Winnebago, Winneconne, the Embarrass River (a Wolf River tributary) and Butte des Morts are included in this system (Becker 1983).

In the Menominee River, along the Wisconsin-Michigan border, lake sturgeon occur in the stretch of river between the White Rapids and Grand Rapids dams (Thuemler 1985). Lake sturgeon are also distributed between the Grand Rapids and Upper Scott dam in the city of Marinette (Thuemler 1985).

In summary, lake sturgeon enjoy a wide distribution throughout North America. Few studies have been reported that comment on the status or uniqueness of individual populations within these systems. Exceptions include Ferguson et al. (1993), who studied the mitochondrial DNA variation in populations of lake sturgeon in the Moose River basin, Ontario. Substantial gene flow among sampling locations was recorded, which suggests the possibility of future population segregation with further habitat alterations due to dam construction or other barriers to migration (Ferguson et al. 1993).

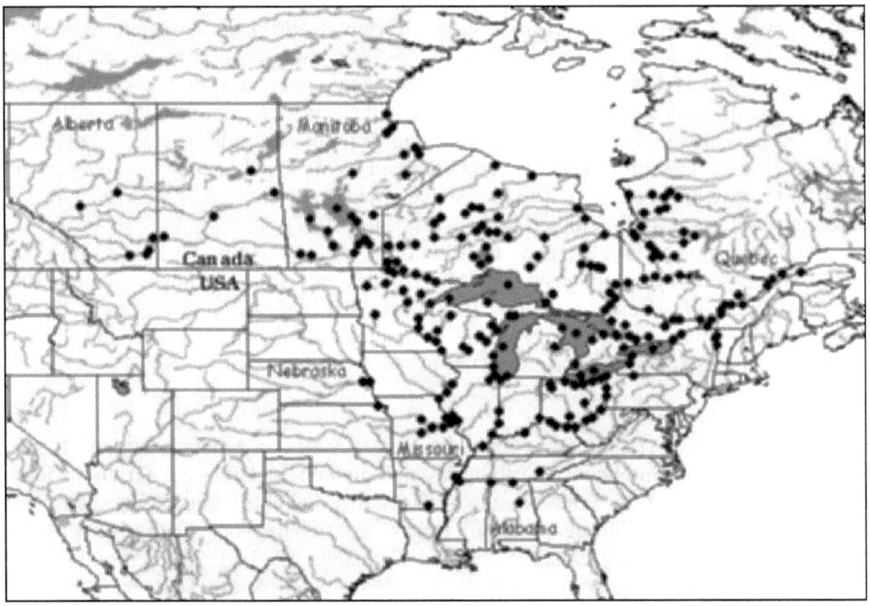

Fig. 2-1. Approximate historical distribution of lake sturgeon in North America.

2.2.1.3 Status

Lake sturgeon were previously extensive throughout its range, although today large numbers are only found in a fragment of its original range (Chiasson et al. 1997, Noakes et al. 1999). Although exact population estimates are difficult to obtain, a continuous decline in annual yield indicates the reduction in lake sturgeon populations throughout Canada and the United States. Population estimates for lake sturgeon in various river systems are provided in Table 2-2. Williams et al. (1989) lists the lake sturgeon as threatened throughout its range in Canada and the United States, as designated by the American Fisheries Society.

Table 2-2 Population estimates for lake sturgeon in various river systems. Average length of fish (cm, FL-fork length, TL-total length where known) are shown in parentheses. 95% confidence intervals (CI) are provided with appropriate references.

River System	Location	Estimate	CI	Reference
Abitibi, Ontario	Island Falls – Abitibi Canyon Dam	994 (94.5 - FL)	830-1213	Payne 1987
Frederick House, Ontario	Neeland's Rapids to Carter Rapids	186 (113.9 - FL)	Unknown	Payne 1987
Groundhog & Mattagami, Ontario	N/A	8429 (105.1 -FL)	6260-11654	Nowak & Jessop 1987
Kenogami, Ontario	Kapeeswatan Lake	10 (114 - TL)	5-18	Sandilands 1987
Kenogami, Ontario	Kapeeswatan Lake. - Ogahalla Rapids	101	82-123	Sandilands 1987
Kenogami, Ontario	Ogahalla Rapids	48 (114 - TL)	35-63	Sandilands 1987
Kenogami, Ontario	Mammamattawa	1314 (62.4 - TL)	1244-1387	Sandilands 1987
Mattagami, Ontario	Sturgeon Falls to Loon Rapids	114 (81.6 - FL)	74-182	Payne 1987
Mattagami, Ontario	Island Falls to Smooth Rock Falls	33 (79.9 - FL)	16-72	Payne 1987
Menominee, Wisconsin	White Rapids dam-Koss bridge	2543 (25-107); 206 (>107)	1882-3204; 115-297	Thuemler 1985
Lake Winnebego, Wisconsin	Lake Winnebago	34100 (>114)	22506-45694	Folz and Meyers 1985

2.2.2 PALLID STURGEON (SCAPHIRHYNCHUS ALBUS)

Pallid sturgeon are distributed within the Missouri and lower Mississippi River basins (Birstein 1993). This species is found in the lower Mississippi River near New Orleans upstream to the Missouri River, and also in the Kansas River upstream to Lawrence, Kansas (Kallemeyn 1983, Keenlyne and Jenkins 1993, Mayden and Kuhajda 1997a). This species is found only occasionally in the Platte, Kansas and St. Francis rivers, and is commonly located in the Yellowstone River in Montana and North Dakota (Kallemeyn 1983). The approximate historical North American distribution of the pallid sturgeon is illustrated in Figure 2-2.

Fig. 2-2. Approximate historical distribution of pallid sturgeon in North America.

2.2.2.1 Status

The pallid sturgeon is classified as an endangered species throughout its range in the United States (Williams et al. 1989, Keenlyne and Jenkins 1993).

2.2.3 SHOVELNOSE STURGEON (SCAPHIRHYNCHUS PLATORYNCHUS)

2.2.3.1 United States

The freshwater shovelnose sturgeon, *Scaphirhynchus platorynchus*, is widely distributed within the inland rivers of the United States (Figure 2-3). This species is found throughout the Mississippi, Missouri and Ohio River drainage basins (Carlson et al. 1985, Keenlyne 1997). It has been reported in the Arkansas, Mississippi, Missouri, St. Francis, Tennessee and Ohio rivers (Keenlyne 1997). It also appears in the Wisconsin River, upstream to the Eau Claire Dam and the Red Cedar River upstream to the Menominee Dam (Becker 1983). This species has also been observed in the White, Illinois, Platte, Yellowstone, Powder and Chippewa rivers (Keenlyne 1997).

2.2.3.2 Status

Williams et al. (1989) lists the shovelnose sturgeon as an endangered species in the United States. Keenlyne (1997) summarizes population estimates from various authors for shovelnose sturgeon in miscellaneous rivers. The population estimates for the Missouri, Mississippi, Red Cedar (Wisconsin) and Tongue (Montana) rivers were cited to be 2500, 1030, 100 and 405 - 537 fish per km, respectively.

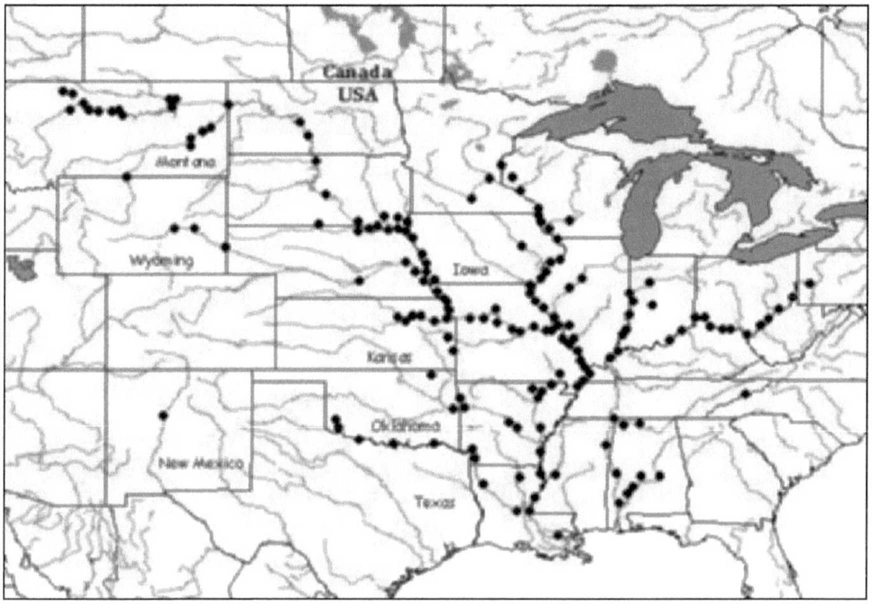

Fig. 2-3. Approximate historical distribution of the shovelnose sturgeon in North America.

2.2.4 ALABAMA SHOVELNOSE STURGEON (SCAPHIRHYNCHUS SUTTKUSI)

The Alabama shovelnose sturgeon, *S. suttkusi*, is found in the Mobile River basin, in Alabama and Mississippi (Keenlyne 1997). Historically, it occurred in the Alabama, Cahaba, Coosa, Mississippi and Tombigbee rivers, although recently it is only reported in the Alabama River, Alabama (Mayden and Kuhajda 1997b). The Alabama shovelnose sturgeon may be considered one of the most endangered species in the United States (Mayden and Kuhajda 1997b).

2.2.5 PADDLEFISH (POLYODON SPATHULA)

The paddlefish, *P. spathula*, is found in the Mississippi River system and smaller rivers that flow into the Gulf of Mexico (Birstein 1993). In Wisconsin, paddlefish inhabit the Mississippi, St. Croix, Chippewa, Red Cedar and Wisconsin rivers (Becker 1983). The approximate historical distribution of paddlefish is illustrated in Figure 2-4.

2.2.5.1 Status

Paddlefish populations have declined significantly over the past 100 years in the Mississippi, Missouri, Ohio and Red rivers (Graham 1997). The AFS classifies the paddlefish as vulnerable throughout its range (Williams et al. 1989) and COSEWIC has declared paddlefish to be extirpated in Canada (Campbell 1991). Extirpated species refers to species of flora or fauna that do not exist in the wild in Canada, but are found elsewhere (Cook and Muir 1984).

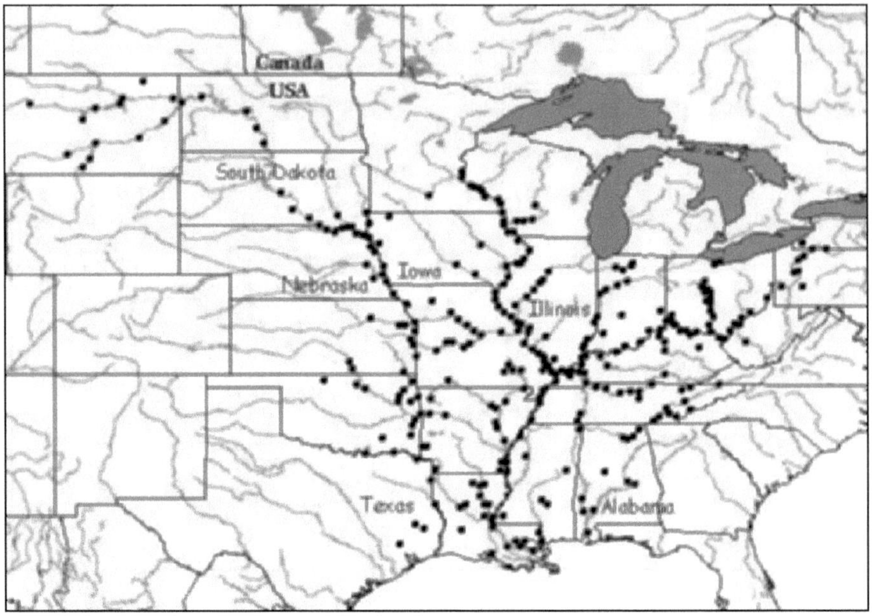

Fig. 2-4. Approximate historical distribution of the paddlefish in North America.

2.2.6 WHITE STURGEON (ACIPENSER TRANSMONTANUS)

The largest of all freshwater fish in North American, *A. transmontanus*, is located along the pacific coast of North America from the Aleutian Islands of Alaska south to Monterey, California (Scott and Crossman 1973, Lane 1991, Parsley et al. 1993). The historical North American distribution of this species is presented in Figure 2-5. Large populations of the white sturgeon are confined to the Sacramento, Columbia and Fraser River systems (Lane 1991).

2.2.6.1 Canada

White sturgeon populations in Canada occur throughout the Fraser River and upper Columbia River systems (Scott and Crossman 1973, Lane 1991). In the Fraser River system, it is distributed in the Harrison, Lower Pitt and Stellako, Taku, Kootenay and Columbia rivers (Scott and Crossman 1973). It occurs in Stuart, Fraser, Kootenay and Duncan Lakes, and possibly in Okanagan Lake (Scott and Crossman 1973).

2.2.6.2 United States

The white sturgeon is found in two major river drainages in Idaho (Cochnauer et al. 1985). It occurs downstream from Shoshone Falls in the Snake River, and in the Clearwater and Salmon rivers, two major tributaries of the Snake River (Cochnauer et al. 1985). Another isolated population exists in the Kootenai River, from Kootenay Lake in Canada, south through Idaho to Kootenai Falls, Montana (Cochnauer et al. 1985).

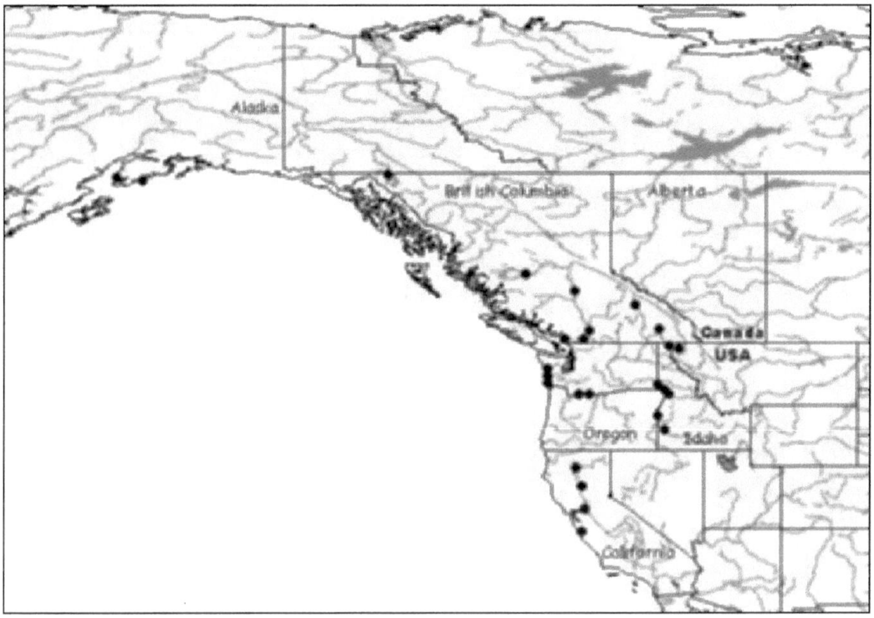

Fig. 2-5. Approximate historical distribution of the white sturgeon in North America.

2.2.6.3 Status

COSEWIC lists the white sturgeon as vulnerable in Canada (Campbell 1991). Although this species is not classified as endangered or threatened in Canada, its distribution appears to be restricted to two river systems, with a substantial population likely occurring only in the Fraser River (Lane 1991).

2.2.7 SHORTNOSE STURGEON (ACIPENSER BREVIROSTRUM)

Distribution of the shortnose sturgeon, *A. brevirostrum*, is confined to the coastal rivers and estuaries along the East coast of North America (Figure 2- 6). It ranges from the Saint John River in New Brunswick, south to the Saint Johns River in Florida (Scott and Crossman 1973, Taubert 1980).

2.2.7.1 Canada

The shortnose sturgeon in Canada is confined to the Saint John River, upstream to Gagetown (Scott and Crossman 1973). The absence of large rivers with warm water estuaries is a limiting factor in the distribution of shortnose sturgeon within Canada (Dadswell 1984). The Saint John River is the northern boundary of the shortnose sturgeon range and is the probable thermal limit for reproductive populations (Dadswell 1984). Estuaries around the Bay of Fundy or the Miramichi River may contain populations of shortnose sturgeon, although none have been identified, possibly due to limited sampling or incorrect classification as Atlantic sturgeon (Dadswell 1984).

Fig. 2-6. Approximate historical distribution of the shortnose sturgeon in North America.

2.2.7.2 United States

Hall et al. (1991) summarize documented shortnose sturgeon distribution in southern rivers of the United States. They range throughout North Carolina, the Winyah Bay system and the Pee Dee and Edisto rivers of South Carolina, in the Altamaha, Ogeechee, and Savannah rivers of Georgia and in the Indian and Saint Johns rivers, Florida (Dadswell 1984, Hall et al. 1991, Birstein 1993). Hastings et al. (1987) records the shortnose sturgeon in the Delaware River from the lower bay upstream to Lambertville, New Jersey. Two populations occur in the Connecticut River, one landlocked between Turners Falls Dam and Holyoke Dam, and one below Holyoke Dam to Long Island Sound (Buckley and Kynard 1985a). Presently, populations are not known to exist from the Delaware River, New Jersey, to the Cape Fear River, North Carolina (Kynard 1997).

2.2.7.3 Status

The AFS lists this species as threatened in the United States and Canada (Williams et al. 1989). COSEWIC and the IUCN classify the shortnose sturgeon as a vulnerable species in Canada (Campbell 1991, IUCN 2000). Populations in the Saint John River, New Brunswick are abundant, and may possibly extend into other Maritime rivers, and therefore the shortnose sturgeon should not be considered endangered in Canada (Moser and Ross 1995). American populations are also larger than originally recorded suggesting the species may be removed from the endangered species list in the near future (Moser and Ross 1995).

2.2.8 GREEN STURGEON (ACIPENSER MEDIROSTRIS)

The green sturgeon, *A. medirostris*, occurs along the Pacific coast in the estuarine areas of larger rivers from the Gulf of Alaska (Aleutian Islands) to Ensenada Mexico (Houston 1988), as illustrated in Figure 2-7. In Canada, this species is found in the Fraser and Skeena rivers, at Victoria off Kyuoquot Sound, and along the west coast of Vancouver Island (Scott and Crossman 1973).

2.2.8.1 Status

COSEWIC and the IUCN have classified the green sturgeon as vulnerable (Campbell 1991, IUCN 2000).

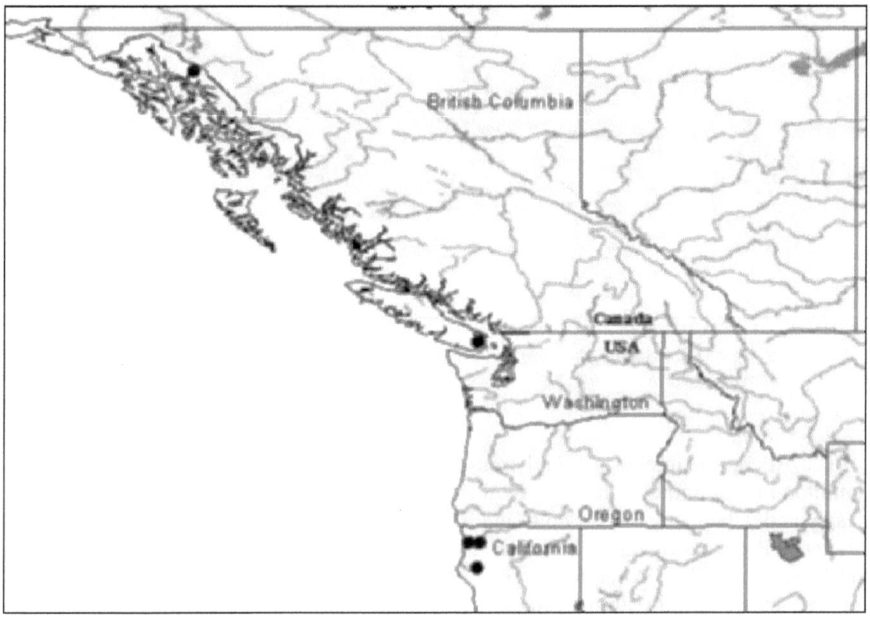

Fig. 2-7. Approximate distribution of the green sturgeon in North America.

2.2.9 ATLANTIC STURGEON (ACIPENSER OXYRINCHUS OXYRINCHUS)

The Atlantic sturgeon is divided into two subspecies, the northern *A. oxyrinchus*, and the southern Gulf sturgeon, *A. oxyrinchus desotoi* (Smith 1985b, Smith and Clugston 1997). *A. o. desotoi* will be discussed separately. *A. o. oxyrinchus* is broadly distributed along the eastern coast of North America in major rivers, estuaries and marine waters (Smith and Clugston 1997). They are found in Hamilton Inlet, Labrador, and George River, Ungava Bay, south to the St. Johns River, Florida (Scott and Crossman 1973, Wooley and Crateau 1985, Musick et al. 2000). The approximate historical distribution of the Atlantic sturgeon is shown in Figure 2-8.

Fig. 2-8. Approximate historical distribution of the Atlantic sturgeon and the sub-species Gulf sturgeon in North America.

2.2.9.1 Canada

The Atlantic sturgeon is found from the George River, Ungava Bay, south in the Hamilton Inlet, Labrador, through the Strait of Belle Isle, into the Gulf of St. Lawrence, Gaspe and Chapleur Bay (Scott and Crossman 1973). It occurs through the St. Lawrence River upstream to Trois-Rivieres, and occasionally into Lac St. Pierre (Scott and Crossman 1973). Atlantic sturgeon are rarely recorded off Newfoundland, and no populations are found within freshwater (Scott and Crossman 1973). This species does occur in Nova Scotia and New Brunswick, within Cheticamp, Passamaquoddy and Aspy bays, the Bay of Fundy, Minas Basin, and the Annapolis, Avon and St. John rivers (Scott and Crossman 1973, Smith and Clugston 1997). It has also been recorded in Oromocto and Grand Lakes and in the Miramichi estuary (Scott and Crossman 1973).

2.2.9.2 United States

Atlantic sturgeon in the United States are found along the Atlantic coast south to the St. Johns River on the east coast of Florida (Scott and Crossman 1973). During the winter, they have been recorded as far south as Port Canaveral and Hutchinson Island, Florida, when cold fronts along the coast induce southward movement (Wooley and Crateau 1985).

2.2.9.3 Status

The Atlantic sturgeon is listed as vulnerable in the United States by the American Fisheries Society (Williams et al. 1989). This species is listed as conservation dependent by the International Union for Conservation of Nature Red List (IUCN 2000).

2.2.10 GULF STURGEON (ACIPENSER OXYRINCHUS DESOTOI)

2.2.10.1 United States

The southern Atlantic sturgeon subspecies, *A.o. desotoi*, is less widely distributed compared to the northern subspecies (Figure 2-8). The Gulf sturgeon is restricted to the Gulf of Mexico in coastal waters from Tampa Bay, Florida, west to the mouth of the Mississippi River (Smith and Clugston 1997). The current range of the Gulf sturgeon appears to be from the Suwannee River, Florida, to eastern Louisiana (Wooley 1985).

2.2.10.2 Status

The Gulf sturgeon is listed as threatened in the United States by the AFS (Williams et al. 1989). The IUCN Red List classifies this species as vulnerable (IUCN 2000).

Table 2-3. Depth (m), Temperature (°C) and substrate preferences of North American sturgeon and paddlefish for the adult life stage. Refer to the text for the definitions of substrate types where applicable. Information that is not available is indicated by N/A.

Species	Depth (m)	Temp (°C)	Substrate	Reference
Acipenser brevirostrum	1-27	2-13	N/A	McCleave et al. 1977 Dadswell 1979
A. fulvescens	4-9	<23.8	Clay, sand, gravel, rock, mud	Harkness and Dymond 1961 Ono et al. 1983 Chiasson et al. 1997
A. o. oxyrinchus	10-40	N/A	N/A	Moser and Ross 1995
A. o. desotoi	8.4	7.5-15	Limestone	Wooley and Crateau 1985 Smith and Clugston 1997 Fox et al. 2000
A. transmontanus	N/A	0-23	N/A	Scott and Crossman 1973
Scaphirhynchus albus	N/A	N/A	Sand, gravel, rock	Mayden and Kuhajda 1997a
S. platorynchus	4.6-6.1	N/A	Sand	Curtis et al. 1997
S. suttkusi	N/A	N/A	Sand, gravel	Mayden and Kuhajda 1996
Polyodon spathula	2-11	13-24	Mud	Southall and Hubert 1984 Pitman and Parks 1994

2.3 Habitat

A variety of habitats are utilized by sturgeon and paddlefish throughout their life cycles. Each stage of life (feeding, overwintering and spawning of adults, development of eggs and larvae and the growth of juveniles) requires a distinct habitat (Auer 1996a). Sturgeon and paddlefish are found in fresh, brackish and saltwater environments, although all species spawn in freshwater. The distribution of sturgeon and paddlefish within a given habitat is dependent on the physical characteristics of an area, including depth, substrate and water velocity. The characteristics of selected habitats by individual species for the adult life stage is summarized in Table 2-3. The characteristics of the preferred habitats of sturgeon and paddlefish for spawning are listed in Table 2-4.

Table 2-4. Depth (m), temperature (°C) and substrate preferences of North American sturgeon and paddlefish during spawning. Refer to the text for the definitions of substrate types where applicable. Information that is not available is indicated by N/A

Species	Depth (m)	Temp (°C)	Substrate	Reference
Acipenser brevirostrum	3-4	9-15	Gravel, rubble, large boulders	Taubert 1980 Buckley and Kynard 1985b
A. fulvescens	0.6-4.7	13-18	Fine-medium gravel-boulders, cobble	Scott and Crossman 1973 LaHaye et al. 1992
A. medirostris	N/A	8.9-16.7	Rocky	Houston 1988
A. o. oxyrinchus	11-13	13-18	Rock, rubble, hard clay	Scott and Crossman 1973 Smith 1985b
A. o. desotoi	5.5-8.1	18.3	N/A	Smith and Clugston 1997
A. transmontanus	N/A	8.9-16.6	Rocky	Scott and Crossman 1973
S. platorynchus	N/A	16.9-20.5	Hard surfaces	Keenlyne 1997
Polyodon spathula	2-12	10-16.1	Gravel, cobble	Crance 1987 Becker 1983

2.3.1 LAKE STURGEON

One of the few Acipenser species that spends its entire life cycle in freshwater is the lake sturgeon. Although they live almost exclusively in freshwater they are capable of inhabiting brackish water, and have occasionally been observed in the St. Lawrence River, Hudson and James Bays in areas where the salinity does not greatly exceed 15% (Harkness and Dymond 1961, LeBreton and Beamish 1998). Juvenile and smaller adult lake sturgeon are not capable of inhabiting environments with salinities above 25% (LeBreton and Beamish 1998). Lake sturgeon are rarely found in water with temperatures above 23.8°C (Ono et al. 1983). This species typically inhabits large lakes and rivers, preferring highly productive shoal waters (Scott and Crossman 1973, Becker 1983).

Lake sturgeon are typically found in waters from 4 - 9m deep (Harkness and Dymond 1961, Scott and Crossman 1973, Houston 1987). Most lake sturgeon have been caught at depths of 4.6 - 9.2m, but records indicate that, on occasion, they have been found as deep as 42.7m (Scott and Crossman 1973). The greatest proportions of recorded catches occur at depths less than 4.5m (Harkness and Dymond 1961). Similarly, Folz and Meyers (1985) found the greatest number of lake sturgeon in Lake Winnebago at depths of 3.7 - 6.1m. In Lake of the Woods, of the Winnipeg – Nelson river system, overwintering sturgeon were found concentrated in water over 7m deep (Rusak and Mosindy 1997). Fish from the same system that overwintered in the Rainy River were observed in water greater than 6m deep. Rusak and Mosindy (1997) attribute discrepancies in depth partiality to the importance of food in the selection of habitat and the depths at which these foods are located in various bodies of water.

Lake sturgeon are anatomically adapted for feeding on bottom living organisms, equipped with a ventral, protrusible tube-like mouth (Harkness and Dymond 1961, Scott and Crossman 1973). Lake sturgeon are opportunistic predators, therefore habitat selection is directly related to the environments that support organisms that they exploit as food (Chiasson et al. 1997). The composition of sturgeon diet is dependent on the availability of different prey organisms (Harkness and Dymond 1961, Scott and Crossman 1973), rather than the size of the fish. The species and abundance of prey items available is dictated by the substrate (Harkness and Dymond 1961, Chiasson et al. 1997), whether gravel (particles >2 and <16mm in size), clay (<0.125mm), mud or sand (>0.125 and <2mm). Lake sturgeon feed on molluscs, crustaceans, insect larvae (*Hexagenia*, *Diptera* and *Trichoptera*) and small fish (Harkness and Dymond 1961, Houston 1987, Chiasson et al. 1997).

Chiasson et al. (1997) reported lake sturgeon stomach contents to contain 86.3% *Hexagenia* (Mayflies) and that *Hexagenia* and *Trichoptera* abundance was greatest in areas with a high percentage of clay substrate. Accordingly, Chiasson et al. (1997) noted an increase in lake sturgeon abundance with an increase in clay substrate. Smith (1985a) documented that lake sturgeon avoided muddy areas and preferred substrates of sand, gravel or rock, where food was most abundant. In contrast, Harkness and Dymond (1961) declared mud to be the prevalent substrate over which sturgeon were located, followed by gravel with some mud. Harkness and Dymond (1961) also reported sturgeon to feed on mayfly nymphs more than any other prey items. This is in accordance with McKinley et al. (1993) who found that *Hexagenia* was the most abundant prey item in stomachs of many lake sturgeon. Mayflies inhabit areas with mud, sand or clay substrate (Harkness and Dymond 1961). Essentially, lake sturgeon are most abundant over bottoms which support their forage base (Harkness and Dymond 1961, Houston 1987).

2.3.1.1 Spawning

Lake sturgeon spawn in the spring, from early May to late June, depending on their geographical location (Scott and Crossman 1973). During spawning, lake sturgeon migrate into smaller streams (Smith 1985a). Generally, spawning does not commence until water temperatures reach 13°C (Harkness and Dymond 1961). Scott and Crossman (1973) report optimum spawning temperatures to be between 13 and 18°C. In Des Prairies River, Quebec, spawning occurred when temperatures rose from 11.6 to 15.4°C (LaHaye et al. 1992). Similarly, in L'Assomption River, spawning occurred when water temperatures increased from 11 to 18°C (LaHaye et al. 1992). An increase in temperature to 11.7°C initiated spawning in the Wolf River system, Wisconsin (Folz and Meyers 1985). In the Fox River, of the same system, spawning occurred at temperatures of 14-16°C, and was postulated to cease if temperatures were to drop below 11.7°C (Folz and Meyers 1985). Harkness and Dymond (1961) maintain that when temperatures fall below 13°C, or exceed 21°C, spawning will be inhibited.

Spawning occurs at depths of 0.6 - 4.7m, in fast-flowing waters, and frequently at the base of low waterfalls that prevent further migration upstream (Harkness and

Dymond 1961, Scott and Crossman 1973). In rivers that do not have suitable spawning habitat, lake sturgeon spawn over rocky ledges where wave action supplies sufficient oxygen for the eggs (Scott and Crossman 1973, Houston 1987). Eggs that are released in rocky areas are protected from predation and other factors that would contribute to egg loss (Houston 1987). In L'Assomption and Des Prairies rivers, spawning occurred primarily in areas with coarse gravel substrates (LaHaye et al. 1992). In both locations, eggs were located over substrates varying from fine-medium gravel to boulders, and were absent in areas with fine sand, clay and silt, and bedrock (LaHaye et al. 1992). Physical characteristics of a spawning site in the Groundhog River, Ontario, included a water velocity over $0.5 m \cdot s^{-1}$, and a substrate of cobble and bedrock (McKinley et al. 1998).

2.3.1.2 Juvenile

Documentation of habitat utilisation by juvenile lake sturgeon is limited (Auer 1996a). Harkness and Dymond (1961) located small sturgeon in 3 - 4.5m of water over rocky, gravelly shores within the mouths of rivers. Juveniles in the Mattagami and Groundhog rivers of Ontario were located in areas of high prey item abundance (Chiasson . 1997). Habitat selection by young lake sturgeon has been related to stock adaptation, suggesting juveniles from different stocks may prefer different habitats (Auer 1999b).

2.3.2 PALLID STURGEON

The pallid sturgeon inhabits areas of rapid current and prefers turbid water conditions (Kallemeyn 1983) that are necessary to conceal it from prey species (Mayden and Kuhajda 1997a). These conditions are historically found in the Missouri and lower Mississippi rivers (Mayden and Kuhajda 1997a). The pallid sturgeon is associated with habitats characterized by sand, gravel or rocky substrates (Mayden and Kuhajda 1997a).

2.3.3 SHOVELNOSE STURGEON

This potamodromous species is frequently located in deep channels of large rivers in areas characterized by increased water flow (Becker 1983). They are primarily located in main channels or along main channel borders, in pools downstream from sandbars, and in association with wing dams (Curtis et al. 1997, Keenlyne 1997). In the Missouri River, this species has been reported to congregate in deeper waters behind sandbars in the spring and fall, and disperse when velocities increase during the summer (Modde and Schmulbach 1977). Wing dams in the Mississippi River may provide velocities similar to pools located behind sandbars in the unchannelised Missouri River (Hurley et al. 1987). Optimum velocities for shovelnose sturgeon in the upper Mississippi River were observed to be $0.2 - 0.4 m \cdot s^{-1}$ near the bottom and $0.4 - 0.7 m \cdot s^{-1}$ at the surface (Hurley et al. 1987). Curtis et al. (1997) observed sturgeon in the Mississippi River most frequently in areas with surface velocities of $0.2 - 0.64 m \cdot s^{-1}$. Quist et al. (1999) located sturgeon in the Kansas River in areas with surface velocities between 0.3 and $0.6 m \cdot s^{-1}$, and bottom velocities from $0.2 - 0.3 m \cdot s^{-1}$.

Shovelnose sturgeon have been reported to prefer stable substrates dominated by gravel and sand (Carlson et al. 1985). Shovelnose sturgeon in the Chippewa River were reported over gravel and sand, and in the Red Cedar River over sand, rubble-gravel, and occasionally bedrock (Becker 1983). Telemetry data for fish tagged in the Mississippi River reveal sand as the predominant substrate where fish were located (Curtis et al. 1997). Shovelnose sturgeon were also located primarily over sand substrates in the Kansas River, although rocky areas were available (Quist et al. 1999). During spring and summer, fish in the Mississippi River were primarily located over sand, but were often found near rocky areas (Hurley et al. 1987). Quist et al. (1999) suggest that shovelnose sturgeon utilize rocky habitats in the summer and fall, which can be attributed to the abundance of benthic invertebrates available in these substrates. Shovelnose sturgeon are opportunistic foragers that feed on a variety of invertebrates (Modde and Shmulbach 1977, Carlson et al. 1985).

The depth at which shovelnose sturgeon are located varies between river systems. Fish have been cited at depths of 4.3 - 10.1m and 1.2 - 5.8m in the upper Missouri River and in the Yellowstone River at 0.9 - 8.8m (Quist et al. 1999). In the Kansas River, Quist et al. (1999) observed fish at depths of 1.0 - 2.0m. Radio-tagged fish in the Mississippi were located at depths ranging from 2.7 - 8.3m, but were most often found in waters 4.6 - 6.1m deep (Curtis et al. 1997). Due to the variability in preferred depth, the primary habitat characteristic which affects shovelnose sturgeon distribution appears to be water velocity, while depth and substrate type are only secondary factors (Hurley et al. 1987, Curtis et al. 1997). More specifically, Quist et al. (1999) suggest bottom water velocity is the most important factor in shovelnose habitat selection.

2.3.3.1 Spawning

Spawning habitat of shovelnose sturgeon is not well described, but it is believed that spawning occurs in tributary streams or along the borders of main river channels over hard bottoms (Keenlyne 1997). Shovelnose sturgeon have been reported to spawn from April to June at water temperatures from 16.9 - 20.5°C (Keenlyne 1997). In the Red Cedar and Chippewa rivers of Wisconsin, spawning occurred during May and June when water temperatures reached 19.5 - 21.2°C (Becker 1983).

2.3.4 ALABAMA SHOVELNOSE STURGEON

The Alabama shovelnose inhabits high velocity areas of river channels preferring stable substrates of sand and gravel, although it has been found over softer sediments (Mayden and Kuhajda 1996). Insect larvae are the primary food preference of Alabama shovelnose, although it has also been reported to feed on snails, mussels, fish eggs and fish (Mayden and Kuhajda 1996).

2.3.5 PADDLEFISH

Paddlefish occur in deeper waters of large rivers in the Mississippi River system (Becker 1983, Southall and Hubert 1984). A variety of habitats are occupied by paddlefish within these rivers being found in main channel, main channel borders, tailwater, secondary channel and backwater areas (Zigler et al. 1999). In the upper Mississippi River, paddlefish are most frequently located in tailwater and main channel borders with wing dams (Southall and Hubert 1984, Moen et al. 1992). In contrast to these findings, Zigler et al. (1999) found paddlefish to inhabit a secondary channel of the upper Mississippi River, the lower Black River, which is similar to a backwater habitat due to low current velocities measured in this area.

Habitat selection by paddlefish is related to variations in current velocity. Paddlefish prefer areas with reduced velocity. Constructed wing dams and naturally occurring sandbars create eddies and scour holes which reduce current, creating ideal conditions for this species (Southall and Hubert 1984, Moen et al. 1992). In the upper Mississippi River, telemetry observations indicate paddlefish selected tailwaters as their preferred habitat, with an average current velocity of $0.32m \cdot s^{-1}$ (Southall and Hubert 1984). Fish also showed a preference for main channel borders, having an average velocity of $0.30m \cdot s^{-1}$ (Southall and Hubert 1984). Moen et al. (1992) located radio-tagged paddlefish most frequently in tailwaters below Lock and Dam 12 of the Mississippi River, which contained a series of wing dams. Fish were observed in areas with water velocity ranging from 0.1 - $0.5m \cdot s^{-1}$, although average velocity was $0.1m \cdot s^{-1}$ (Moen et al. 1992). Rosen et al. (1982) found habitat selection to vary with season. Throughout the spring and early fall paddlefish were located downstream from sandbars, occurring in velocities of 0 - $0.3m \cdot s^{-1}$ (Rosen et al. 1982). From fall to spring, fish were located in deeper habitats.

Deep waters provide ideal habitats for paddlefish. During the winter, fish entered areas with depth greater than 3m (Rosen et al. 1982). Pitman and Parks (1994) located paddlefish in the Neches River, Texas, from 3.4 - 11m during the winter, with an average depth of 7.6m. In the summer, fish were found at depths of 2 - 3m in the Missouri River (Rosen et al. 1982) and 2 - 10m in the Neches River, with an average depth of 4.4m (Pitman and Parks 1994). Paddlefish that occupy river or tailwater sections are regularly found in deeper water (Pitman and Parks 1994).

2.3.5.1 Spawning

Spawning occurs when water temperatures reach 10°C, and water levels increase (Becker 1983). Becker (1983) cited a study on paddlefish spawning in the Osage River, Missouri. Fish were observed spawning over a gravel bottom when water temperatures reached 16.1°C and water levels had increased by 2.7m. The overall depth at which paddlefish prefer to spawn was cited to range from 2 - 12m and may depend on the size of the individual fish (Crance 1987). In the upper Mississippi River, paddlefish located in wing dam and tailwater habitats in the spring were presumed to be searching for spawning sites (Moen et al. 1992). These areas are characterized by rocky substrates ideal for paddlefish spawning (Moen et al. 1992).

2.3.6 WHITE STURGEON

The white sturgeon is a semi-anadromous species which has been caught in salt, brackish and freshwater (Scott and Crossman, 1973, Musick et al. 2000). This species has been found to occur in waters with temperatures ranging from 0 - 23°C (Scott and Crossman 1973). Smaller white sturgeon are bottom feeders, feeding primarily on chironomids (Scott and Crossman 1973). Larger fish feed on crayfish (Scott and Crossman 1973) and some have been observed feeding on salmonids (North et al. 1993). White sturgeon typically inhabit shallow waters, but have been recorded at depths of 30m (Scott and Crossman 1973). White sturgeon in the Columbia and Fraser rivers were observed primarily in large pools within the main channels (Lane 1991).

2.3.6.1 Spawning

White sturgeon spawning occurs when water temperatures are between 8.9 and 16.6°C, in areas with rocky substrate and swift currents (Scott and Crossman 1973).

White sturgeon in the Columbia River spawned during spring and early summer, which coincided with increased currents that may have aided larval dispersion downstream (Parsley et al. 1993). Spawning occurred at a median water temperature of 14°C, which is ideal for the development of white sturgeon eggs, and in areas with increased velocity, over coarse substrates (Parsley et al. 1993).

2.3.6.2 Juvenile

Young-of-the-year white sturgeon in the lower Columbia River were located in deep areas with reduced water velocity (Parsley et al. 1993). These fish have been caught at depths of 9 - 57m, over bottoms of clay (particles of 0.00024 - < 0.004mm), cobble (64 -< 250mm), gravel (2 - <64mm), sand (0.062 - <2mm) and mud (0.004 - <0.062mm) with the majority of captures in areas dominated by sand substrates (Parsley et al. 1993). Juvenile white sturgeon were found at depths of 2 - 58m, primarily over sand (Parsley et al. 1993).

2.3.7 SHORTNOSE STURGEON

The shortnose sturgeon is considered a semi-anadromous species as it is found mainly in estuarine environments (Dadswell 1979, Musick et al. 2000). This species primarily occurs around the mouths of large rivers in estuarine or nearshore marine areas (Dadswell 1984, Bain 1997). Habitat selection appears to be influenced by latitude and the physical characteristics of specific river systems (Dadswell 1984). Shortnose sturgeon in the Saint John River, New Brunswick, overwintered in lower saltwater portions of the estuary and deeper regions of estuarine lakes (Dadswell 1979). Shortnose sturgeon in Georgia were observed entering rivers only during the spring to spawn, and spent the remainder of the year in the estuary (Dadswell 1984). Fish in Montsweag Bay, Maine were most frequently located in fresh and lower salinity waters of the Kennebec River estuary during the summer (McCleave et al. 1977). Similarly, Buckley and Kynard (1985a) found shortnose sturgeon in the Connecticut River remained in more riverine habitats, only periodically entering saltwater.

Shortnose sturgeon move throughout all areas of a river system, but may remain at important resting and/or feeding aggregation sites for longer periods of time (Hastings et al. 1987). Aggregation areas are often related to food (Kieffer and Kynard 1993). Shortnose sturgeon in the Merrimack River, Massachusetts used only a few discrete areas of the river, suggesting only a few locations are characterized by resources necessary for optimum foraging (Kieffer and Kynard 1993).

Depth may also be an important characteristic of aggregation areas (Hastings et al. 1987). Fish in the Hudson River estuary remained in deep, channel habitats from late spring through early fall. In the late fall as water temperatures decrease, adult shortnose concentrated in a few overwintering locations (Bain 1997). In the St. John River estuary, this species overwintered in portions of estuarine lakes greater than 10m deep (Dadswell 1979). McCleave et al. (1977) found sturgeon in a Maine estuary at depths ranging from 1 - 27m.

2.3.7.1 Spawning

Shortnose sturgeon spawn in the spring, from April to early June, in rivers or areas of fresh water within estuaries (Scott and Crossman 1973, Dadswell 1984). Spawning is initiated at water temperatures ranging from 9 - 15°C (Taubert 1980, Dadswell 1979). Radio-tagged shortnose sturgeon were observed spawning in the Savannah

River at water temperatures of 9 - 12°C (Hall et al. 1991), and in the Connecticut River when temperatures increased to 11.5 - 14°C (Buckley and Kynard 1985b).

The ideal substrate composition for shortnose sturgeon spawning grounds includes gravel, rubble and large boulders (Taubert 1980, Buckley and Kynard 1985b). Buckley and Kynard (1985b) found suitable spawning substrate throughout a 2km reach of the Connecticut River and concluded depth and water velocity may be the primary factors determining spawning locations. Spawning occurred in areas 3 - 4m deep in a hole in the river bottom (Buckley and Kynard 1985b). A wider range of depths, and therefore velocities, are available to fish spawning in deep depressions.

Hall et al. (1991) found shortnose sturgeon in the Savannah River spawned in areas of increased water velocity with river velocity at the time of spawning 0.52 - 1.04$m \cdot s^{-1}$ at the surface, and 0.82$m \cdot s^{-1}$ near the bottom. Similarly, Buckley and Kynard (1985b) found the highest concentration of fish in velocities of 0.7 - 1$m \cdot s^{-1}$ during the spawning period. A specific velocity is necessary for the survival of eggs after being released from the spawning fish (Buckley and Kynard 1985b). Eggs may lose their adhesiveness before having the opportunity to attach to the substrate if they are released into a river with excessive flow (Buckley and Kynard 1985b). However, if eggs are released in extremely reduced velocity areas, individual egg survival may decrease due to factors including respiratory stress, fungal growth and predation (Buckley and Kynard 1985b). Survival is also related to water velocity, as adequate flow is necessary for the transport of larvae downstream (Buckley and Kynard 1985b). Buckley and Kynard (1985b) concluded water velocity is the determining factor in environmental threshold limits for spawning fish. Radio tagged shortnose in the Connecticut River departed from the spawning area in 1983 before they had spawned (Buckley and Kynard 1985b). Telemetry data suggested environmental conditions were not favourable and reproduction was impeded.

2.3.7.2 Juvenile

Juvenile shortnose sturgeon have been located in the same deep estuarine habitats used by adults (Collins and Smith 1997, Bain 1997). The first years of juveniles are spent in deep tidal freshwater or low salinity estuarine habitats (Dadswell 1984, Collins and Smith 1997). Juvenile shortnose sturgeon in the Savannah River remained in riverine or brackish areas within 2 - 5km of saltwater termination (Hall et al. 1991). Similarly, juveniles in the Hudson River occupied areas of the tidal reach (Bain 1997). It is not evident if juveniles move out of the estuary into coastal marine waters (Bain 1997). Juvenile shortnose sturgeon feed over sand, mud and small gravel substrates (Hall et al. 1991).

2.3.8 GREEN STURGEON

This anadromous species only enters freshwater to spawn, but can be found in brackish waters near river mouths along the Pacific coast (Houston 1988). Specific habitat selections are thought to be similar to those of the white sturgeon (Houston 1988). Adults feed mainly on benthic species, including chironomids, mysids, copepods and other invertebrates (Houston 1988).

2.3.8.1 Spawning

It is thought the spawning characteristics of green sturgeon are similar to those of the white sturgeon (Houston 1988). White sturgeon spawn in rapid flowing waters, over rocky substrates, at temperatures ranging from 8.9 - 16.7°C (Scott and Crossman 1973).

2.3.9 ATLANTIC STURGEON

Although Atlantic sturgeon reproduce in freshwater, and young may spend up to four years in the river, they eventually migrate to the sea and are therefore considered anadromous (Scott and Crossman 1973, Bain 1997, Musick et al. 2000). Details of adult Atlantic sturgeon habitat preferences are not well documented (Scott and Crossman 1973). They are a benthic species, but have been observed jumping out of the water on occasion (Scott and Crossman 1973). Sonically tagged fish in the lower Cape Fear River, North Carolina were located at depths in excess of 10m throughout the year, with little movement in the summer (Moser and Ross 1995). Based on these observations, Moser and Ross (1995) infer Atlantic sturgeon may be restricted to a few deep, freshwater holes that offer thermal protection.

2.3.9.1 Spawning

Atlantic sturgeon migrate into rivers to spawn in freshwater from February to June, depending on the river system (Smith 1985b). Adults arrived on spawning sites in February in the Suwannee River, Florida, St. Mary's River, Georgia and Winyah Bay, South Carolina. In the Chesapeake Bay and Delaware systems fish were observed at spawning sites in April, and in the Gulf of Maine and Gulf of St. Lawrence systems in May-June (Smith 1985b). Spawning has been reported to occur at water temperatures of 13 - 19°C in South Carolina and 13.3 - 17.8°C in the Delaware River (Smith 1985b). Atlantic sturgeon spawn over substrates consisting of rocks, rubble and hard, clay bottoms, in fast flowing water from 10 - 12m deep (Smith 1985b, Scott and Crossman 1973).

2.3.9.2 Juvenile

Juvenile Atlantic sturgeon essentially inhabit estuarine regions, moving into higher salinity waters during late fall and winter (Collins and Smith 1997). In the fall when temperatures dropped below 20°C, juveniles in the Hudson River moved into brackish water to overwinter (Bain 1997). Lazzari et al. (1986) also reported juveniles to overwinter in deep waters of the lower estuary, with an upstream movement in the spring in response to increasing water temperatures. In contrast to these observations, some juvenile sturgeon were located in the upper tidal portion of the Delaware River through December at water temperatures close to 0.5°C (Lazzari et al. 1986). Juvenile sturgeon remained in the Hudson River for 2 - 6 years before migrating to marine waters (Bain 1997).

2.3.10 GULF STURGEON

The Gulf sturgeon is an anadromous sub-species, inhabiting saltwater as an adult, and reproducing in freshwater (Musick et al. 2000). Telemetry observations in the Apalachicola River, Florida, revealed late spring and summer habitat preferences to include areas with sand and gravel substrate, at an average depth of 8.4m, with an average water velocity of $0.64 m \cdot s^{-1}$ (Wooley and Crateau 1985). A fall staging area in the Brothers River was characterized by substrates of sand and clay at depths of 11m in velocities of $0.62 m \cdot s^{-1}$ (Wooley and Crateau 1985). Fish overwintered in high velocity areas over 14m deep in water temperatures of 7.5 - 15°C (Smith and Clugston 1997).

2.3.10.1 Spawning

In the Suwannee River, Gulf sturgeon spawned in the upper reaches of the river when temperatures reached 18.3°C (Smith and Clugston 1997). Fish in the upper Apalachicola River are speculated to spawn at water temperatures of 22.5 - 23°C (Wooley and Crateau 1985).

2.4 Migratory Behaviour

Sturgeon and paddlefish migrate to avoid adverse conditions, to ensure successful reproduction and to optimize feeding (Auer 1996a, Bemis and Kynard 1997). As outlined earlier, sturgeon and paddlefish exhibit three migratory patterns. Paddlefish, lake sturgeon, pallid, shovelnose and Alabama shovelnose sturgeon remain in freshwater throughout their lives; white and shortnose sturgeon migrate to estuarine and brackish water after spawning, and green, Atlantic and Gulf sturgeon out-migrate to the sea after spawning (Auer 1996a, Boreman 1997). Little information exists on pallid, Alabama shovelnose and green sturgeon movements, and therefore these species' migrations are not discussed. Future studies using telemetry are necessary to examine the behaviour of these species to determine their movement patterns and subsequent habitat use.

2.4.1 LAKE STURGEON

Lake sturgeon spend the majority of their life cycle within freshwater. Their movements generally involve a spring migration to potential spawning areas, a post-spawning dispersal to feeding grounds and a fall migration to overwintering sites (McKinley et al. 1998).

Only sturgeon that are ready to spawn make a migration in the spring to the spawning grounds (Rusak and Mosindy 1997). Movements begin when the ice has melted from the rivers, although some fish have been observed travelling under the ice (Scott and Crossman 1973). Water temperature appears to be the dominating factor in signalling the fish. In 1990, two male radio-tagged lake sturgeon that overwintered in the Rainy River along the United States-Canada border, began upstream migrations when temperatures reached 5 - 6°C (Rusak and Mosindy 1997). In the same year, two males from the same system that overwintered in the Lake of the Woods commenced upstream movements at 12°C. In a previous year, six fish began moving upstream at temperatures of 12.5 - 13.5°C. Rusak and Mosindy (1997) concluded temperature was the environmental signal that initiates spawning migrations, as spawning dates and the timing of peak flow rates varied. The differences observed between the commencement of migration of river and lake fish was attributed to the river experiencing an earlier rise in the ambient water temperature than the lake.

Lake sturgeon spawning migrations are extensive. This species is often recorded migrating as much as 128km to suitable spawning grounds, however total movements have not been reported to exceed 400km (Scott and Crossman 1973). Based on the average length of lake sturgeon, Auer (1996a) concluded this species should be capable of travelling 1000 - 1800km, although these distances are not typically observed. Auer (1996a) summarizes the total distances travelled by lake sturgeon within various natural river systems that have not been altered by human influences. Her findings included lake sturgeon in the Bad River, Wisconsin, travelling 32km to spawn near natural waterfalls and rapids. In the Grand River, Michigan, fish migrated 64km before reaching an impassable set of rapids. Fish in the Sturgeon River, Michigan travelled 69km upstream to a set of rapids below a hydroelectric facility. Finally, a total spawning migration of 228km across Lake Winnebago and up the Wolf River was observed in the Lake Winnebago system, Wisconsin.

Lake sturgeon disperse quickly downstream after spawning. Rapid downstream movements have been noted for fish in the Wisconsin, Sturgeon, Wolf and Mattagami rivers (Auer 1996a, Auer 1999a, McKinley et al. 1998). Lake sturgeon in the Mattagami River left the spawning grounds at Cypress Falls before temperatures increased above 13°C (McKinley et al. 1998). Auer (1999a) cited a study on fish in the Wolf River in which sturgeon travelled 456km in 31 days to arrive at Lake Winnebago. Downstream movements may be an avoidance response to unfavourable conditions, including stranding and sun exposure due to decreasing river flows (Auer 1999a). In addition, fish may be migrating in search of food, as they do not feed during spawning, relying instead on lipid stores (McKinley et al. 1998).

In the fall, fish migrate short distances to previously inhabited overwintering sites. Lake sturgeon in the Rainy River and Lake of the Woods had relocated to overwintering sites by the end of October, when flows had decreased to a yearly low, and water temperatures were between 4 and 9°C (Rusak and Mosindy 1997). Movements during the winter are often reduced, with fish gathering in distinct regions of river systems (Harkness and Dymond 1961, Rusak and Mosindy 1997). Sturgeon overwintering in Lake of the Woods, Ontario limited their movement to an area of approximately 5 - 6km near the mouth of the Rainy River (Rusak and Mosindy 1997). Fish in the river had an individual average movement of 5.2km (Rusak and Mosindy 1997).

Apart from spawning migrations, lake sturgeon movements are not extensive (Scott and Crossman 1973). Threader and Brousseau (1986) studied lake sturgeon movements in the Moose River, Ontario. Eighty percent of marked fish were recaptured in the original tagging area, and the remaining fish had moved less than 5km. This pattern was also documented for fish observed in the Menominee River, Wisconsin-Michigan where only 2% of fish tagged in the White Rapids section of the river had moved to another area of the river and only 1.6% of fish tagged in the Grand Rapids section had moved (Thuemler 1997). Mark-recapture methods in the St. Lawrence and Ottawa rivers revealed fish in this system are also relatively sedentary, as a high proportion of fish were recaptured within 5km of the original tagging site (Fortin et al. 1993). Although the majority of fish in this system were sedentary, a small number of fish were located at the opposite end of the study site, a distance of approximately 225km. Despite the observed sedentary habits of most sturgeon within these systems, it is reported that fish may travel long distances if necessary to find ideal habitats (Houston 1987).

Rusak and Mosindy (1997) observed spring and summer movements of non-spawning fish in the Rainy River and Lake of the Woods. Fish that had overwintered in the river moved into Lake of the Woods at the beginning of May, while fish that had remained in the lake overwinter began moving within the lake in mid-May. Movements were initiated by increasing water temperatures and velocities.

2.4.1.1. Movement Rates

The rate of lake sturgeon movements in the Groundhog and Mattagami rivers decreased in the summer months when temperatures increased (McKinley et al. 1998). During July and August, water temperatures in these rivers reached 17 - 23°C. McKinley and Power (1991) also recorded a decline in movement rates as water temperatures approached 19°C in the Mattagami River. Lake sturgeon were calculated to swim $75 cm \cdot s^{-1}$ at water temperatures of 5°C, and only $30 cm \cdot s^{-1}$ when water temperatures were increased to 19°C. Contrary to these findings, Rusak and Mosindy (1997) found lake sturgeon in the Lake of the Woods and the Rainy River, to have a decreased rate of movement when temperatures decreased in the winter. Fish moved an average of $1.10 km \cdot d^{-1}$ during the winter. The highest rates of movement were observed in the spring, as water temperatures increased and long-distance movement was initiated (Rusak and Mosindy 1997). Spring rates averaged $0.84 km \cdot d^{-1}$. Seasonal discrepancies in rates of movement may be attributed to monitoring frequencies, as fish in the Lake of the Woods and Rainy River were only tracked once a month during the winter, but twice per week in the spring (Rusak and Mosindy 1997).

2.4.2 SHOVELNOSE STURGEON

Although shovelnose sturgeon are capable of migrating long distances, telemetry data suggests they are predominantly a sedentary species (Hurley et al. 1987, Curtis et al. 1997). Shovelnose sturgeon in the Mississippi and Missouri rivers typically exhibit limited movement (Hurley et al. 1987). Fish movements were reduced during the winter, moving on average less than a total of 2km (Quist et al. 1999). However, migrations upstream by shovelnose sturgeon are made for the purposes of spawning (Becker 1983). Seasonal migrations in the Missouri River have been extensive, with total movements cited to be in excess of 480km from the tagging site (Modde and Schmulbach 1977).

2.4.2.1 Movement Rates

The greatest rate of movement by radio-tagged fish in the upper Mississippi River occurred in May, at a speed of $0.60 km \cdot d^{-1}$ (Hurley et al. 1987). The rate of movement decreased in June, July and August, at 0.21, 0.28 and $0.17 km \cdot d^{-1}$ respectively. These observations are similar to those made in a study of shovelnose sturgeon in the Missouri River, in which movement rates of fish were greatest during the spawning season, and lower in the spring and summer (Hurley et al. 1987).

2.4.3 PADDLEFISH

Paddlefish travel extensively within river systems (Moen et al. 1992, Zigler et al. 1999). Paddlefish are mobile throughout the year, and extensive movements are not exclusively related to spawning migrations (Rosen et al. 1982, Zigler et al. 1999). In the Missouri River, Gavins Point Dam restricts upstream movement (Rosen et al. 1982). Upstream movement was limited in fish released within 71km of the dam, averaging 20km. Downstream movement was unlimited, and fish travelled an average of 147km (Rosen et al. 1982). Five fish were reported to move over 200 km within 3 -

8 months. Paddlefish within the upper Mississippi River travelled up to 11.5km within a day, with speeds estimated up to 1.55km·h^{-1}, although movement among individuals varied greatly (Zigler et al. 1999).

Rosen et al. (1982) discusses the importance of paddlefish mobility in relation to habitat requirements. Slow flowing water is required for paddlefish growth and development, while spawning requires habitats that contain fast-flowing, deep water with gravel substrates (Rosen et al. 1982). The distance between these suitable habitats necessitates long distance movements by paddlefish.

2.4.4 WHITE STURGEON

White sturgeon are capable of travelling long distances within short periods of time (Haynes et al. 1978, North et al. 1993). Between June and October, radio-tagged white sturgeon in the mid-Columbia River moving upstream generally travelled 3 - 12km per week, while fish moving downstream travelled 16 or more kilometres per week (Haynes et al. 1978). The total distance travelled by these fish ranged from 8.0 - 87.0km (Haynes et al. 1978). North et al. (1993) found the average distance moved by white sturgeon in three lower Columbia River reservoirs to be 8.1km, based on mark-recapture methods. Some fish were reported to have travelled up to 152 km. Although white sturgeon are capable of long distance movements, Haynes et al. (1978) found 14 of 26 white sturgeon in the unimpounded mid-Columbia River to move less than 0.8 km from their release points. The majority of movements occurred from June to October, at which point water temperatures increased to 15°C and movement ceased (Haynes et al. 1978). A dormant period extended from November to May during which sturgeon movement was less than 0.2km (Haynes et al. 1978).

Dams restrict movement within reservoirs on the lower Columbia River but on occasion fish were observed passing dams (North et al. 1993). Only 27 of 635 recaptured fish had bypassed dams. Ninety-six percent of the sturgeon able to pass dams were travelling downstream (North et al. 1993). Navigational locks, spillways, fishladders and turbines are available to facilitate the downstream passage of dams (North et al. 1993). Navigational locks and fishladders designed for salmonid use are available for upstream movements of sturgeon (North et al. 1993). Impoundments on the Snake and Columbia rivers were not constructed with adequate fishways, barring any upstream or downstream movement between various sections (Cochnauer et al. 1985).

2.4.5 SHORTNOSE STURGEON

Shortnose sturgeon exhibit complex migratory strategies that vary depending on the river system. Migration patterns of this species involves movement between distinct feeding, spawning and overwintering sites.

Spawning migrations generally occur in the spring, originating from overwintering locations (Kieffer and Kynard 1993, O'Herron et al. 1993) and are initiated by changes in water temperature. Pre-spawning fish in the Merrimack River, Massachusetts began an upstream migration to spawning grounds at an average temperature of 7.1°C (Kieffer and Kynard 1993). Movements in the Connecticut River began when temperatures reached 2.5°C (Buckley and Kynard 1985a). The spawning migration of fish in the Connecticut River was unique, as 30 - 40% of radio-tagged fish migrated in the fall, overwintering in the spawning area below Holyoke Dam (Buckley and Kynard

1985a). Fall migrations began from Agawam, a summering area, with an increase in discharge and decrease in temperature from 19 to 11°C. Only 17% of fish migrated to Holyoke Dam in the spring from overwintering sites.

Post-spawning fish in the Connecticut and Merrimack rivers moved downstream when water temperatures increased to approximately 14°C (Buckley and Kynard 1985a, Kieffer and Kynard 1993). Within two weeks of spawning, shortnose sturgeon in the Savannah River had moved downstream into brackish water (Hall et al. 1991). Shortnose sturgeon in the Merrimack River exhibited two patterns of dispersal. Over three years, ten fish moved to Amesbury reach, a freshwater portion of the river, while the remaining ten travelled to saline areas after spawning (Kieffer and Kynard 1993). In the Connecticut River, fish moved to three areas after spawning, the freshwater portions at Agawan and Hartford, and the estuarine areas of the lower river (Buckley and Kynard 1985a). Fish in the Delaware River travelled downstream to the Philadelphia area, remaining in freshwater (O'Herron et al. 1993). Essentially, the final destination of post-spawning migrations varies with individuals within a river system.

In late spring, fish congregate at feeding areas. During the summer months, shortnose sturgeon moved extensively throughout Montsweag Bay, a Maine estuary (McCleave et al. 1977) with movements presumably being related to feeding behaviours. In the fall, fish move to overwintering sites. Sturgeon in the Delaware River relocated upstream to an area off Duck Island to overwinter in freshwater (O'Herron et al. 1993). Fish in the Merrimack River overwintered at various locations throughout the Amesbury reach (Kieffer and Kynard 1993). Movements during the winter months are limited.

2.4.5.1 Movement Rates

Shortnose sturgeon have the ability to move quickly (Buckley and Kynard 1985a). Telemetry observations indicated shortnose sturgeon in the Connecticut River were moving at an average rate of $16 km \cdot d^{-1}$ upriver, and $10.5 km \cdot d^{-1}$ downriver (Buckley and Kynard 1985a). A similar rate of movement was recorded for shortnose sturgeon in a Maine estuary, where fish were reported to travel at an average of $22 km \cdot d^{-1}$ (McCleave et al. 1977). Contrary to these findings, Dadswell (1979) found shortnose sturgeon in a Saint John River estuary to move an average of $4 km \cdot d^{-1}$. Discrepancies between movement rates were attributed to variations in river flow and sampling techniques. Fish in the Connecticut River and in the Maine estuary were tracked daily using telemetry, while fish in the St. John River estuary were sampled over extended periods of time by mark and recapture methods.

2.4.6 ATLANTIC STURGEON

Information on Atlantic sturgeon movements is limited. This species is reported to migrate extensively along the Atlantic coast (Bain 1997, Smith and Clugston 1997). Atlantic sturgeon tagged in the Hudson River were recovered in marine waters and in the mouths of rivers from North Carolina north to Massachusetts (Bain 1997). Adult migrations are likely similar, if not more extensive, than juveniles (Smith et al. 1984).

Although juvenile Atlantic sturgeon remain in their natal riverine system for 1 - 6 years, they eventually migrate to the ocean over time. Movement of juveniles to deeper portions of rivers has been reported in the fall and winter months when temperatures decrease (Smith 1985b). A decrease in temperature has also initiated movements of juveniles into saline and occasionally coastal waters (Smith 1985b). Immature fish are capable of travelling large distances within short periods. Juvenile Atlantic sturgeon have been recorded to migrate 645km, from North Carolina to Long Island, New York (Smith 1985b). This distance was travelled in 65 days, with an average rate of speed of 9.9 $km \cdot d^{-1}$ (Smith 1985b). In South Carolina, juveniles migrated 466km to Virginia and 595km to North Carolina (recapture 326 days after initial tagging) and juvenile fish in Quebec travelled 559km to Nova Scotia (Smith et al. 1984, Smith 1985b).

2.4.7 GULF STURGEON

Gulf sturgeon begin upstream movements into coastal rivers in February when temperatures increase to 16 - 19°C (Smith and Clugston 1997). In the Suwannee River, peak movements occur in late March or early April when water temperatures have increased to 20°C (Smith and Clugston 1997). As summer temperatures drop below 20°C in November, downstream migrations back to the Gulf of Mexico begin (Smith and Clugston 1997).

Gulf sturgeon radio-tagged in the upper Apalachicola River below the Jim Woodruff Lock and Dam began downstream migrations when water temperatures dropped from 32 to 23°C in late September (Wooley and Crateau 1985). Sturgeon travelled 152km over an average of 8 days (average 24.3$km \cdot d^{-1}$) to an Apalachicola tributary, the Brothers River, which is considered a fall staging area (Wooley and Crateau 1985). Sturgeon remained in this section through November until temperatures dropped to between 16 and 19°C (Wooley and Crateau 1985). Fish travelled along the Apalachicola River into Apalachicola Bay, with some continuing into the Gulf of Mexico to overwinter. Two tagged fish remained in the Apalachicola River throughout the year. Movement of fish upstream began in early April, and by the end of May all radio-tagged fish had migrated upstream to the original tagging site below the dam.

Gulf sturgeon have been observed migrating south to Florida Bay during exceptionally cold winters (Wooley and Crateau 1985). During normal winter temperatures, this movement is not observed (Wooley and Crateau 1985). It is proposed that these movements are related to the availability of new shallow-water feeding habitats generated by cooler water temperatures (Wooley and Crateau 1985).

2.5 Management

Sturgeon and paddlefish are increasingly vulnerable to anthropogenic influences. Dams, human exploitation and pollution are reducing populations of these species to dangerous levels. Furthermore, sturgeon and paddlefish are slow-growing, late-maturing fish with non-annual spawning, and therefore populations that have declined due to human impacts are slow to recover (Ono et al. 1983).

The construction of dams has numerous detrimental impacts on sturgeon and paddlefish. They obstruct the passage of fish to ideal habitats, modify the genetic variability of fish populations, and alter water levels and rates of flow necessary for various life stages of individual species (Ferguson et al. 1993, Boreman 1997). The blockage of migratory routes to ideal habitats, especially to spawning grounds, is of primary concern. Shortnose sturgeon migrations in rivers without barriers are extensive, with total migrations generally exceeding 200km, depending on the river system (Kynard 1997). Dams interrupt this naturally extensive migratory behaviour in this species, and fish are not able to spawn on historical spawning grounds. Lake sturgeon in the Wolf River, Wisconsin, are forced to spawn below the Shawano dam, 201km upstream of Lake Winnebago. Fish in this system were observed spawning below the dam, depositing large amounts of eggs, some in masses 15cm thick (Kempinger 1988). Auer (1996a) suggests eggs concentrated in such large quantities may be detrimental to this species, as the success of hatching may be poor, survival may be reduced, and there may be increased vulnerability to predation or disease.

An additional concern with the construction of dams is the impact on the genetic integrity of fish populations. Ferguson et al. (1993) observed considerable gene flow among sampling locations for lake sturgeon in the Moose River basin. Dams essentially segment fish into isolated populations, with the potential result of increased inbreeding and random genetic drift, which may lead to a loss of genetic variability for fish in this and similar systems (Ferguson and Duckworth 1997). Populations with reduced genetic variability are more vulnerable to environmental stresses, and therefore have a decreased chance of survival.

A final concern related to the construction of dams is the alteration of water levels and the effect of varying flow regimes on essential habitats. Varying water levels and velocity has been reported to affect the spawning success of lake sturgeon in various systems. Daily fluctuations in river flow velocity had a negative impact on the reproductive success of lake sturgeon in the Sturgeon River, Michigan, with greater success at flows near run of the river (Auer 1996b). Alterations in flow regimes will affect the substrate composition necessary for successful spawning, egg incubation and prey availability. The abundance of macroinvertebrates as a food source is directly related to the composition of the substrate. An alteration in substrate composition will affect food availability, which could result in a reduction of the nutritional status of fish (McKinley et al. 1993). In addition, under conditions of reduced flow, embryos may be exposed to varying water temperatures, oxygen concentrations and desiccation (Ferguson and Duckworth 1997). Furthermore, juveniles risk predation or oxygen and temperature stress if confined to shallow pools while mortality of adults may occur due to anoxic conditions in shallow pools (Ferguson and Duckworth 1997).

Populations of lake sturgeon in Ontario have been reduced substantially due to human exploitation (Brousseau 1987). Brood stocks are removed quickly by commercial fisheries under current regulations. Consequently, young, immature fish are caught, leading to a reduction in the standing stock (Brousseau 1987). The harvest of sturgeon and paddlefish must be restricted to ensure protection of viable populations. Stocks of lake sturgeon in the White Rapids section of the Menominee River are currently maintained by harvest regulations, however further restrictions may be necessary if exploitation rates increase beyond the present level (Thuemler 1997). In the Grand Rapids section of the Menominee River, the standing stock of sturgeon is lower (Thuemler 1997). The number of legal size sturgeon is decreasing, and exploitation rates are high. Harvest restrictions must be re-evaluated in this section of the river (Thuemler 1997). The highest sturgeon standing stock in the Menominee occurs in the lowest section of the river (Thuemler 1997). Although an improvement in water quality enabled a stock expansion, the exploitation rate is high and therefore harvest should be restricted (Thuemler 1997).

Habitat protection is a primary factor in the conservation and regeneration of sturgeon and paddlefish populations. For future conservation efforts, upstream and downstream fishways are required for sturgeon and paddlefish to migrate to spawning and feeding grounds. The provision of artificial spawning habitats and the establishment of minimum flow regimes are also necessary for the successful recovery of these unique fish.

2.6 References

Auer, N. A. 1996a. Importance of habitat and migration to sturgeons with emphasis on lake sturgeon. Canadian Journal of Fisheries and Aquatic Science 53: 152-160.

Auer, N. A. 1996b. Response of spawning lake sturgeons to changes in hydroelectric facility operation. Transactions of the American Fisheries Society 125: 66-77.

Auer, N. A. 1999a. Population characteristics and movements of lake sturgeon in the Sturgeon River and Lake Superior. Journal of Great Lakes Research 25: 282-293.

Auer, N. A. 1999b. Lake sturgeon: a unique and imperilled species in the Great Lakes. Pages 515-536 in W. W. Taylor and C. P. Ferreri, editors. Great Lakes Fisheries Policy and Management. Michigan State University Press, Michigan.

Bain, M. B. 1997. Atlantic and shortnose sturgeons of the Hudson River: common and divergent life history attributes. Environmental Biology of Fishes 48: 347-358.

Becker, G. C. 1983. Fishes of Wisconsin. Madison: University of Wisconsin Press.

Bemis, W. E. and Kynard, B. 1997. Sturgeon rivers: an introduction to acipenseriform biogeography and life history. Environmental Biology of Fishes 48: 167-183.

Birstein, V. J. 1993. Sturgeons and paddlefishes: Threatened fishes in need of conservation. Conservation Biology 7: 773-787.

Bogue, M. B. 2000. Fishing the Great Lakes: an environmental history. University of Wisconsin Press. Madison, Wisconsin.

Boreman, J. 1997. Sensitivity of North American sturgeons and paddlefishes to fishing mortality. Environmental Biology of Fishes 48: 399-405.

Brousseau, C. S. 1987. The lake sturgeon (*Acipenser fulvescens*) in Ontario. Proceedings of a Workshop on the Lake Sturgeon (*Acipenser fulvescens*) , Ont.Fish.Tech.Rep.Ser.No.23 2-9.

Buckley, J. and Kynard, B. 1985a. Yearly movements of shortnose sturgeons in the Connecticut River. Transactions of the American Fisheries Society 114: 813-820.

Buckley, J. and Kynard, B. 1985b. Habitat use and behaviour of pre-spawning and spawning shortnose sturgeon, *Acipenser brevirostrum*, in the Connecticut River. In North American sturgeons: biology and aquaculture potential. Edited by F.P.Binkowski and S.I.Doroshov.Dr.W.Junk Publishers, Dordrecht, The Netherlands. pp. 111-117.

Campbell, R. R. 1991. Rare and endangered fishes and marine mammals of Canada: COSEWIC Fish and Marine Mammal Subcommittee Status Reports: VII. The Canadian Field-Naturalist 105: 151-156.

Carlson, D. M., Pflieger, W. L., Trial, L., and Haverland, P. S. 1985. Distribution, biology and hybridization of *Scaphirhynchus albus* and *S. platorynchus* in the Missouri and Mississippi rivers. Environmental Biology of Fishes 14: 51-59.

Carlson, D. M. 1995. Lake sturgeon waters and fisheries in New York State. Journal of Great Lakes Research 21: 35-41.

Chiasson, W. B., Noakes, D. L. G., and Beamish, F. W. H. 1997. Habitat, benthic prey, and distribution of juvenile lake sturgeon (*Acipenser fulvescens*) in northern Ontario rivers. Canadian Journal of Fisheries and Aquatic Science 54: 2866-2871.

Cochnauer, T. G., Lukens, J. R., and Partridge, F. E. 1985. Status of white sturgeon, *Acipenser transmontanus*, in Idaho. In North American sturgeons: biology and aquaculture potential. Edited by F.P.Binkowski and S.I.Doroshov.Dr.W.Junk Publishers, Dordrecht, The Netherlands. pp. 127-133.

Collins, M. R. and Smith, T. I. J. 1997. Distributions of shortnose and Atlantic sturgeons in South Carolina. North American Journal of Fisheries Management 17: 995-1000.

Cook, F. R. and Muir, D. 1984. The committee on the status of endangered wildlife in Canada (COSEWIC): history and progress. Canadian Field-Naturalist 98: 63-70.

Crance, J. H. 1987. Habitat suitability index curves for paddlefish, developed by the Delphi technique. North American Journal of Fisheries Management 7: 123-130.

Curtis, G. L., Ramsey, J. S., and Scarnecchia, D. L. 1997. Habitat use and movements of shovelnose sturgeon in Pool 13 of the upper Mississippi River during extreme low flow conditions. Environmental Biology of Fishes 50: 175-182.

Dadswell, M. J. 1979. Biology and population characteristics of the shortnose sturgeon, *Acipenser brevirostrum* LeSuer 1818 (Osteichthyes: Acipenseridae), in the Saint John River Estuary, New Brunswick, Canada. Canadian Journal of Zoology 57: 2186-2210.

Dadswell, M. J. 1984. Status of the shortnose sturgeon, *Acipenser brevirostrum*, in Canada. Canadian Field-Naturalist 98: 75-79.

Ferguson, M. M., Bernatchez, L., Gatt, M., Konkle, B. R., Lee, S., Malott, M. L., and McKinley, R. S. 1993. Distribution of mitochondrial DNA variation in lake sturgeon (*Acipenser fulvescens*) from the Moose River basin, Ontario, Canada. Journal of Fish Biology 43: 91-101.

Ferguson, M. M. and Duckworth, G. A. 1997. The status and distribution of lake sturgeon, *Acipenser fulvescens*, in the Canadian provinces of Manitoba, Ontario and Quebec: a genetic perspective. Environmental Biology of Fishes 48: 299-309.

Folz, D. J. and Meyers, L. S. 1985. Management of the lake sturgeon, *Acipenser fulvescens*, population in the Lake Winnebago system, Wisconsin. North American sturgeons: biology and aquaculture potential. 135-146.

Fortin, R., Mongeau, J.-R., Desjardins, G., and Dumont, P. 1993. Movements and biological statistics of lake sturgeon (*Acipenser fulvescens*) populations from the St. Lawrence and Ottawa River system, Quebec. Canadian Journal of Zoology 71: 638-650.

Fox, D. A., Hightower, J. E., and Parauka, F. M. 2000. Gulf sturgeon spawning migration and habitat in the Choctawhatchee River system, Alabama-Florida. Transactions of the American Fisheries Society 129: 811-826.

Graham, K. 1997. Contemporary status of the North American paddlefish, *Polyodon spathula*. Environmental Biology of Fishes 48: 279-289.

Hall, J. W., Smith, T. I. J., and Lamprecht, S. D. 1991. Movements and habitats of shortnose sturgeon, *Acipenser brevirostrum* in the Savannah River. Copeia 1991: 695-702.

Harkness, W. J. K. and Dymond, J. R. 1961. The lake sturgeon: The history of its fishery and problems of conservation. Ontario Department of Lands and Forests. Fish and Wildlife Branch, Maple, Ontario, Canada.

Hastings, R. W., O'Herron, I. J. C., Schick, K., and Lazzari, M. A. 1987. Occurrence and distribution of shortnose sturgeon, *Acipenser brevirostrum*, in the upper tidal Delaware River. Estuaries 10: 337-341.

Haynes, J. M., Gray, R. H., and Montgomery, J. C. 1978. Seasonal movements of white sturgeon (*Acipenser transmontanus*) in the Mid-Columbia River. Transactions of the American Fisheries Society 107: 275-280.

Houston, J. J. 1987. Status of the lake sturgeon, *Acipenser fulvescens*, in Canada. The Canadian Field-Naturalist 101: 171-185.

Houston, J. J. 1988. Status of the green sturgeon, *Acipenser medirostris*, in Canada. The Canadian Field-Naturalist 102: 286-290.

Hurley, S. T., Hubert, W. A., and Nickum, J. G. 1987. Habitats and movements of shovelnose sturgeon in the upper Mississippi River. Transactions of the American Fisheries Society 116: 655-662.

International Union for Conservation of Nature. 2000. 2000 IUCN Red List of Threatened Animals. [Online] Available: http://www.redlist.org [January 2, 2001].

Kallemeyn, L. 1983. Status of the pallid sturgeon. Fisheries 8: 3-9.

Keenlyne, K. D. and Jenkins, L. G. 1993. Age at sexual maturity of the pallid sturgeon. Transactions of the American Fisheries Society 122: 393-396.

Keenlyne, K. D. 1997. Life history and status of the shovelnose sturgeon, *Scaphirhynchus platorynchus*. Environmental Biology of Fishes 48: 291-298.

Kempinger, J. J. 1988. Spawning and early life history of lake sturgeon in the Lake Winnebago system, Wisconsin. American Fisheries Society Symposium 5: 110-122.

Kieffer, M. C. and Kynard, B. 1993. Annual movements of shortnose and Atlantic sturgeons in the Merrimack River, Massachusetts. Transactions of the American Fisheries Society 122: 1088-1103.

Kynard, B. 1997. Life history, latitudinal patterns, and status of the shortnose sturgeon, *Acipenser brevirostrum*. Environmental Biology of Fishes 48: 319-334.

LaHaye, M., Branchard, A., Gendron, M., Verdon, R., and Fortin, R. 1992. Reproduction, early life history, and characteristics of the spawning grounds of the lake sturgeon (*Acipenser fulvescens*) in Des Prairies and L'Assomption rivers, near Montreal, Quebec. Canadian Journal of Zoology 70: 1681-1689.

Lane, E. D. 1991. Status of the white sturgeon, *Acipenser transmontanus*, in Canada. The Canadian Field-Naturalist 105: 161-168.

Lazzari, M. A., O'Herron, I. J. C., and Hastings, R. W. 1986. Occurrence of juvenile Atlantic Sturgeon, *Acipenser oxyrinchus*, in the upper tidal Delaware River. Estuaries 48: 356-361.

LeBreton, G. T. O. and Beamish, F. W. H. 1998. The influence of salinity on ionic concentrations and osmolarity of blood serum in lake sturgeon, *Acipenser fulvescens*. Environmental Biology of Fishes 52: 477-482.

Lee, D. S., Gilbert, C. R., Hocutt, C. H., Jenkins, R. E., McAllister, D. E., and Stauffer, J. R., Jr. 1980. Atlas of North American Freshwater Fishes. North Carolina State Museum of Natural History Publication Number 1980-12. North Carolina Biological Survey. Raleigh, N.C.

Mayden, R. L. and Kuhajda, B. R. 1996. Systematics, taxonomy, and conservation status of the endangered Alabama shovelnose sturgeon, *Scaphirhynchus suttkusi* Williams and Clemmer (Actinopterygii, Acipenseridae). Copeia 1996: 241-273.

Mayden, R. L. and Kuhajda, B. R. 1997a. Threatened fishes of the world: *Scaphirhynchus albus* (Forbes and Richardson, 1905) (Acipenseridae). Environmental Biology of Fishes 48: 420-421.

Mayden, R. L. and Kuhajda, B. R. 1997b. Threatened fishes of the world: *Scaphirhynchus suttkusi* Williams and Clemmer, 1991 (Acipenseridae). Environmental Biology of Fishes 48: 418-419.

McCleave, J. D., Fried, S. M., and Towt, A. K. 1977. Daily movements of shortnose sturgeon, *Acipenser brevirostrum*, in a Maine Estuary. Copeia 1977: 149-157.

McKinley, R. S. and Power, G. 1991. Measurement of activity and oxygen consumption for adult lake sturgeon (*Acipenser fulvescens*) in the wild using radio-transmitted EMG signals. Wildlife Telemetry, remote monitoring and tracking of animals 307-318.

McKinley, R. S., Singer, T. D., Ballanytne, J. S., and Power, G. 1993. Seasonal variation in plasma nonesterified fatty acids of lake sturgeon (*Acipenser fulvescens*) in the vicinity of hydroelectric facilities. Canadian Journal of Fisheries and Aquatic Science 50: 2440-2447.

McKinley, R. S., Van Der Kraak, G., and Power, G. 1998. Seasonal migrations and reproductive patterns in the lake sturgeon, *Acipenser fulvescens*, in the vicinity of hydroelectric stations in northern Ontario. Environmental Biology of Fishes 51: 245-256.

Modde, T. and Schmulbach, J. C. 1977. Food and feeding behaviour of the shovelnose sturgeon, *Scaphirhynchus platorhynchus*, in the unchannelized Missouri River, South Dakota. Transactions of the American Fisheries Society 106: 602-608.

Moen, C. T., Scarnecchia, D. L., and Ramsey, J. S. 1992. Paddlefish movements and habitat use in Pool 13 of the upper Mississippi River during abnormally low river stages and discharges. North American Journal of Fisheries Management 12: 744-751.

Moser, M. L. and Ross, S. W. 1995. Habitat use and movements of shortnose and Atlantic sturgeons in the lower Cape Fear River, North Carolina. Transactions of the American Fisheries Society 124: 225-234.

Musick, J. A., Harbin, M. M., Berkeley, G. H., Burgess, G. H., Eklund, A. M., Findley, L., Gilmore, R. G., Golden, J. T., Ha, D. S., Huntsman, G. R., McGovern, J. C., Parker, S. J., Poss, S. G., Sala, E., Schmidt, T. W., Sedberry, G. R., Weeks, H., and Wright, S. G. 2000. Marine, estuarine, and diadromous fish stocks at risk of extinction in North America (Exclusive of Pacific salmonids). Fisheries 25: 6-30.

Noakes, D. L. G., Beamish, F. W. H., and Rossiter, A. 1999. Conservation implications of behaviour and growth of the lake sturgeon, *Acipenser fulvescens*, in northern Ontario. Environmental Biology of Fishes 55: 135-144.

North, J. A., Beamesderfer, R. C., and Rien, T. A. 1993. Distribution and movements of white sturgeon in three lower Columbia River reservoirs. Northwest Science 67: 105-111.

Nowak, A. M. and Jessop, C. S. 1987. Biology and management of the lake sturgeon (*Acipenser fulvescens*) in the Groundhog and Mattagami rivers, Ontario. Proceedings of a Workshop on the Lake Sturgeon (*Acipenser fulvescens*) , Ont.Fish.Tech.Rep.Ser.No.23 20-32.

O'Herron, I. J. C., Able, K. W., and Hastings, R. W. 1993. Movements of shortnose sturgeon (*Acipenser brevirostrum*) in the Delaware River. Estuaries 16: 235-240.

Ono, D., Williams, J. D., and Wagner, A. 1983. Vanishing fishes of North America. Stone Wall Press. Washington.

Parsley, M. J., Beckman, L. G., and McCabe, Jr. G. T. 1993. Spawning and rearing habitat use by white sturgeons in the Columbia River downstream from McNary Dam. Transactions of the American Fisheries Society 122: 217-227.

Payne, D. A. 1987. Biology and population dynamics of lake sturgeon (*Acipenser fulvescens*) from the Frederick House, Abitibi and Mattagami rivers, Ontario. Proceedings of a Workshop on the Lake Sturgeon (*Acipenser fulvescens*) , Ont.Fish.Tech.Rep.Ser.No.23 10-19.

Pitman, V. M. and Parks, J. O. 1994. Habitat use and movement of young paddlefish (*Polyodon spathula*). Journal of Freshwater Ecology 9: 181-189.

Quist, M. C., Tillma, J. S., Burlingame, M. N., and Guy, C. S. 1999. Overwinter habitat use of shovelnose sturgeon in the Kansas River. Transactions of the American Fisheries Society 128: 522-527.

Rosen, R. A., Hales, D. C., and Unkenholz, D. G. 1982. Biology and exploitation of paddlefish in the Missouri River below Gavins Point Dam. Transactions of the American Fisheries Society 111: 216-222.

Rusak, J. A. and Mosindy, T. 1997. Seasonal movements of lake sturgeon in Lake of the Woods and the Rainy River, Ontario. Canadian Journal of Zoology 74: 383-395.

Sandilands, A. P. 1987. Biology of the lake sturgeon (*Acipenser fulvescens*) in the Kenogami River, Ontario. Proceedings of a Workshop on the Lake Sturgeon (*Acipenser fulvescens*), Ont.Fish.Tech.Rep.Ser.No.23 33-46.

Scott, W. B. and Crossman, E. J. 1973. Freshwater fishes of Canada. Fish.Res.Board Can.Bull.184 966pp.

Smith, C. L. 1985a. The inland fishes of New York State. New York State Department of Environmental Conservation. Albany, NY.

Smith, T. I. J., Marchette, D. E., and Ulrich, G. F. 1984. The Atlantic sturgeon fishery in South Carolina. North American Journal of Fisheries Management 4: 164-176.

Smith, T. I. J. 1985b. The fishery, biology and management of Atlantic sturgeon, *Acipenser oxyrinchus*, in North America. Environmental Biology of Fishes 14: 61-72.

Smith, T. I. J. and Clugston, J. P. 1997. Status and management of Atlantic sturgeon, *Acipenser oxyrhinchus*, in North America. Environmental Biology of Fishes 48: 335-346.

Southall, P. D. and Hubert, W. A. 1984. Habitat use by adult paddlefish in the upper Mississippi River. Transactions of the American Fisheries Society 113: 125-131.

Taubert, B. D. 1980. Reproduction of shortnose sturgeon (*Acipenser brevirostrum*) in Holyoke Pool, Connecticut River, Massachusetts. Copeia 1980: 114-117.

Threader, R. W. and Brousseau, C. S. 1986. Biology and management of the lake sturgeon in the Moose River, Ontario. North American Journal of Fisheries Management 6: 383-390.

Thuemler, T. F. 1985. The lake sturgeon, *Acipenser fulvescens*, in the Menominee River, Wisconsin-Michigan. Environmental Biology of Fishes 14: 73-78.

Thuemler, T. F. 1997. Lake sturgeon management in the Menominee River, a Wisconsin-Michigan boundary water. Environmental Biology of Fishes 48: 311-317.

Williams, J. E., Johnson, J. E., Hendrickson, D. A., Contreras-Balderas, S., Williams, J. D., Navarro-Mendoza, M., McAllister, D. E., and Deacon, J. E. 1989. Fishes of North America endangered, threatened, or of special concern: 1989. Fisheries 14: 2-20.

Wooley, C. M. 1985. Evaluation of morphometric characters used in taxanomic separation of Gulf of Mexico sturgeon, *Acipenser oxyrinchus desotoi*. North American sturgeons: biology and aquaculture potential. 97-103.

Wooley, C. M. and Crateau, E. J. 1985. Movement, microhabitat, exploitation, and management of Gulf of Mexico sturgeon, Apalachicola River, Florida. North American Journal of Fisheries Management 5: 590-605.

Zigler, S. J., Dewey, M. R., and Knights, B. C. 1999. Diel movement and habitat use by paddlefish in Navigation Pool 8 of the upper Mississippi River. North American Journal of Fisheries Management 19: 180-187.

CHAPTER 3

ENVIRONMENTAL REQUIREMENTS, PREFERENCES, AND TOLERANCE LIMITS OF NORTH AMERICAN STURGEONS

JOSEPH J. CECH JR.[1] AND SERGE I. DOROSHOV[2]
[1]Department of Wildlife, Fish, and Conservation Biology; and Centre for Aquatic Biology and Aquaculture
[2]Department of Animal Science; and Centre for Aquatic Biology and Aquaculture
University of California, 1 Shields Avenue, Davis, CA 95616 USA

3.1 Introduction

Sturgeons are "living fossils" (Gardiner 1984) that have enormous value for studying vertebrate evolution, including physiological adaptations to the environment. Regrettably, human impacts have reduced their stocks to levels near extinction (Birstein et al. 1997). A more complete understanding of sturgeon environmental requirements and preferences will help protect their natural reproduction, properly manage the existing small fisheries, and develop sustainable aquaculture. The three genera (including paddlefish) of North American Acipenseriformes live in an amazing variety of habitats. Although all of them spawn in fresh water, the various species (n = 8-9), depending on where taxonomic lines are drawn) can be exposed to a wide range of light, environmental temperature, dissolved gas (oxygen and carbon dioxide), salinity, depth, and velocity conditions (Table 3-1). This chapter reviews these species' environmental requirements, preferences, and tolerance limits.

3.2 North American Species

Eight species of Acipenseriformes inhabit rivers, lakes, estuaries, and coastal waters of the North America between the sub arctic and subtropics (Table 3-1). Four species are amphihaline (live in fresh and seawater), including anadromous Atlantic and green sturgeons and semi-anadromous shortnose and white sturgeons. The Atlantic and green sturgeons use freshwater habitat for spawning and for the first years of life (they do not have resident, freshwater populations); white and shortnose sturgeons reside mainly in the river estuaries and adjacent coastal areas, although the white sturgeon can migrate up to 1,000km along the Pacific coast (Kohlhorst et al. 1991). Four other North American species: lake, pallid, shovelnose sturgeons, and paddlefish, are inland freshwater fish, although lake sturgeon may occur in a tidal zone of the lower river

G.T.O. LeBreton et al. (eds.), Sturgeons and Paddlefish of North America, 73–86.
© 2004 *Kluwer Academic Publishers. Printed in the Netherlands.*

reaches. Sturgeons are long-lived and some species can reach very large sizes (Table 3-1).

Table 3- 1. Distributions and characteristics of North American sturgeon and paddlefish. Habitat: seawater (SW), brackish water (BW), freshwater (FW). Maximum age, length, and weight from Carey & Judge (2000, lake sturgeon), Cailliet et al. *(2001, white sturgeon), and the FishBase (for other species)*

Species	Habitat	Age (y)	Length (cm)	Weight (kg)	Status
Atlantic Coast (from Hudson Bay to Mexican Gulf)					
Atlantic & Gulf sturgeon *Acipenser oxyrinchus & A. o. desotoi*	SW, BW, FW	60	430	368	Fishery moratorium Gulf -*threatened*
Shortnose sturgeon *A. brevirostrum*	BW, FW	67	143	23	*Endangered*
Pacific Coast (from Alaska to California)					
White sturgeon *A. transmontanus*	SW, BW, FW	104	610	816	Harvested, One stock *endangered*
Green sturgeon *A. medirostris*	SW, BW, FW		213	159	Harvested
Hudson-James Bay, St. Lawrence-Great Lakes, Mississippi-Missouri drainages					
Lake sturgeon *A. fulvescens*	FW, BW	152	274	125	Harvested
Shovelnose *Scaphirhynchus platorynchus*	FW		100	5	Harvested
Pallid sturgeon *S. albus*	FW	41	168	45	*Endangered*
Paddlefish *Polyodon spathula*	FW	30	221	91	Harvested

Despite their high fecundity (up to 2 million eggs in Atlantic sturgeon), they have low, overall reproductive rates due to late sexual maturity, e.g. 10-25 years in females, and long breeding intervals (Boreman 1997). All the species spawn in rivers broadcasting their gametes on the river current, and they require specific water velocities for egg dispersal, and a hard spawning substrate for the egg attachment. Unregulated commercial fisheries for caviar, and the dams that changed hydraulic regimes of sturgeon rivers rendered the rivers unsuitable for spawning fragmented sturgeon stocks and reduced their abundance to small fractions of historic levels (Waldman and Secor 1998).

3.2.1 LIGHT

Light intensity and day length (photoperiod) influence behaviour, growth, and reproduction of fish, yet light is the least studied environmental factor in sturgeon. Sturgeons generally prefer dimly lit, moderately turbid water. Cultured white sturgeon held outdoors in clean water seek shaded tank areas when exposed to sunlight, but swim undisturbed over the entire tank area during cloudy days or during early morning and late afternoon hours. Green sturgeon are more photophobic and exhibit a pronounced nocturnal behaviour, i.e. high swimming activity at night and sluggish, demersal behaviour during the day. Lankford et al. (2003) reported a more pronounced stress response (high plasma cortisol and lactate levels) in green sturgeon stressed at night, when compared with those stressed during the daytime. Studies on retinal photoreceptors of white sturgeon revealed one type of rods and three types of cones, with blue-sensitive, green-sensitive, and red-sensitive visual pigments (Loew and Sillman 1993). Shovelnose sturgeon and paddlefish retinas had a different type of long and tightly packed rods, which may have value for light intensity discrimination and orienting behaviour in dimly lit environments (Sillman et al. 1999).

Photoperiodicity appears to regulate sturgeon growth and reproduction, in a manner similar to that shown in salmonids. Juveniles of the Eurasian beluga (*Huso huso*) grew faster under a cyclic photoperiod (16L: 8D) compared to constant light (24L) condition (Trenkler and Semenkova 1995). White sturgeon, reared in outdoor tanks and raceways under year-round, constant temperature and feeding regime conditions, grew significantly faster during spring and summer compared to fall and winter (S. Doroshov, University of California, and R. Schneider, Clear Springs Food Co., Idaho, unpublished observations). All of the North American Acipenseriformes, studied so far, are long-day breeders, with a spawning season approximately between the spring equinox and summer solstice with the spawning run peak often coinciding with the maximal river flow. The majority of the Eurasian species also spawn during spring and summer (Dettlaff et al. 1993). However, there are ecotypes (stellate sturgeon in Kura and Sefid-Rud rivers, South Caspian Sea) and species (Chinese sturgeon in the Yangtze River) that have well-documented fall spawning during the September-November period, associated with river flooding (Berg 1948, Wei et al. 1997).

3.2.2 ENVIRONMENTAL TEMPERATURE

Environmental temperature controls metabolism, growth and reproduction in ectothermic fish (Brett 1979). Fishes have preferred and optimal temperature ranges (Neill et al. 1972, Neill and Magnuson 1974) which often coincide (Jobling 1981). While the distribution range of North American Acipenseriformes extends over a zone with the temperature variation up to 30°C, they generally prefer and perform optimally under cool (e.g., <25°C) temperature conditions. McKinley et al. (1998) studied lake sturgeon movements in the Mattagami River system (northern Ontario, Canada), using radio-telemetered fish. The lake sturgeon started an upstream migration in January (water temperature: 3°C) to suspected spawning sites in May, when temperatures reached 8-10°C. They dispersed downstream as temperatures approached 13°C (McKinley et al. 1998). Gulf of Mexico sturgeon (subspecies of Atlantic sturgeon) adults and large juveniles swim upriver from the Gulf in the spring when the water temperature is 15–20°C (Chapman and Carr 1995, Sulak and Clugston 1998, Fox et al. 2000) and return to the Gulf in the fall when water temperatures range from 18 to 23°C. Adult white sturgeon migrate up the Sacramento River from San Francisco Bay during late fall and remain at low river temperatures (7-12°C) before spring spawning (optimal spawning temperature range: 14-16°C, Kohlhorst 1976). Adult female white sturgeon require exposure to cold (ca. 10°C) temperatures for oocyte development and ovulation to proceed normally (Webb et al. 1999, Linares-Casenave et al. 2002). Semi-anadromous shortnose sturgeon juveniles (<56 cm TL) moved down the Savannah River (Georgia - South Carolina, USA) when temperatures decreased to < 22°C, to Savannah Harbour and the mouths of the Front and Middle Rivers (Collins et al. 2002).

Activity and growth of young sturgeons generally increase with temperature increases until an optimal temperature is reached, usually below 25°C. Juvenile lake sturgeon tend to be more active at 19°C than at 6°C (Peake 1999) and juvenile white sturgeon were more active at 16°C than at 10 or 20°C, (near-air-saturated dissolved O_2 levels) (Table 3-2). Juvenile white sturgeon grew faster at 20-25°C than at 15°C under normoxic (Table 3-1) conditions (Cech et al. 1984), and juvenile Atlantic sturgeon reared for six months in a hatchery grew faster at 17°C, reaching up to 323mm TL, than when reared at 10°C, reaching a maximum of 156mm mean TL (Welsh et al. 2002). Finally, green sturgeon (mean wet weight: 150g) from the Klamath River grew faster at 15-19°C, compared with those at 11°C, when fed either full (repletion) or 50% of repletion rations (Mayfield and Cech 2004). Cultured white sturgeon reared in warm water of 18-22°C grew and matured much faster compared with their wild counterparts. Fully matured, cultured males and females reached 10 and 32kg live weights at ages 4 and 8 years, respectively (Doroshov et al. 1997).

Young life history stages may be the most temperature sensitive within sturgeon species. Laboratory studies show that Gulf of Mexico sturgeon eggs, embryos, and larvae have the highest survival rates in the 15 - 20°C range, and that survival decreased significantly at temperatures >25°C (Chapman and Carr 1995). The 14 - 17°C temperature range is optimal for the development and survival of white and lake sturgeons' embryos; hatching rates of both species decreased at 20°C and complete arrest of development occurred at 23°C in white sturgeon (Wang et al. 1985). Embryos of green sturgeon incubated over the 11-26°C range had high hatching rates at 11-19°C, low hatching rate at 20°C, high rates of abnormal embryos at 22°C, and did not hatch at 23°C (Van Eenennaam, Linares and Doroshov, unpublished observation). Another laboratory experiment showed that the development rate and dry matter losses during yolk resorption in larval white sturgeon increased directly with temperature within the 11-20°C range, and that larval survival decreased dramatically at temperatures >20°C (Wang et al. 1987). Larval paddlefish exhibited low survival to metamorphosis at 24°C and no survival at 28°C (Kroll et al. 1992). In general, the results of laboratory studies on early life stages are in good agreement with the field observations on spawning temperature ranges of several stocks (Kohlhorst 1976, McCabe and Tracy 1994, Bruch and Binkowski 2002, Perrin et al. 2003).

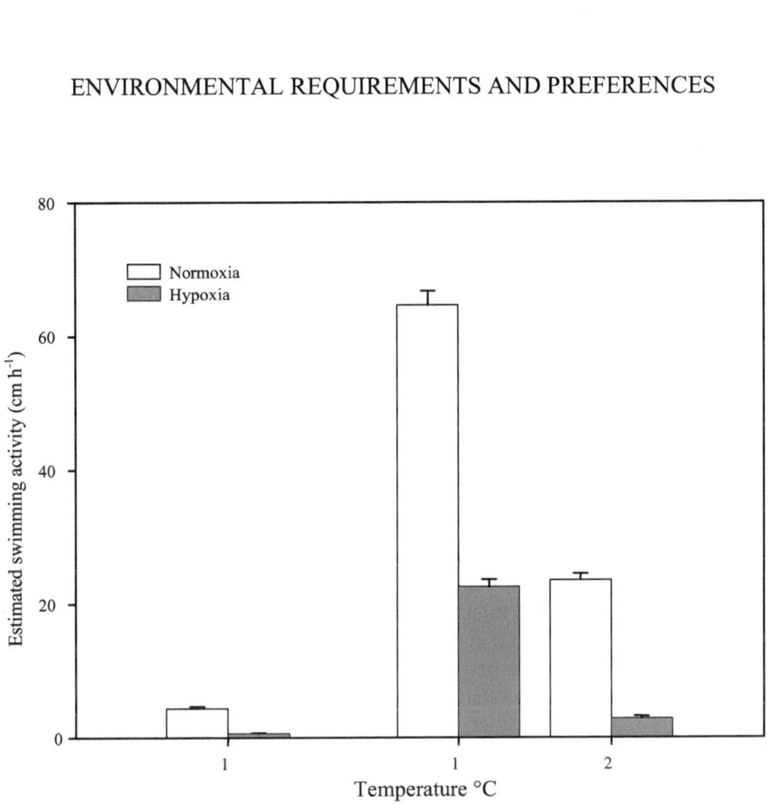

Fig. 3-1. White sturgeon estimated swimming activity at various temperatures, under normoxic and hypoxic conditions.

3.2.3 DISSOLVED OXYGEN (0_2)

Low dissolved O_2 conditions (hypoxia) typify eutrophic, nutrient-rich aquatic systems, where a high respiring biomass overwhelms O_2 inputs from inflowing water, photosynthesis, and atmospheric diffusion. Sturgeon are typically sensitive to dissolved O_2 decreases and hypoxic conditions impair their respiratory metabolism, foraging activity, and growth rates. Expressed as mg O_2 consumed per gram body weight per hour "small" (0.2g body weight at 10°C), "medium" (1.9g at 16°C), and "large" (63g at 20°C) juvenile white sturgeon all displayed significant O_2 consumption rate decreases with exposure to mild hypoxia, representing 51% of air-saturation levels, at temperatures typical for these life stages (Crocker and Cech 1997). This level of hypoxia also significantly decreased activity at these temperatures and at 25°C (Figure 3-2) (Cech et al. 1984) at least partially explaining their decreased O_2 consumption rates (Crocker and Cech 1997). In turn, the decreased activity level may account for decreased food consumption rates and/or decreased energy storage (although these were not quantified), and thus the significantly slower growth in juvenile white sturgeon found to occur under mild hypoxia (58% of air-saturated conditions) in comparison to the growth rate under normoxic conditions at 15, 20, and 25°C, (Cech et al. 1984). This hypometabolic response, seen also in adults (Burggren and Randall 1978), may benefit white sturgeon in natural habitats where decreased activity, at least in the juveniles, would decrease oxygen demand, thereby conserving oxygen resources in hypoxic habitats until conditions improved.

In contrast, activity of Atlantic and shortnose sturgeon did not change with exposure to moderate hypoxia (3mg O_2 l^{-1}) at 15°C (Daniel Baker, Dept. Biology, Univ. New Brunswick, Saint John, N.B. Canada; pers. comm.). Gill ventilatory frequencies increased by 50% in both species during this hypoxic exposure. Finally, plasma lactate concentrations, indicating a partial shift to anaerobic metabolism, increased in both species after exposure to 2mg O_2 l^{-1}) at 15°C, which is close to the minimum dissolved O_2 concentration (1mg O_2 l^{-1}) measured in the Saint John River (Daniel Baker, Dept. Biology, Univ. New Brunswick, Saint John, N.B. Canada; pers. comm.). Shortnose sturgeon died, regardless of the life stage, at oxygen concentrations < 2mg l^{-1} (Jenkins et al. 1993).

3.2.4 DISSOLVED CARBON DIOXIDE (CO_2)

Sturgeon are exposed to higher-than-atmospheric CO_2 (hypercapnic) conditions in some natural environments such as poorly circulated sloughs, and in high-density culture systems that employ O_2 injection (Wedemeyer 1996). In these culture systems, pure O_2 is injected carefully into the inflowing water, allowing the culture of higher densities of fish, which produce proportionately more CO_2. Because CO_2 solubility in water is 25-30 times greater than that of O_2, the CO_2 tends to stay dissolved, producing the hypercapnic conditions. Further, because much CO_2 hydrates in water, producing carbonic acid, water pH decreases with increasing CO_2 additions. Crocker and Cech (1996) measured significant decreases in the growth, apparently due to decreased food consumption, of juvenile white sturgeon, which were exposed to sturgeon-farm-production-tank CO_2 conditions (45-75mg CO_2 l^{-1}, pH 7.0). The decreased growth rates were attributable to the CO_2 additions, rather than to the resulting pH decreases, from experiments using mineral acid (HCl) additions to simulate the "low" pH environment (pH 7.1) without the CO_2 additions (Crocker and Cech 1996).

Hypercapnia also affected white sturgeon blood gas, cardiovascular, and acid-base characteristics. In vitro studies showed that increasing CO_2 decreased white sturgeon blood O_2 affinity. The partial pressure of dissolved O_2 at which blood was half saturated ("arterial" P50) decreased by 21mm Hg at 15°C and by 28mm Hg at 20°C) at the "high-hypercapnic" levels associated with the high-density white sturgeon culture, compared with normocapnic levels (Crocker and Cech 1998). Thus, higher O_2 partial pressures (PO_2s) are needed to adequately saturate white sturgeon blood with oxygen for respiratory purposes under high hypercapnia, especially at the higher (20°C) temperature. In vivo studies showed that hypercapnic exposure induced stress as detected by significantly increased plasma norepinephrine, epinephrine, and cortisol concentrations, in white sturgeon (Crocker and Cech 1998). Furthermore, hypercapnic exposure increased white sturgeon struggling, when compared with those exposed to normocapnia. Although hypercapnic exposure, per se, did not affect gut blood flow, the struggling events significantly decreased it, possibly affecting nutrient absorption (Crocker et al. 2000). White sturgeon arterial CO_2 conditions directly reflected those in the water, while arterial PO_2 showed either no difference or a slight increase in the hypercapnic fish, presumably due to increased gill ventilation. As expected, arterial pH significantly decreased in the hypercapnic fish. However, after two days, arterial pH started to compensate, increasing in pH towards the level found in normocapnic sturgeon, due to significant accumulation of bicarbonate ion. This response was more limited (ca. 35% of pH restoration) and slower than the response observed in most teleosts studied so far (Crocker and Cech 1998).

3.2.5 SALINITY

All bony fishes with jaws must maintain quite a tight control of total salt and water concentrations (osmoregulation) and specific ion concentrations (ionic regulation) for efficient physiological functioning (LeBreton and Beamish 1998). Regulatory mechanisms are similar in sturgeon and teleost fishes (reviewed by Cech 2000). Depending on their life history characteristics, sturgeon have been characterized as freshwater, semi-anadromous, or anadromous (Doroshov 1985). The lake sturgeon, although generally considered to be a freshwater species of the Great Lakes, Mississippi River, and Hudson Bay regions of North America, has been observed in elevated salinity regions of the St. Lawrence River and Hudson Bay (Harkness and Dymond 1961, Vladykov and Greeley 1963). Because salinities in these systems may approach 28ppt (Martini 1986), it was hypothesized that lake sturgeon possessed hyperosmoregulatory capabilities (LeBreton and Beamish 1998). Juvenile lake sturgeon showed no signs of osmoregulatory distress at salinities below 23ppt. However, after 24h exposure to 2ppt, swimming and feeding activity decreased, and smaller individuals (640-690g live weight) died with exposure to 25ppt (LeBreton and Beamish 1998). Mean plasma osmolarity and electrolyte concentrations showed sharp increases in 25-ppt-acclimated sturgeon, compared with those acclimated to lower salinities, indicating poor regulation (Table 3-2). In contrast, anadromous white sturgeon adults acclimated to freshwater or to 22-26ppt salinity showed few osmolarity or plasma electrolyte concentration differences (McEnroe and Cech 1985, Table 2). McEnroe and Cech (1985) also demonstrated that juvenile white sturgeon show increased percent survival with increased size when abruptly transferred to hypertonic water of 15ppt salinity. Whereas none of the 0.4 - 0.9g (live weight) white sturgeon survived the transfer, half of the 0.9 - 1.8g fish and all of the 4.9-9.5g fish survived the transfer (McEnroe and Cech 1985).

Juveniles of anadromous Gulf of Mexico sturgeon, of Gulf of Mexico coastal water rivers such as the Suwannee River in Florida, also showed body size-related abilities to tolerate more saline waters. At age 13 months, only sturgeon of greater than approximately 200g live weight could survive direct transfer into 25ppt salinity water, with half of the larger juveniles (>450g) surviving at 30ppt (Altinok et al. 1998). After acclimation to brackish (20ppt) and to sea water (34ppt), gulf sturgeon increased the number and size of mitochondria-rich, chloride cells in their gills (Altinok et al. 1998).After two weeks acclimation to brackish water, all of the size classes showed significantly increased plasma osmolality, and plasma Na^+ and Cl^- concentrations after subsequent acclimation to seawater conditions, compared with those acclimated to freshwater (0ppt) conditions (Altinok et al. 1998, Table 3-2).

The semi-anadromous shortnose sturgeon inhabits coastal rivers and estuaries of North America's Atlantic coast. Acoustically tagged juveniles (<56cm TL) mostly inhabited Savannah River (Georgia - South Carolina, USA) salinities between 0.1ppt and 5.4ppt except for brief high tide events when they were exposed to salinity increases to 17.6ppt (Collins et al. 2002). Comparative growth rates of age-16-month, shortnose sturgeon juveniles (mean live weight: 274g), acclimated to four salinities, showed that mean weight gain was significantly lower at 20ppt, compared with those measured at lower (0, 5, 10ppt) salinities (Jarvis et al. 2001).

Table 3- 2. Lake sturgeon (LS), Gulf sturgeon (GS), and white sturgeon (WS) sample sizes, mean (SD) weights, serum osmolarities (or osmolalities in mOsm kg^{-1} for GS) and electrolyte concentrations, in various salinities. Data are from LeBreton and Beamish (1998, LS), Altinock et al. *(1998, GS) and McEnroe and Cech (1985, WS).*

Species	Salinity	Sample size	Mass (g)	Osm (mOsm l^{-1})	Na^+ (mmol l^{-1})	Cl^- (mmol l^{-1})
LS	0	5	1195 (284)	247 (9)	126 (5)	117 (6)
LS	10	5	2275 (511)	273 (10)	141 (5)	133 (4)
LS	15	5	1225 (185)	301 (26)	155 (14)	152 (16)
LS	25	5	1228 (316)	451 (56)	224 (26)	223 (19)
GS	0	6	110-170	261	152	90
GS	34	8	110-170	306	160	158
GS	0	6	230-290	266	131	118
GS	34	8	230-290	286	149	145
GS	0	6	460-700	255	125	112
GS	34	8	460-700	291	147	145
WS	0	7	> 10,000	236 (7)	132 (4)	111 (4)
WS	22-26	7	> 10,000	249 (14)	125 (8)	122 (3)

3.2.6 SUBSTRATE/DEPTH/VELOCITY

North American sturgeon substrate, depth and velocity preferences vary with species. Peake (1999) found that juvenile lake sturgeon significantly preferred a sand substrate, compared with rock, gravel, or smooth plastic substrates in a laboratory flume. Shortnose sturgeon preferred sand and cobble substrates in lower velocity, curved reaches and deeper channels (avoiding the higher velocity, straight runs of the Connecticut and Merrimack Rivers (Massachusetts, USA), presumably to save energy by not requiring strenuous swimming (Kynard et al. 2000). Using radio and ultrasonic telemetry, Bramblett and White (2001) determined that pallid sturgeon preferred sandy substrates while shovelnose sturgeon preferred gravel and cobble substrates in the Yellowstone-Missouri River system. Although pallid sturgeon used greater depths significantly more often than the shovelnose sturgeon, their depth ranges (0.6 - 14.5m for pallids, 0.9-10.1m for shovelnose) overlapped. Water velocity preferences also overlapped for these two species, but the mean bottom velocity at pallid sturgeon locations (0.65m s^{-1}) was significantly less than that for shovelnose sturgeon (0.78m s^{-1}, Bramblett and White 2001). During low-water conditions of the Mississippi River in 1988, radio-tagged shovelnose sturgeon were found in the highest-available-velocity areas: with 0.23m s^{-1} water velocity at 5.8m depth and a sand bottom (Curtis et al. 1997). During winter in the Kansas River (Kansas, USA), radio-tagged shovelnose sturgeon preferred 1.0 - 2.0-m depths with 0.02 - 0.79m s^{-1} currents over sand substrates (Quist et al. 1999). Potential spawning areas are suggested by spring and early summer aggregations of pallid sturgeon in the lowermost 14km of the Yellowstone River. In contrast, the infrequent use of the Missouri River reach between the Fort Peck Dam and the confluence with the Yellowstone River suggest that the dam's influences (alterations of hydrograph, temperature, sediment and organic matter transport) have shifted the habitat characteristics to be similar to the less turbid conditions upstream of the pallid sturgeon's native range (Bramblett and White 2001). Missouri River fishes evolved under high-turbidity conditions (Pflieger and Grace 1987). Both pallid and shovelnose sturgeon avoid the clear water, impounded areas of the system, where visually oriented, non-native species are favoured (Bramblett and White 2001).

Substantial river flows, following natural hydrographs, typically constitute optimal sturgeon spawning habitat. Changes in hydroelectric operations below the Prickett Dam on the Sturgeon River (Michigan, USA) to near run-of-the-river flows attracted 74% more lake sturgeon (68% more females) with increased reproductive readiness to spawning sites (Auer 1996). Spawning Sacramento River white sturgeon upstream movements (up to 25km d^{-1}) were stimulated by small increases in river flow (Schaffter 1997). Similarly, Parsley and Beckman (1994) showed that decreased spring and summer river discharges from Columbia River (Washington and Oregon, USA) hydropower system operations decreased the quantity and quality of white sturgeon spawning habitat. Gulf of Mexico sturgeon eggs were found on bedrock limestone thinly overlain with fine sand and densely distributed elliptical gravel in the southern half of the Suwanee River at 2-4m depth (Sulak and Clugston 1998). Laboratory flume studies on age-0, Gulf of Mexico sturgeon revealed a preference shift from sand substrates at low velocities to cobble substrates at moderate velocities and no preferences at high velocities (Chan et al. 1997). Preferred spawning velocities and substrates vary for different species and locations, with a typical velocity range of 1-2.5m s^{-1} and a hard substrate, such as cobble, gravel, boulders and sand (McCabe and Tracy 1994, Bruch and Binkowski 2002, Perrin et al. 2003). Variations in thickness and the adhesive properties of the outer-egg chorion layer in different species may be associated with spawning habitat preferences (Deng et al. 2002).

3.2.7 NUTRITION

All fishes require sufficient quantities of high-quality food in their diets to develop, grow, and reproduce. First-feeding larval fish are the most susceptible to food deprivation. Resting-routine, O_2 consumption rates of green sturgeon from hatching to 31 days post hatch increased 5-fold under normoxic conditions, before yolk sac reserves became limiting and metabolic rates declined (Gisbert et al. 2003). These increased O_2 consumption rates were associated with organogenesis, acquisition of organ functions, and the conversion of yolk sac into new metabolizing larval biomass. If food-deprived, larvae showed body mass decreases, presumably from tissue resorption, and a 3-fold decrease in O_2 consumption rates within 15-31 days post hatch, during which most larvae died (Gisbert et al. 2001).

Sturgeon rely heavily on chemosensory mechanisms, olfaction and gustation, for finding food (Kasumyan 2002). Shortnose sturgeon fed on crustacea, insects, and small molluscs in the Hudson River estuary (Carlson and Simpson 1987). Subadult Gulf of Mexico sturgeon fed on lancelets, brachiopods, amphipods, grass shrimp, polychaetes, and gastropods (Mason and Clugston 1993). Two-year-old lake sturgeon, which were starved for 60 days, showed decreased plasma glucose, but only to day 10, followed by increased plasma concentrations of 11 free amino acids. The maintenance or increased plasma levels of glucogenic amino acids in combination with the maintenance of blood glucose concentrations indicated active gluconeogenic processes in the liver supported by muscle proteolysis (Gillis and Ballantyne 1996). Consequently, amino acids are available for the sturgeon's oxidative metabolic needs under those conditions (Gillis and Ballantyne 1996).

Juvenile white sturgeon specific growth rate increased linearly with increasing ration size, when fed tubificid worms (Cui et al. 1996). When first-feeding, these fish preferred chopped tubificid worms to a prepared diet, showing increased survival and growth rates when either consistently fed on the Tubifex diet or switched to it from the prepared diet (Lindberg and Doroshov 1986). Juvenile white sturgeon generally digest carbohydrates quite well (Herold et al. 1995) without adverse effects on growth and liver functions (Fynn-Aikens et al. 1993). To determine the specific dietary requirements of juvenile white sturgeon, Hung et al. (1987) found that purified diets containing 42.9 - 43.8% crude protein, 15.5% lipids (including the phospholipid, lecithin), 7.4 - 8.1% moisture, and 4.1 - 4.2% minerals produced the best growth rates. Sturgeon dietary requirements are reviewed by Hung and Deng (2002).

3.2.8 OUTLOOK

This review summarizes the environmental requirements, preferences, and tolerance limits of North American sturgeons. The recent declines of sturgeon populations, world-wide, due to habitat alterations and overharvesting (i.e., for their roe or flesh), has made the acquisition of such knowledge about these ancient fishes even more urgent. For example, more information is needed about these species' passage around barriers such as dams in rivers. Data from radio and ultrasonic-tagged lake sturgeon in the upper Mississippi River showed that navigation dams can be barriers to their upstream movements (Knights et al. 2002). With appropriate data for management and modelling efforts (e.g., concerning river fragmentation by dams, Jager et al. 2001), the best, science-based decisions can be made to make sure that these magnificent, North American fishes do not slip into extinction.

Acknowledgments

We thank our graduate students for their technical and intellectual contributions to our research, the UC Davis Center for Aquatic Biology and Aquaculture (CABA) staff (especially, P. Lutes and E. Hallen) for logistical support, N. Kondakov (artist and zoologist) for the Atlantic sturgeon drawing in Figure 1 (from "Commercial Fishes of the USSR", 1949, Pishchepromizdat, Moscow), C. Woodley for drawing Figure 2, two anonymous reviewers for manuscript improvement, T. Blake for manuscript editing, and the CALFED Ecosystem Restoration Program, the Anadromous Fish Restoration Program (of the US Fish and Wildlife Service and the US Bureau of Reclamation), and the University of California Agricultural Experiment Station for research support.

3.3 References

Altinok, I., S.M. Galli, and F.A. Chapman. 1998. Ionic and osmotic regulation capabilities of juvenile Gulf of Mexico sturgeon, *Acipenser oxyrinchus desotoi*. Comp. Biochem. Physiol. 120A:609-616.

Auer, N.A. 1996. Response of spawning lake sturgeons to change in hydroelectric facility operation. Trans. Am. Fish. Soc. 125:66-77.

Berg, L.S. 1948. The freshwater fishes of the USSR and adjacent countries. 4th ed., Part 1. Akademia Nauk USSR, Moscow/Leningrad (in Russian, English translation published by Israel Program for Scientific Translations, 1965).

Birstein, V.J., J.R. Waldman, and W.E. Bemis (ed.). 1997. Sturgeon Biodiversity and Conservation. Kluwer Academic Publishers, Dordrecht.

Boreman, J. 1997. Sensitivity of North American sturgeons and paddlefish to fishing mortality. Env. Biol. Fish. 48:399-405.

Bramblett, R.G. and R.G. White. 2001. Habitat use and movements of pallid and shovelnose sturgeon in the Yellowstone and Missouri rivers in Montana and North Dakota. Trans. Am. Fish. Soc. 130:1006-1025.

Bruch, R.M. and F.P. Binkowski. 2002. Spawning behaviour of lake sturgeon (*Acipenser fulvescens*). J. Appl. Ichthyol. 18: 570-579.

Burggren W.W., and Randall, D.J. 1978. Oxygen uptake and transport during hypoxic exposure in the sturgeon, *Acipenser transmontanus*. Resp. Physiol. 34: 171-183.

Cailliet, G.M., A.H. Andrews, E.J. Burton, D.L. Waters, D.E. Kline, L.A. Ferry-Graham. 2001. Age determination and validation studies of marine fishes: do deep-dwellers live longer? Exp. Gerontol. 36: 739-764.

Carey J.R. and D.S. Judge. 2000. Longevity records: lifespans of mammals, birds, amphibians, reptiles, and fish. Odense University Press.

Carlson D.M. and K.W. Simpson. 1987, Gut contents of juvenile shortnose sturgeon in the upper Hudson Estuary. Copeia 1987(3):796-802.

Cech, J.J., Jr. 2000. Osmoregulation in bony fishes. pp. 614-622. In: R.R. Stickney (ed.) Encyclopaedia of Aquaculture. John Wiley & Sons, Inc. (New York).

Cech, J.J., Jr., S.J. Mitchell, and T.E. Wragg, 1984. Comparative growth of juvenile white sturgeon and striped bass: effects of temperature and hypoxia. Estuaries 7:12-18.

Chan, M.D., E.D. Dibble, and K.J. Kilgore. 1997. A laboratory examination of water velocity and substrate preference by age-0 gulf sturgeons. Trans. Am. Fish. Soc. 126:330-333.

Chapman, F.A. and S.H. Carr. 1995. Implications of early life stages in the natural history of the Gulf of Mexico sturgeon, *Acipenser oxyrinchus desotoi*. Env. Biol. Fish 43:407-413.

Collins, M.R., W.C. Post, D.C. Russ, and T.I.J. Smith. 2002. Habitat use and movements of juvenile shortnose sturgeon in the Savannah River, Georgia-South Carolina. Trans. Am. Fish. Soc. 131:975-979.

Crocker, C.E. and J.J. Cech, Jr. 1996. The effects of hypercapnia on the growth of juvenile white sturgeon, *Acipenser transmontanus*. Aquaculture. 147:293-299.

Crocker, C.E. and J.J. Cech, Jr. 1997. Effects of environmental hypoxia on oxygen consumption rate and swimming activity in juvenile white sturgeon, *Acipenser transmontamus*, in relation to temperature and life intervals. Env. Biol. Fish. 50:383-389.

Crocker, C.E. and J.J. Cech, Jr. 1998. Effects of hypercapnia on blood gas and acid-base status of the white sturgeon, *Acipenser transmontanus*. J. Comp. Physiol. B168:50-60.
Crocker, C.E., A.P. Farrell, A.K. Gamperl, and J.J. Cech, Jr. 2000. Cardiorespiratory responses of white sturgeon to environmental hypercapnia. Am. J. Physiol.: Reg., Int. And Comp. Physiol.279:R617-628.
Cui, Y., S.S.O. Hung, and X. Zhu. 1996. Effect of ration and body size on energy budget of juvenile white sturgeon. J. Fish Biol.49:863-876.
Curtis, G.L., J.S. Ramsey, and D.L. Scarnecchia. 1997. Habitat use and movements of shovelnose sturgeon in pool 13 of the upper Mississippi River during extreme low flow conditions. Env. Biol. Fish. 50:175-182.
Deng, X, J.P. Van Eenennaam, and S.I. Doroshov. 2002. Comparison of early life stages and growth of green and white sturgeon. Am. Fish. Soc. Symp. 28:237-248.
Dettlaff, T.A., A.S. Ginsburg, O.I. Schmalhausen. 1993. Sturgeon Fishes: developmental biology and aquaculture. Springer-Verlag, Berlin.
Doroshov, S.I. 1985. The biology and culture of sturgeon. pp. 251-274 In: J. Muir and R. Roberts (eds.) Recent advances in aquaculture. Croon Helm. Publ. (London).
Doroshov, S.I., G.P. Moberg, and J.P. Van Eenennaam. 1997. Observations on the reproductive cycle of cultured white sturgeon, *Acipenser transmontanus*. Env. Biol. Fish 48:265-278.
Fox , D.A., J.E. Hightower, and F.M. Parauka. 2000. Gulf sturgeon spawning migration and habitat in the Choctshatchee River system, Alabama-Florida. Trans. Am. Fish. Soc. 129:811-826
Fynn-Aikens, K., S.S.O. Hung, and S.G. Hughes. 1993. Effects of feeding a high level of D-glucose on growth and liver function in juvenile white sturgeon. Fish Physiol. Biochem.12:317-325.
Gardiner, B.G. 1984. Sturgeons as living fossils. pp. 148-152 in: N. Eldredge and S.M. Stanley (ed.). Living Fossils. Springer-Verlag, New York.
Gillis, T.E. and J.S. Ballantyne. 1996. The effects of starvation on plasma free fatty acid and glucose concentrations in lake sturgeon. J. Fish Biol. 49:1306-1316.
Gisbert, E., J.J. Cech, Jr. and S.I. Doroshov. 2001. Routine metabolism of larval green sturgeon (*Acipenser medirostris* Ayres). Fish Physiol. Biochem. 25:195-200.
Harkness, W.J.K. and J.R. Dymond. 1961. The lake sturgeon. Ontario Department of Lands and Forests, Fish and Wildlife Branch. Maple, Ontario. 97 pp.
Herold, M.A., S.S.O. Hung, and K. Fynn-Aikens. 1995. Apparent digestibility coefficient of carbohydrates for white sturgeon. Prog. Fish-Cult.57:137-140.
Hung, S.S.O. and D.F. Deng. 2002. Sturgeon, Acipenser pp. 344-357. In: C.D. Webster and C. Lim (eds.) Nutrient Requirements and Feeding of Finfish for Aquaculture. CABI Publ. (New York).
Hung, S.S.O., B.J. Moore, C.E. Bordner, and F.S. Conte. 1987. Growth of juvenile white sturgeon (*Acipenser transmontanus*) fed different purified diets. J. Nutr. 117:328-334.
Jager, H.I., J.A. Chandler, K.B. Lepla, and W. Van Winkle. 2001. A theoretical study of river fragmentation by dams and its effects on white sturgeon populations. Env. Biol. Fish. 60:347-361.
Jarvis, P.L., J.S. Ballantyne, and W.E. Hogans. 2001. The influence of salinity on the growth of juvenile shortnose sturgeon. N. Am. J. Aquacult. 63:272-276.
Jenkins, W.E., T. Smith, L. Heyward & D.M. Knott. 1993. Tolerance of shortnose sturgeon, *Acipenser brevirostrum*, juveniles to different salinity and dissolved oxygen concentrations. Proc. Ann. Conf. SE Assoc. Fish & Wildl. Agen. 47:476-484.
Jobling, M. 1981. Temperature tolerance and the final preferendum: rapid methods for the assessment of optimal growth temperatures. J. Fish Biol. 19:439-455.
Kasumyan, A.O. 2002. Sturgeon food searching behaviour evoked by chemical stimuli: a reliable sensory mechanism. J. Appl. Ichthyol. 18:685-690.
Knights, B.C., J.M. Vallazza, S.J. Zigler, and M.R. Dewey. 2002. Habitat and movement of lake sturgeon in the upper Mississippi River system, USA. Trans. Am. Fish. Soc. 131:507-522.
Kohlhorst, D.W. 1976. Sturgeon spawning in the Sacramento River in 1963 as determined by distribution of larvae. Calif. Fish and Game 62: 32-40.
Kohlhorst, D.W., L.W. Botsford, J.S. Brennan, and G.M. Cailliet. 1991. Aspects of the structure and dynamics of an exploited Central California population of white sturgeon (*Acipenser transmontanus*). pp. 277-294 in: P. Williot (ed.). Acipenser. CEMAGREF Publ., Bordeaux.
Kroll, K.J., J.P. Van Eenennaam, S.I. Doroshov, J.E. Hamilton, T.R. Russell. 1992. Effect of water temperature and formulated diets on growth and survival of larval paddlefish. Trans. Amer. Fish. Soc. 121: 538-543.
Kynard, B., M. Horgan, M. Kieffer, and D. Seibel. 2000. Habitats used by shortnose sturgeon in two Massachusetts rivers, with notes on estuarine Atlantic sturgeon: a hierarchical approach. Trans. Am. Fish. Soc. 129:487-503.
Lankford, S.E., T.E. Adams, J.J. Cech, Jr. 2003. Time of day and water temperature modify the physiological stress response in green sturgeon, *Acipenser medirostris*. Comp. Biochem. Physiol. Part A 135:291-302.

LeBreton, G.T.O. and F.W.H. Beamish 1998. The influence of salinity on ionic concentrations and osmolarity of blood serum in lake sturgeon, *Acipenser fulvescens*. Env. Biol. Fish. 52:477-482.
Linares-Casenave, J., J.P. Van Eenennaam and S.I. Doroshov. 2002. Ultrastructural and histological observations on temperature-induced follicular ovarian atresia in the white sturgeon. J. Appl. Ichthyol. 18:382-390.
Lindberg, J.C. and S.I. Doroshov. 1986. Effect of diet switch between natural and prepared foods on growth and survival of white sturgeon juveniles. Trans. Am. Fish. Soc. 115:166-171.
Loew, E.R. and A.J. Sillman. 1993. Age-related changes in the visual pigments of the white sturgeon (*Acipenser transmontanus*). Can. J. Zool. 71:1552-1557.
Martini , I.P. 1986. Canadian inland seas. Elsevier Science Publishers Co., New York. 494 pp.
Mason, W.T., Jr., and J.P. Clugston. 1993. Foods of the Gulf sturgeon in the Suwannee River, Florida. Trans. Am. Fish. Soc. 12:378-385.
Mayfield , R.B. and J.J. Cech, Jr. 2004. Temperature effects on green sturgeon bioenergetics. Trans. Am. Fish. Soc. (in press)
McCabe, G.T., Jr. and C.A. Tracy. 1994. Spawning and early life history of white sturgeon, *Acipenser transmontanus*, in the lower Columbia River. Fish. Bull. 92: 760-772.
McEnroe, M. and J.J. Cech, Jr. 1985. Osmoregulation in juvenile and adult white sturgeon, *Acipenser transmontanus*. Env. Biol. Fish. 14:23-30.
McKinley, S., G. Van Der Kraak, and G. Power. 1998. Seasonal migrations and reproductive patterns in the lake sturgeon, *Acipenser fulvescens*, in the vicinity of hydroelectric stations in northern Ontario.
Neill, W.H. and J.J. Magnuson. 1974. Distributional ecology and behavioural thermoregulation of fishes in relation to heated effluent from a power plant at Lake Monona, Wisconsin. Trans. Am. Fish. Soc. 103:663-710.
Neill, W.H. J.J. Magnuson, and G.G. Chipman. 1972. Behavioural thermoregulation by fishes: a new experimental approach. Science 176:1443-1445.
Parsley, M.J. and L.G. Beckman. 1994. White sturgeon spawning and rearing habitat in the lower Columbia River. N. Am. J. Fish. Manage. 14:812-827.
Peake, S. 1999. Substrate preferences of juvenile hatchery-reared lake sturgeon, *Acipenser fulvescens*. Env. Biol. Fish. 56:367-374.
Perrin, C.J., L.L. Rempel, and M.L. Rosenau. 2003. White sturgeon spawning habitat in an unregulated river: Fraser River, Canada. Trans. Am. Fish. Soc. 132: 154-165.
Pflieger, W.L., and T.B. Grace. 1987. Changes in the fish fauna of the Lower Missouri River, 1940-1983. pp. 166-177 in W.J. Matthews and D.C. Heins, editors. Community and evolutionary ecology of North American stream fishes. University of Oklahoma Press, Norman.
Quist, M.C., J.S. Tillma, M.N. Burlingame, and C.S. Guy. 1999. Overwinter habitat use of shovelnose sturgeon in the Kansas River. Trans. Am. Fish. Soc. 128:522-527.
Schaffter, R.G. 1997. White sturgeon spawning migrations and location of spawning habitat in the Sacramento River, California. Calif. Fish Game 83:1-20.
Sillman, A.J., C.J. O'Leary, C.D. Tarantino, E.R. Loew. 1999. The photoreceptors and visual pigments of two species of Acipenseriformes, the shovelnose sturgeon (*Scaphirhynchus platorynchus*) and the paddlefish (*Polyodon spathula*). J. Comp. Physiol. A 184:37-47.
Sulak, K.L. and J.P. Clugston 1998. Early life history stages of gulf sturgeon in the Suwannee River, Florida. Trans. Am. Fish. Soc. 127:758-771.
Trenkler, I.V., T.B. Semenkova. 1995. The influence of photoperiod, pinealectomy and pharmacological preparations on growth rate and metabolism in young sturgeons. Pp. 34-42 in: A.D. Gershanovich and T.I.J. Smith (ed.). Proceedings of the International Symposium on Sturgeons. VNIRO Publ., Moscow.
Vladykov, V.D. and J.R. Greeley. 1963. Order Acipenseroidei. pp. 24-60. In Fishes of the Western North Atlantic, Mem. Sears Found. Mar. Res. 1:24-60.
Waldman, J.R. and D.H. Secor. 1998. Caviar trade in North America: an historical prospective. pp. 77-89 in: D.F. Williamson, G.W. Benz, and C.M. Hoover (ed.). Proceedings of the Symposium on the Harvest, Trade and Conservation of North American Paddlefish and Sturgeon. TRAFFIC North America/World Wildlife Fund, Washington DC.
Wang, Y. L., F.P. Binkowski, and S.I. Doroshov. 1985. Effect of temperature on early development of white and lake sturgeon, *Acipenser transmontanus* and *A. fulvescens*. pp. 43-50 in F.P. Binkowski and S.I Doroshov (ed.). North American sturgeons: biology and aquaculture potential. Dr. W. Junk Publishers, Dordrecht, The Netherlands.
Wang, Y.L., R.K. Buddington, and S.I. Doroshov. 1987. Influence of temperature on yolk utilization by the white sturgeon, *Acipenser transmontanus*. J. Fish Biol. 30: 263-271.

Webb, M.A.H., J.P. Van Eenennaam, S.I. Doroshov, G.P. Moberg. 1999. Preliminary observations on the effects of holding temperature on reproductive performance of female white sturgeon, *Acipenser transmontanus* Richardson. Aquaculture 176:315-329.
Wedemeyer, G.A., 1996. Physiology of fish in intensive culture systems. Chapman and Hall. New York.
Wei, Q., F. Ke, J. Zhang, P. Zhuang, J. Luo, R. Zhou and W. Yang. 1997. Biology, fisheries, and conservation of sturgeons and paddlefish in China. Env. Biol. Fish. 48:241-255.
Welsh, S.A., S.M. Eyler, M.F. Mangold, and A.J. Spells. 2002. Capture locations and growth rates of Atlantic sturgeon in the Chesapeake Bay. Amer. Fish. Soc. Symp. 28:183-194.

CHAPTER 4

THE ECOLOGY AND FUNCTIONAL MORPHOLOGY OF FEEDING OF NORTH AMERICAN STURGEON AND PADDLEFISH

MICHAEL J. MILLER
Ocean Research Institute, The University of Tokyo, Nakano, Tokyo 164-8639, Japan

4.1 Introduction

Sturgeons and paddlefishes of the order Acipenseriformes are two of the largest freshwater fishes and are unique in having an unusual combination of sensory systems, jaw structures, and feeding modes. The basic body form of these fishes and possibly the general nature of their feeding mechanism may date back 200 million years based on fossil records (Bemis et al. 1997). The four extant genera of sturgeon of the family Acipenseridae (Acipenser, Huso, Pseudoscaphirhynchus, Scaphirhynchus) and the two genera of paddlefishes of the family Polyodontidae (Polyodon, Psephurus) share a variety of characteristics such as their cartilaginous endoskeleton and similar jaw suspension, but they have contrasting head shapes and feeding mechanisms that enable some species to have very different feeding modes.

The unique combination of morphological features associated with feeding in sturgeons and paddlefishes appear to have made them successful foragers that have withstood the test of time. The common features of all four genera of sturgeon include a ventrally flattened snout or rostrum with four barbels in front of their ventrally protruding jaws. In addition to their eyes and olfactory rosettes, they have electroreceptors and a lateral line canal system on the underside of the snout. In contrast, the two genera of paddlefish have a long pointed dorsoventrally flattened rostrum that is covered with electroreceptors, only two barbels, and in the case of the North American paddlefish, *Polyodon spathula*, jaws that open to a forwardly opening mouth.

Based on their similar head shapes and the position of their jaws, all Acipenser and Scaphirhynchus species appear to use the same powerful suction feeding mechanism that enables them to feed on aquatic animals on or within the upper sediment layer or to prey on small fishes (Miller, 1987; Bemis et al. 1997; Carroll and Wainwright, 2003). However, the adults of the beluga sturgeon, *Huso huso*, appear to feed more on fishes (Vecsei et al. 2002) and like the Chinese paddlefish, *Psephurus gladius*, have more forward pointing jaws (Grande and Bemis, 1991) that suggest they are more specialized for feeding on fishes than on benthic prey (Bemis et al. 1997). In contrast, the North American paddlefish has a completely different feeding mode that consists of using its

G.T.O. LeBreton et al. (eds.), Sturgeons and Paddlefish of North America, 87–102.
© 2004 *Kluwer Academic Publishers. Printed in the Netherlands.*

larger, more forward pointing mouth and modified gill rakers to filter feed on small planktonic animals (Rosen and Hales, 1981).

The feeding ecology of almost all eight species of North American sturgeons and paddlefish has been studied to some extent. Along the west coast of North America there is an abundant anadromous species, the white sturgeon, *Acipenser transmontanus*, whose food habits and feeding in aquacultural environments have been studied fairly extensively in recent years. The other west coast anadromous species is the green sturgeon, *A. medirostris*. It has recently been artificially spawned and raised in captivity (Van Eenennaam, 2001). The habitat use and food habits of the river sturgeons, *Scaphirhynchus platorynchus*, and *S. alba* (Carlson et al. 1985; Keenlyne, 1997), and the paddlefish (Jennings and Zigler, 2000), have also been studied in various portions of their ranges in the Mississippi River basin. Another freshwater species, the lake sturgeon, *Acipenser fulvescens*, has been studied in some of the lakes and rivers within its range in the north central region of North America (Chiasson et al. 1997). Other anadromous sturgeon are the Gulf, *Acipenser oxyrinchus desotoi*, and the Atlantic sturgeon, *Acipenser oxyrinchus oxyrinchus*, found in rivers and coastal areas in the northern Gulf of Mexico and along the east coast of North America and parts of eastern Florida, respectively (Smith and Clugston, 1997). The shortnose sturgeon, *Acipenser brevirostrum*, is considered to be amphidromous because its use of the marine environment is limited to the estuaries of its natal rivers along the east coast of North of America (Kynard, 1997).

This chapter provides an overview of the feeding ecology of North American sturgeons and paddlefish by reviewing the literature on the food habits of each species, and by describing the functional basis of their feeding mechanisms in terms of their sensory systems that may be used in feeding and their basic jaw structure and feeding behaviour.

4.2 Feeding Ecology of North American Sturgeons and Paddlefish

4.2.1 FOOD HABITS OF STURGEONS

Sturgeon feed on a variety of prey items including molluscs, insects, crustaceans and fishes depending on the availability of food and on the size of sturgeon. The habitats of the seven species North America sturgeons vary widely from rivers and estuaries, to lakes, to the coastal marine environment, and the diets of the various species appear to reflect the different types of prey species found in each area.

Perhaps the most studied species of North America sturgeon is the white sturgeon, whose feeding habits have been studied in all different sizes of juveniles and adults. White sturgeon as small as 15mm TL in the lower Colombia River fed primarily on the gammarid amphipods *Corophium* spp. as did fish ranging in size up to 290mm, but the length of *Corophium* spp. eaten by larval and young-of-the-year white sturgeon increased with increasing fish length (Muir and McCabe 2000). Copepods (Cyclopoida), Ceratopogonidae larvae, and Diptera pupae and larvae (primarily chironomids) were also consumed, especially at the onset of exogenous feeding. McCabe et al. (1993) found that the tube-dwelling amphipod *Corophium salmonis* was the most important prey for two size classes of white sturgeon (144-350 and 351-724mm FL), but that the smaller size class preyed more heavily on *C. salmonis* than did the larger individuals. Additional temporally important prey for the smaller white sturgeon included *Corophium spinicorne*, *Neomysis* sp., chironomid larvae, and eulachon eggs, *Thaleichthys pacificus*. Other temporally important prey for the larger individuals included the bivalve *Corbicula fluminea*, *Corophium spinicorne*, chironomid larvae, and eulachon eggs. *Corophium spinicorne* was also suggested to be the most important food item of young-of-the-year and larger juvenile white sturgeon in the Sacramento-San Joaquin Delta in California, where *Tendipidae* larvae and Neomysis sp. were also eaten (Schreiber, 1962; Radtke, 1966). However, adults in the San Pablo and Suisun Bays of the Sacramento-San Joaquin River estuary were found to feed on several species of clams throughout the year, with barnacles, crabs, shrimp, herring eggs, and several fish species being seasonally important in their diets (Mckechnie and Fenner, 1971). Similarly, in addition to invertebrates such crayfish, chironomid and stonefly larvae and mysids, fishes were found in half of the stomachs of adult white sturgeon during the summer in the lower Fraser River (Semakula and Larkin, 1968). In May, eulachon spawn in the Fraser River and were the main food item.

The food habits of the species of North American sturgeon that live exclusively in freshwater also have been studied and their diets appear to reflect the availability of food items in the various habitats in which they live. The lake sturgeon feeds primarily on aquatic insects such as Ephemeroptera, Diptera, and Trichoptera, with molluscs, crustaceans, fishes and fish eggs sometimes being eaten depending on the availability of food in a particular area (Harkness and Dymond, 1961; Chiasson et al. 1997; Beamish et al. 1998). This species has recently been observed to feed upon the introduced zebra mussels (Jackson et al. 2002). The shovelnose sturgeon are opportunistic feeders that prey primarily on aquatic insects, with insect larvae of the Trichoptera, Diptera, and Ephemeroptera being their primary food items (Held, 1969; Modde and Schmulbach, 1977; Keenlyne, 1997). Carlson et al. (1985) found a similar utilization of aquatic insects by both shovelnose and pallid sturgeons in the Mississippi and Missouri Rivers. However, fishes were also consumed in considerable quantities by the pallid sturgeon and individuals thought to be hybrids between these two species.

The other anadromous or potamodromous North American sturgeons also appear to be opportunists that feed on the available food items in the various environments that they inhabit. Juvenile shortnose sturgeon mostly remain in the freshwater reaches of coastal rivers of the east coast of North America until they are about 45 cm in length. They feed on benthic crustaceans such as gammarid amphipods, isopods and mysids, or insects such as dipteran and chironomid larvae. Larger and older individuals then move into estuaries and feed primarily on any molluscs that are available (Dadswell, 1979; Dadswell, et al. 1984; Carlson and Simpson, 1987). Atlantic sturgeon off the coast of New Jersey fed primarily on polychaetes, isopods, and amphipods with their importance showing seasonal variation (Johnson et al. 1997). The isopod *Politolana concharum* was the most important individual prey eaten. In the Gulf of Mexico, large juveniles and adults of the threatened Gulf sturgeon fed primarily on lancelets, brachiopods, amphipods and other crustaceans, polychaetes, and gastropods during spring in estuarine regions of the Suwannee River, Florida (Mason and Clugston, 1993). Small individuals near the mouth of the river fed on amphipods, grass shrimp, isopods, oligochaetes, polychaetes, and chironomid and ceratopogonid larvae found in the intertidal zone. Juveniles and adults in upstream freshwater areas of the river ate relatively little during the summer and fall, despite apparently abundant food supplies.

It appears that these Gulf sturgeon use spring fed parts of the river as temperature refuges during summer and apparently do not feed much in freshwater based on their stomach contents and on the nitrogen isotope ratios of their tissue (Mason and Clugston, 1993; Gu et al. 2001).

In addition to studies of sturgeon food habits in the wild, considerable information about feeding tendencies and influences on their growth have been obtained from aquaculture studies. The culture techniques for North American sturgeons were pioneered with white sturgeon, and early studies on their feeding and growth were critical to the development of the sturgeon aquaculture industry because of the importance of learning how to get young sturgeon to accept artificial diets (Monaco et al. 1981; Buddington and Doroshov, 1984; Buddington and Christofferson, 1985). This was found to be the biggest challenge because sturgeon that initially fed on natural foods usually had difficulty in switching to artificial diets (Lindberg and Doroshov, 1986). Research on feeding and growth of sturgeon has continued with white sturgeon (see Hung et al. 1993; Deng et al. 2003) and has been expanded to include the Lake sturgeon (Dilauro et al. 1998; Fajfer et al. 1999), Atlantic sturgeon (Kelly and Arnold, 1999), Gulf sturgeon (Mohler et al. 1996, 2000; Bardi et al. 1998: Barrows, 1998), shortnose sturgeon (Jarvis et al. 2001), and green sturgeon (Van Eenennaam et al. 2001). In addition, various studies have examined the characteristics of the digestive systems of sturgeons in an attempt to gain a better understanding of their feeding biology and to enhance aquaculture efforts (e.g. Buddington and Doroshov, 1986; Gawlicka et al. 1995; Domeneghini et al. 2002).

4.2.2 FOOD HABITS OF PADDLEFISH

The American paddlefish is found only in the Mississippi River drainage and is unique among the acipenseriform fishes because in the juvenile and adult stage it is a filter feeding planktivore that feeds primarily on small zooplankton. However, the earliest stages of paddlefish appear to actively feed upon individual prey items such as zooplankton and insects until they reach about 120–250mm TL when the gill rakers develop enough to be used as a filter (Ruelle and Hudson, 1977; Rosen and Hales, 1981; Michaletz et al. 1982). When they are preying on individual zooplankton, larger species such as *Daphnia* may be actively selected (Michaletz et al. 1982). Once they make the transition to filter feeding, paddlefish presumably forage exclusively by filtering water through their gill rakers. Larger paddlefish are ram suspension filter feeders (Sanderson et al. 1994) and they feed mostly on zooplankton and occasionally small insects, insect larvae, and small fish (Ruelle and Hudson, 1977; Rosen and Hales, 1981; Michaletz et al. 1982).

In studies on the culture of paddlefish, young have been raised on both brine shrimp, *Artemia salina*, and krill, *Euphausia superba*, and on artificial commercial diets. It was found that the fry were particulate feeders and that they would readily accept both types of diets at the onset of exogenous feeding, but did not exhibit the imprinting reported for some sturgeon species when reared on natural diets (Kroll et al. 1992, 1994).

4.2.3 FEEDING BEHAVIOUR AND JAW MECHANISMS OF STURGEONS AND PADDLEFISH

The opportunistic feeding behaviour of sturgeon and paddlefish that can be inferred from food habits studies suggests that these fish will prey on whatever suitable types of food items are available in a particular area. Both sturgeon and paddlefish are often found in relatively deep or turbid environments where vision would be of relatively little use, so their ability to detect and capture prey in these environments is due to their unique combinations of sensory systems and their unusual jaw mechanisms. In fact, it is these characteristics that make sturgeon and paddlefish so unusual in comparison to other fishes, because no other fishes have the combination of a ventrally positioned mouth and highly protrusible jaws preceded by both barbels and numerous electroreceptors in the case of sturgeon, or the long rostrum covered with numerous electroreceptors preceding the large mouth opening for filter feeding as in paddlefish. In this section, the sensory systems, feeding behaviour and the general structure of the jaw mechanisms of the North American species of sturgeon and paddlefish will be outlined and the functional implications of their jaw mechanisms will be described.

4.2.4 FEEDING BEHAVIOUR AND SENSORY SYSTEMS OF STURGEONS

Larval sturgeon quickly develop the morphological features associated with their primary sensory systems that are used in feeding. White sturgeon larvae hatched in the laboratory and raised at 16°C showed that the major features of the head were formed by day 9 (Figure 4-1). The development of these features proceeds in conjunction with the absorption of their yolk reserves during a period of hiding within the substrate, which occurs after a brief swim-up period immediately after hatching (Brewer, 1987; Miller 1987). During this swim-up period, the undeveloped larvae swim randomly over the substrate and up in the water column. This behavioural pattern likely functions to enable the larvae to be transported by currents away from the hatching area and to become more widely distributed downstream to reduce predation. Similar behaviour has been observed in other species of sturgeon, although the timing and duration of this period of swimming in the water column appears to vary among species (Kynard et al. 2002).

Once white sturgeon larvae emerged from hiding they were ready to initiate feeding, and this occurs on day 10 at 18-20°C and on day 13 at 16°C (Miller 1987). This is similar to shortnose sturgeon larvae that begin to feed after 8-10 days at 15-17°C when they are about 15mm TL (Kynard, 1997). When these tiny larvae begin feeding, their electroreceptors also appear to be developed (Figure 4-1). In larger juveniles these ampullary electroreceptors (Jørgensen, 1980) are quite apparent as seen in the white sturgeon juvenile in Figure 4-2. These ampullary receptors are able to detect weak electrical fields in some species of sturgeon (Teeter et al. 1980; New and Bodznick, 1985) and are likely to be used in feeding in much the same way as has been observed in elasmobranches (see Wilkens et al. 2001). On the underside of the snout of sturgeon there also is a lateral line canal system, and other canals are located on the lateral and dorsolateral surface of the head (Norris, 1924). This basic pattern of sensory systems in relation to the position of the mouth is the same for all North American sturgeon species, except that the two species of Scaphirhynchus have larger fleshy lobes on their lips, which are presumably taste buds, and their barbels are fringed and not smooth as in Acipenser (Figure 4-2; Berstein et al. 1997, p. 290)

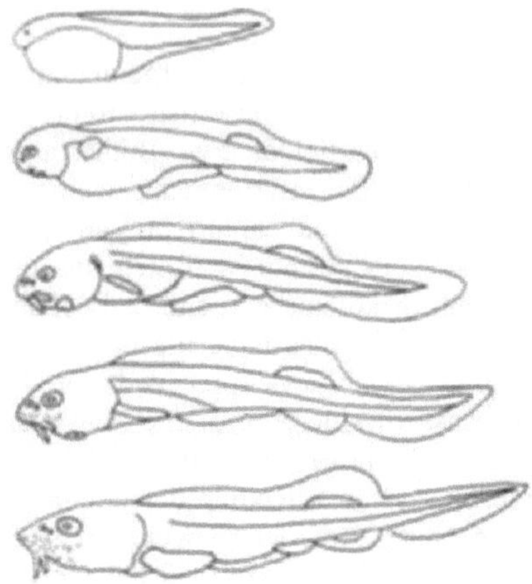

Fig. 4-1. Lateral views of developmental stages of white sturgeon, Acipenser transmontanus, raised at 16°C, showing the days post-hatching and total length of each specimen.

Even though sturgeon have relatively large eyes when they are small, there is presently no indication that vision is used in feeding at any stage of their lives. In fact the relative size of the eye appears to diminish as sturgeon grow larger. Early studies with sturgeon found that they are non-visual feeders (Sbikin, 1974). A more recent study on the feeding behaviour of young white sturgeon suggested that olfactory cues are important in detecting the general presence of food and that their barbels are important for detecting prey items, because their foraging behaviour in the presence of benthic food generally consists of continuous movement over the substrate with their barbels positioned to detect food (Miller, 1987). However, when food is not present they show an increased tendency to move into the water column and swim to another area before returning to the substrate. When white sturgeon initiate feeding they will readily feed on chopped Tubifex worms, and they prefer this type of food over artificially prepared food (Lindberg and Doroshov, 1986). Small juveniles also are clearly able to capture small fish in total darkness (Miller, 1987). This and the dorsolateral position of the eyes of sturgeon relative to their ventrally located mouth, suggest that the eyes of sturgeon may be used for other purposes such as predator avoidance or distinguishing between light and dark in their typically low light environments as discussed by Sillman et al. (1999). It is most likely that a combination of olfactory and tactile chemosensory cues are used by sturgeon for feeding along with their electroreceptors and possibly lateral line canals in some circumstances, but more research is needed to verify the specific sensory systems used in prey detection by sturgeon.

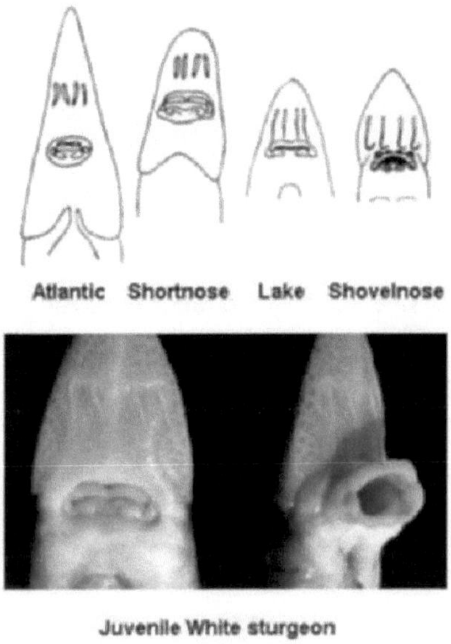

Fig. 4-2. Ventral views of the heads of various sizes of four species of North American sturgeon showing the locations of their barbels relative to the position of the mouth (top), and ventral and ventrolateral views of the head and jaws of an approximately 400 mm TL juvenile white sturgeon (bottom).

4.2.5 STURGEON JAW MORPHOLOGY AND PROTRUSION

All sturgeon have a streamlined body profile and their highly protrusible jaws are only visible during feeding or if the jaw is forced out of the head 4-2). The jaw structures of several North American species of Acipenser and Scaphirhynchus have been described to some degree and the protrusion of their jaws has been filmed (Jollie, 1980; Miller, 1987; Bemis et al. 1997; Findeis, 1997; Carrol and Wainwright, 2003). The jaw structure and protrusion mechanism of sturgeon is unique among fishes because the jaws are not fixed to the cranium and can therefore be protruded quite far outside of the head. In white sturgeon and other species, the jaws are suspended dorsally by the hyomandibular bone with the symplectic bone (or interhyal; see Findeis 1997) as an intermediate component, and ventrally by the hyoid apparatus through the anterior ceratohyal and the symplectic (Figure 4-3). These dorsal and ventral connections are also supported by two ligaments (not shown). Another unique characteristic of their feeding mechanism is the palatal complex that extends the oral surface of the upper jaw posteriorly. This surface shears across the tongue pad and biting ridges that are formed by the ventral hyobranchial skeleton to enable sturgeon to grip prey within the buccal cavity (Findeis, 1997).

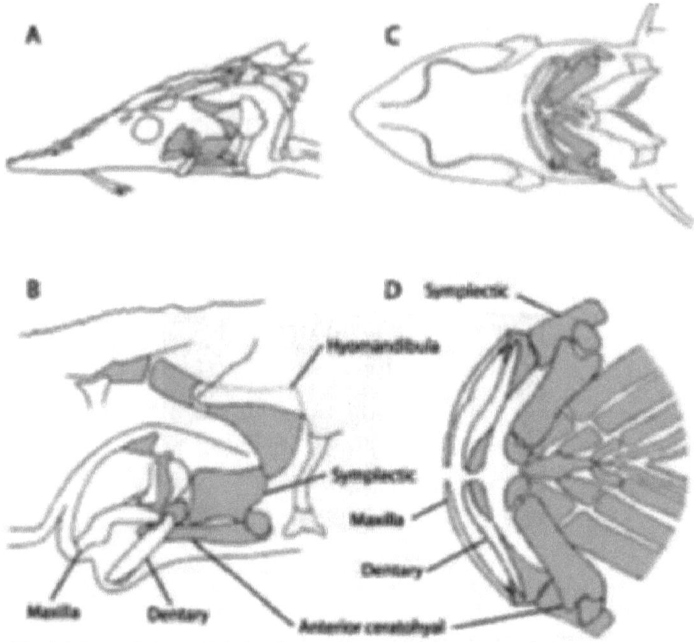

Fig. 4-3. Lateral views of the head of an 81 mm TL white sturgeon, Acipenser transmontanus (A), and a lateral view of the jaws and suspensorium (B), a ventral view of the head (C), and a ventral view of the jaws, ventral suspension, and ventral hyobranchial skeleton (D) of an 83mm individual. Cartilage is shown by the shading, but the cartilage and bone of the palate area is not distinguished in (A) and (B).

The general mechanisms of protrusion and retraction of the jaws and of jaw opening and closing in sturgeon has been deduced from the bone and muscle structure and various other means. Jaw protrusion appears to be caused by forward movement of the hyomandibular bone that results from contraction of the large protractor hyomandibularis muscle that is attached to its anterior margin (Miller, 1987; Bemis et al. 1997; Carrol and Wainwright, 2003). The anterior movement of the hyomandibula causes the jaws and palate to be pushed forward into the cranium which then causes them to be thrust downward and out of the head as shown in Figure 4-4. The jaws are then closed by retraction of the adductor mandibulae muscle that inserts from the top of the palate down to the lower jaw and functions to pull the lower jaw (dentary and Meckel's cartilage) up to the upper jaw (maxillary) causing the mouth to close (Carrol and Wainwright, 2003). The jaws are retracted by contraction of the retractor hyomandibularis muscle that is attached to the posterior margin of the hyomandibula, which would result in the jaws being pulled back into the head. See Carrol and Wainwright (2003) for a detailed discussion of the apparent functional mechanics of jaw protrusion and hyoid depression in sturgeon.

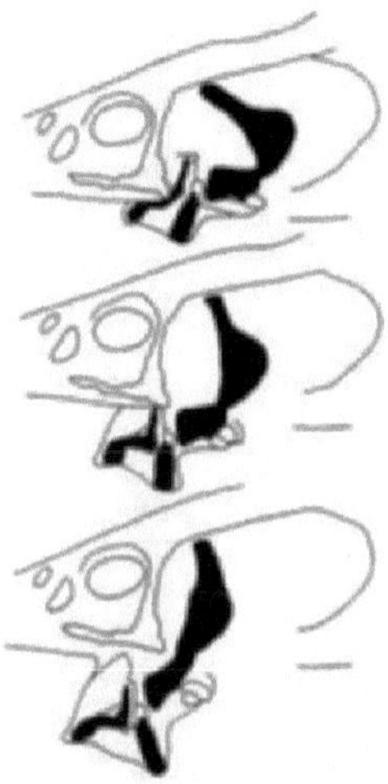

Fig. 4-4. Sequential diagrams of jaw protrusion in a 12 mm TL cleared and stained white sturgeon, Acipenser transmontanus, showing the hyomandibula, symplectic, dentary, maxillary and quadratojugal bones in black. The jaws were induced to be protruded by pulling both of the hyomandibula bones anteriorly using attached threads (not shown).

Observations of feeding events of both Acipenser and Scaphirhynchus have indicated that when the jaws of sturgeons are protruded there is considerable suction pressure created. This suction pressure is caused by the negative pressure created by the expansion of the buccal cavity by various means during the process of jaw protrusion (Carrol and Wainwright, 2003). This powerful suction pressure is illustrated in Figure 4-5A where a 200mm TL white sturgeon was recorded capturing a stunned salmon floating in the water column. The first frame of this sequence shows the mouth opening and the second frame suggests that a powerful suction pressure has then caused the salmon to be sucked into the mouth, resulting in it being folded in half, with both the head and tail protruding from the mouth after the jaws were closed in the third frame. Carrol and Wainwright (2003) also observed the food items being sucked into the mouth of the pallid sturgeon during jaw protrusion. The remarkable extension of the jaws in the white sturgeon is clearly shown in Figure 4-5B where the same 200mm sturgeon is attempting to further ingest a previously captured salmon. In this sequence the jaws are rapidly protruded out over the head of the salmon and then the jaws were closed. The force of the retracting jaws was then used to attempt to force the fry through the esophagus and into the stomach. The same sturgeon also was observed to be able to protrude its jaws asymmetrically more towards one side while attempting to capture a stunned salmon on the left side of its head (Figure 4-6). In this sequence the jaws can be seen protruding out toward the left and then closing without capturing the prey. These observations of the protrusion of the jaws of sturgeon and the apparently very powerful suction pressure generated during protrusion help to explain the ability of sturgeon to capture such a wide range of prey types, as has been found in the food habits studies on the various species found in North America.

Fig. 4-5. Frames of VHS video recordings of feeding events of a 200mm TL white sturgeon, Acipenser transmontanus, capturing (A) and processing (B) stunned salmon. Frames are sequential except for one frame being omitted between the second and third frames of (B).

4.2.6 FEEDING BEHAVIOUR AND SENSORY SYSTEMS OF PADDLEFISH

Paddlefish development differs somewhat in comparison to sturgeon because in contrast to sturgeon, which are essentially the same in their general body form as the adults when they initiate feeding, paddlefish differ considerably from their adults in having a much shorter rostrum when they initiate feeding. The rostrum of sturgeon extends out to some degree during growth, but it does not extend out nearly as far as in paddlefish, which have a remarkably long rostrum (Figure 4-7, 4-8A). In fact, the youngest paddlefish lack a long rostrum and show little resemblance to their adults (Figure 4-7). Young paddlefish also have two relatively prominent barbels preceding their mouth that at 14 days posthatch are proportionally almost as big as in small sturgeon (Bemis and Grande, 1992; Bemis et al. 1997), but the relative size of these barbels becomes greatly reduced with growth. However, the rostrum gets proportionally much longer with growth (Grande and Bemis, 1991) and eventually can make up as much as half the total body length. This long rostrum and the side of the head and operculum are covered with thousands of electroreceptors (Jørgensen et al. 1972).

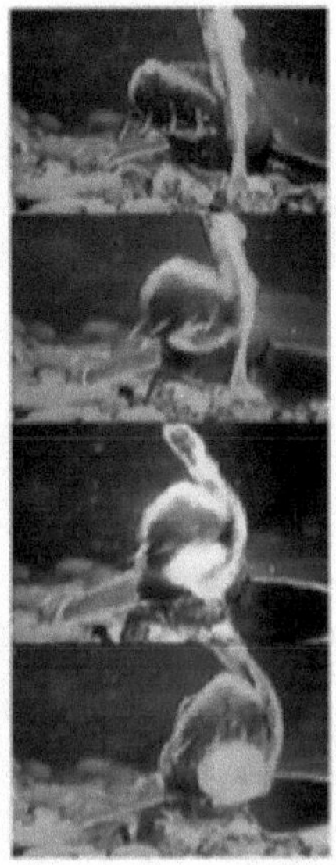

Fig. 4-6. Sequential frames of a VHS video recording of a 200 mm TL white sturgeon, Acipenser transmontanus, attempting to capture a stunned salmon fry floating in the water column.

The feeding behaviour and possible sensory systems that are used by paddlefish for feeding appear to change somewhat with development. When less than about 200mm, paddlefish feed selectively on individual zooplankton (Rosen and Hales, 1981; Michaletz et al. 1982). The prominence of the two barbels in paddlefish larvae suggest the possibility that these could be initially important in feeding, but they also appear to have electroreceptors at this time as well (Bemis and Grande, 1992; Bemis et al. 1997), so it is unclear what senses are used in feeding by young paddlefish. This is especially true when considering the fact that young paddlefish will feed on artificial diets in the water column, but not on the bottom (Kroll et al. 1992, 1994). Kroll et al. (1992) also reported that juveniles larger that 120 mm will feed on floating trout pellets and are not obligate filter feeders. Precisely how paddlefish detect artificial foods that do not generate electrical fields like zooplankton needs to be determined.

Larger juveniles that prey selectively on individual zooplankton have a relatively long rostrum and do clearly appear to use electroreception for locating individual prey, because they are successful at detecting and capturing prey in the absence of other cues such as light and olfactory cues (Wilkens et al. 1997; Wilkens et al. 2001). Paddlefish also show feeding responses to artificially generated electrical fields, which further supports the importance of the electric sense in their feeding behaviour (Wilkens et al. 1997: Wojtenek et al. 2001). Other studies have found evidence that electrical noise such as that from a swarm of *Daphnia* may enhance the ability of paddlefish to detect individual prey using electroreception, through a process called stochastic resonance (Russell et al. 1999).

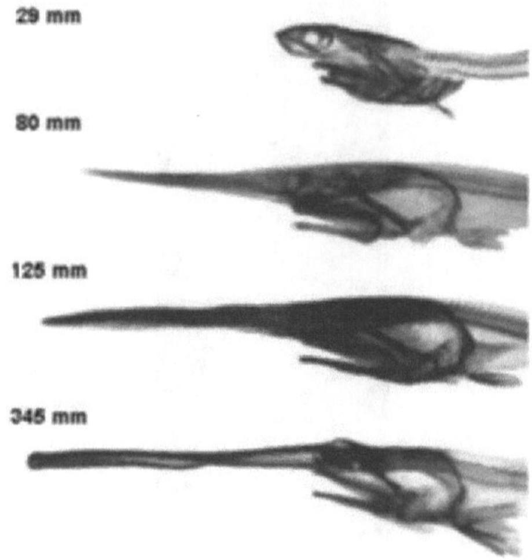

Fig. 4-7. Side views of the heads of various sizes of juvenile paddlefish that were cleared and double stained, which show the rapid growth of the rostrum (modified from Grande and Bemis, 1991).

Recent studies on paddlefish feeding have shown that juveniles use electroreception to detect the electric fields of individual zooplankton, and presumably they use this sense when filter feeding as well. Remarkably, juvenile paddlefish can detect and capture plankton at distances up to 90mm from the surface of the rostrum (Wilkens et al. 2001). The larger paddlefish that are filter feeders may use electroreception to identify larger concentrations of zooplankton or to determine when to switch from ram ventilation (Burggren and Bemis, 1992) to the faster swimming speed that they appear to use during ram suspension feeding (Sanderson et al. 1994). However, there is no evidence of selective behaviour for filter feeding on particular types of zooplankton, because their gut contents contain the same complement of the appropriate sized species that are present in their environment (Michaletz et al. 1982).

4.2.7 PADDLEFISH JAW MORPHOLOGY AND FEEDING

The structure and function of the jaws of the American paddlefish are quite different than the jaws of sturgeon, and their forward pointing mouth can open very wide (Figure 4-8A). The jaws of paddlefish are largely comprised of the same structural components as found in sturgeon, but they are configured in a very different way. The biggest differences are that their mouth is pointed forward (Figure 4-8B) and the upper jaw is firmly fixed to the cranium and is not protrusible (Grande and Bemis, 1991; Bemis et al. 1997). Because of their large mouth opening, the upper and lower jaws are longer and more slender than those of sturgeon (Figure 4-8C), but their jaws are still suspended dorsally through articulations with the symplectic (interhyal) and hyomandibula bones (Grande and Bemis, 1991; Bemis et al. 1997; Findeis, 1997). The mechanics of the opening and closing of the jaws have been described by Bemis (1987).

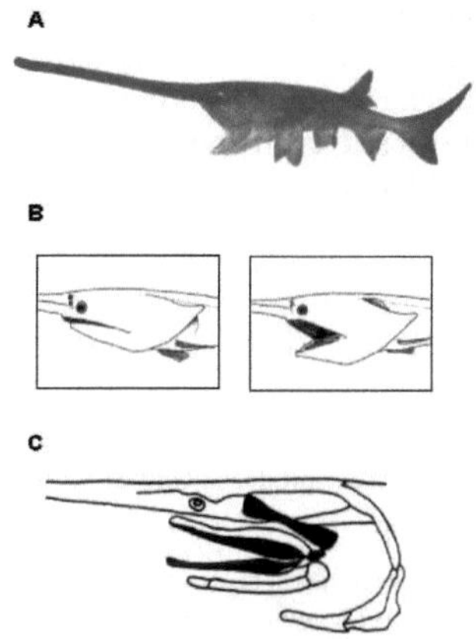

Fig. 4-8. A photograph of a 180 mm paddlefish, Polyodon spathula, (A; modified from Wojtenek et al. 2001), diagrams showing the mouth closed and open in a juvenile paddlefish (B; modified from Bemis et al. 1997), and the jaw structure of an adult paddlefish showing the hyomandibula, symplectic (interhyal), dentary and maxilla in black (C; modified from Traquair, 1877)

4.2.8 THE OPPORTUNISTIC FEEDING MECHANISMS OF STURGEON AND PADDLEFISH

The available information about the feeding ecologies, sensory systems and jaw mechanisms of the North American species of sturgeon and paddlefish indicate that these fish are highly adaptable and are able to feed on available benthic or epibenthic prey in the case of sturgeon, or on whatever type of suitably sized zooplankton are available in the case of paddlefish. The numerous electroreceptors and pronounced barbels of sturgeon appear to enable them to feed on any type of soft-bodied invertebrates or molluscs that are present on or within the substrate. They also are likely to be able to utilize seasonally available food sources such as fish eggs or small fishes. Their powerful suction feeding mechanism gives them the ability to suck these food items rapidly into the mouth and allows them to feed opportunistically on any palatable prey that may be encountered. Similarly, paddlefish appear to have specialized in detecting zooplankton using electroreception and with age they develop a remarkably long rostrum that is covered with electroreceptors that are used in prey detection. Young paddlefish may select larger zooplankton, but once the transition to filter feeding is made, juvenile and adult paddlefish appear to ingest any zooplankton larger than the mesh size of their gill rakers.

The North American species of sturgeon and paddlefish appear to have become specialized to use their unusual sensory systems and jaw mechanisms to feed opportunistically in their respective benthic and pelagic habitats. Sturgeon probably feed primarily in relatively low visibility, benthic habitats, whereas paddlefish feed primarily in the water column of often turbid freshwater rivers or lakes. Both groups are able to feed during both night and day in deep or shallow water, on foods that may be usually unavailable to visually feeding fishes. Their feeding success is probably not influenced much by environmental perturbations that affect water clarity. This would enable sturgeon and paddlefish to be very effective at taking maximum advantage of seasonally or temporally abundant food resources such as the spawning events of other fish species, or zooplankton blooms. The historical reports of the great abundances of sturgeon and paddlefish suggest that their feeding strategies were quite successful before the effects of overfishing, declines in prey populations, and habitat loss, and these strategies might be considered even more successful if measured by the long evolutionary history of the remarkable fishes of the Acipenseriformes.

4.3 References

Bardi, R.W., Chapman F.A., and Barrows, F.T. 1998. Feeding trials with hatchery-produced Gulf of Mexico sturgeon larvae. Progr. Fish. Cult. 60: 25-31.

Barrows, F.T. 1998. Feeding trials with hatchery-produced Gulf of Mexico sturgeon larvae. Progr. Fish Cult . 60: 25–31.

Beamish, F.W.H., Noakes, D.L.G., and Rossiter, A. 1998. Feeding ecology of juvenile lake sturgeon, *Acipenser fulvescens*, in Northern Ontario. Can. Field Nat.. 112: 459-468.

Bemis, W.E. 1987. Functional mechanics of jaw-opening and jaw-closing in the paddlefish, *Polyodon spathula*. Amer. Zool. 27: 87A.

Bemis, W.E. and Grande, L. 1992. Early development of the actinoperygian head. I. External development and staging of the paddlefish *Polyodon spathula*. J. Morphol. 213: 47-83.

Bemis, W.E., Findeis, E.K., and Grande, L. 1997. An overview of Acipenseriformes. Env. Biol. Fish. 48: 25-71.

Birstein, V.J., Waldman, J.R., and Bemis, W.F. (eds.) 1997. Sturgeon biodiversity and conservation. Developments in the environmental biology of fishes Series No. 7. Kluwer Academic Publishers, The Netherlands. 444 pp.

Brewer, S.D. 1987. Behavioural responses and patterns of distribution during the early life history of white sturgeon, *Acipenser transmontanus*. M.S. Thesis, University of Washington. 65 pp.

Buddington, R.K., and Christofferson, J.P. 1985. Digestive and feeding characteristics of the chondrosteans. Env. Biol. Fish. 14: 31-41.

Buddington, R.K., and Doroshov, S.I.. 1984. Feeding trials with hatchery produced white sturgeon juveniles, *Acipenser transmontanus*. Aquaculture 36: 237-243.

Buddington, R.K., and Doroshov, S.I. 1986. Structural and functional relations of the white sturgeon alimentary canal (*Acipenser transmontanus*). J. Morphol. 190: 201-213.

Burggren, W.W., and Bemis, W.E. 1992. Metabolism and ram gill ventilation in juvenile paddlefish, *Polyodon spathula* (Chondrostei, Polyodontidae). Physiological Zoology 65: 515-539.

Carlson, D.M., Pflieger, W.L., Trial, L., and Haverland, P.S. 1985. Distribution, biology, and hybridization of *Scaphirhynchus albus* and *Scaphirhynchus platorynchus* in the Missouri and Mississippi rivers. Env. Biol. Fish. 14: 51–59.

Carlson, D.M., and Simpson, K.W. 1987. Gut contents of juvenile shortnose sturgeon in the upper Hudson estuary. Copeia 1987: 796-802.

Carrol, A.M., and Wainwright, P.C. 2003. Functional morphology of prey capture in the sturgeon, *Scaphirhynchus albus*. J. Morphol. 256: 270-284.

Chiasson, W.B., Noakes, D.L.G., and Beamish, F.W.H. 1997. Habitat, benthic prey, and distribution of juvenile lake sturgeon (*Acipenser fulvescens*) in northern Ontario rivers. Can. J. Fish. Aquat. Sci. 54: 2866-2871.

Dadswall, M.J. 1979. Biology and population characteristics of the shortnose sturgeon, *Acipenser brevirostrum*, in the Saint John River estuary, New Brunswick, Canada. Can. J. Zool. 57: 2186-2210.

Dadswall, M.J., Taubert, B.D., Squiers, T.S., Marchette, D., and Buckley, J.B. 1984. Synopsis of biological data on shortnose sturgeon, *Acipenser brevirostrum* LeSueur 1818. U.S. Dept. of Commerce, NOAA Tech. Rep. NMFS 14, 45 pp.

Deng, D.F., Koshio, S., Yokoyama, S., Bai, S.C., Shao, Q.J., Cui, Y.B., and Hung, S.S.O. 2003. Effects of feeding rate on growth performance of white sturgeon (*Acipenser transmontanus*) larvae. Aquaculture 217: 589-598.

Dilauro, M.N., Krise, F.W., and Flynn-Aikins, K. 1998. Growth and survival of lake sturgeon larvae fed formulated diets. Progr. Fish Cult. 60: 293-296.

Domeneghini, C., Radaelli, G., Bosi, G., Arrighi, S., Di Giancamillo, A., Pazzaglia, M., Mascarello, F. 2002. Morphological and histochemical differences in the structure of the alimentary canal in feeding and runt (feed deprived) white sturgeons (*Acipenser transmontanus*). J. Appl. Ichthyol. 18: 341-346.

Fajfer, S., Meyers, L., Willman, G., Carpenter, T., and Hansen, M.J. 1999. Growth of juvenile lake sturgeon reared in tanks at three densities. N. Am. J. Aquacult. 61: 331-335.

Findeis, E.K. 1997. Osteology and phylogenetic interrelationships of sturgeons (Acipenseridae). Env. Biol. Fish. 48: 73-126.

Gawlicka, A., Teh, S.J., Hung, S.S.O, Hinton, D.E., Delanoue, J. 1995. Histological and histochemical-changes in the digestive-tract of white sturgeon larvae during ontogeny. Fish Physiol. Biochem. 14: 357-371.

Grande, L., and Bemis, W.E. 1991. Osteology and phylogenetic relationships of fossil and recent paddlefishes (Polyodontidae) with comments on the interrelationships of Acipenseriformes. J. Vert. Paleo. 11 (Suppl. 1): 1-121.

Gu, B., Schel,l D.M., Frazer, T., Hoyer, M., Chapman, F.A. 2001. Stable carbon isotope evidence for reduced feeding of Gulf of Mexico sturgeon during their prolonged river residence period. Estuar. Coast. Shelf Sci. 53: 275-280.

Harkness, W.J.K., and Dymond, J.R. 1961. The lake sturgeon: the history of its fishery and problems of conservation. Fish and Wildlife Branch, Ontario Dept. of Land and Forest, Toronto. 121 pp.

Held, J.W. 1969. Some early summer foods of the shovelnose sturgeon in the Missouri River. Trans. Amer. Fish. Soc. 98: 514–517.

Hoxmeier, R.J.H., and DeVries, D.R. 1997. Habitat use, diet, and population size of adult and juvenile paddlefish in the lower Alabama River. Trans. Am. Fish. Soc. 126: 288–301.

Hung, S.S.O., Lutes, P.B., Shqueir, A.A., and Conte, F.S. 1993. Effect of feeding rate and water temperature on growth of juvenile white sturgeon (*Acipenser transmontanus*). Aquaculture. 115: 297–304.

Jackson, J.R., VanDeValk, A.J., Brooking, T.E., vanKeeken, O.A., and Rudstam, L.G. 2002. Growth and feeding dynamics of lake sturgeon, *Acipenser fulvescens*, in Oneida Lake, New York: results from the first five years of a restoration program. J. Applied Ichthyol. 18: 439-443.

Jarvis, P.L., Ballantyne, J.S., and Hogans, W.E. 2001. The influence of salinity on the growth of juvenile shortnose sturgeon. N. Am. J. Aquacult. 63: 272-276.

Jennings, C.A., and Zigler, S.J. 2000. Ecology and biology of paddlefish in North America: historical perspectives, management approaches, and research priorities. Rev. Fish Biol. Fisheries 10: 167–181.

Johnson, J.H., Dropkin, D.S., Warkentine, B.E., Rachlin, J.W, and Andrews, W.D. 1997. Food habits of Atlantic sturgeon off the central New Jersey coast. Trans. Am. Fish. Soc. 126: 166-170.

Jollie, M. 1980. Development of head and pectoral girdle skeleton and scales in Acipenser. Copeia 1980: 226-229.

Jørgensen, J.M., Flock, Å., and Wersäll, J. 1972. The Lorenzian ampullae of *Polyodon spathula*. A. Zellforsch. 130: 362-277.

Jørgensen, J.M. 1980. The morphology of the Lorenzian ampullae of the sturgeon *Acipenser ruthenus* (Pisces: Chondrostei). Acta Zool. 61: 87-92.

Kelly, J.L., Arnold, D.E. 1999. Effects of ration and temperature on growth of age-0 Atlantic sturgeon. N. Am. J. Aquacult. 61: 51–57.

Keenlyne, K.D. 1997. Life history and status of the shovelnose sturgeon, *Scaphirhynchus platorynchus*. Env. Biol. Fish. 48: 291-298.

Kroll, K.J., Van Eenennaam, J.P., Doroshov, S.I., Hamilton, E.J., and Russell, R.R. 1992. Effect of water temperature and formulated diets on growth and survival of larval paddlefish. Tran. Am. Fish. Soc. 121: 538-543.

Kroll, K.J., Van Eenennaam, J.P., Doroshov, S.I., Linares, J., Hamilton, E.J., and Russell, R.R. 1994. Growth and survival of paddlefish fry raised in the laboratory on natural and artificial diets. Progr. Fish. Cult. 56: 169-174.

Kynard, B. 1997. Life history, latitudinal patterns, and status of the shortnose sturgeon, *Acipenser brevirostrum*. Env. Biol. Fish. 48: 319-334.

Kynard, B., Henyey, E., and Horgan, M. 2002. Ontogenetic behaviour, migration, and social behaviour of pallid sturgeon, *Scaphirhynchus albus*, and shovelnose sturgeon, *S. platorynchus*, with notes on the adaptive significance of body colour. Env. Biol. Fish. 63: 389-403.

Lindberg, J.C., and Doroshov, S.I. 1986. Effect of diet switch between natural and prepared foods on growth and survival of white sturgeon juveniles. Tran. Am. Fish. Soc. 115: 166-171.

Mason, W.T., Jr., and Clugston, J.P. 1993. Foods of the Gulf Sturgeon in the Suwannee River, Florida. Trans. Amer. Fish. Soc. 122: 378-375.

McCabe, G.T., Jr., Emmett, R.L., and Hinton, S.A. 1993 Feeding ecology of juvenile white sturgeon (*Acipenser transmontanus*) in the lower Columbia River. N.W. Sci. 67: 170-180.

Mckechnie, R.J., and Fenner, R.B. 1971. Food habits of white sturgeon, *Acipenser transmontanus*, in San Pablo and Suisun Bays, California. Calif. Fish Game 57: 209-212.

Michaletz P.H., Rabeni C.F., Taylor, W.W., and Russell, T.R. 1982. Feeding ecology and growth of young-of-the-year paddlefish in hatchery ponds. Trans. Am. Fish. Soc. 111: 700-709.

Miller, M.J. 1987. Feeding in the White Sturgeon, *Acipenser transmontanus*: Ontogeny, Functional Morphology, and Behaviour. M.S. Thesis, University of Washington. 75 pp.
Modde, T., and Schmulbach, J.C. 1977. Food and feeding behaviour of the shovelnose sturgeon, *Scaphirhynchus platorynchus,* in the unchannelized Missouri River, South Dakota. Trans. Amer. Fish. Soc. 106: 602–608.
Monaco, G., Buddington, R.K., and Doroshov, S.I. 1981. Growth of white sturgeon, *Acipenser transmontanus*, under hatchery conditions. J. World Mar. Soc. 12: 113-121.
Mohler, J.W., Fynn-Aikins, K, and Barrows, R. 1996. Feeding trials with juvenile Atlantic sturgeons propagated from wild broodstock. Progr. Fish. Cult. 58: 173-177.
Mohler, J.W., King, M.K., and Farrell, P.R. 2000. Growth and first-feeding and fingerling Atlantic sturgeon under culture conditions. N. Am. J. Aquacult. 62: 174-183.
Muir, W.D. and McCabe, G.T, Jr. 2000. Diet of first-feeding larval and young-of-the-year white sturgeon in the lower Columbia River. N.W. Sci. 74: 25-33.
New, J.G., and Bosdznick, D. 1985. Segregation of electroreceptive and mechanoreceptive lateral line afferents in the hindbrain of chondrostean fishes. Brain Res. 336: 89-98.
Norris, H. W. 1924. The lateral line organs of the shovelnose sturgeon, distribution and innervation. Proc. Iowa Acad. Sci. 31: 443-444.882
Radtke, L.D. 1966. Distribution of smelt, juvenile sturgeon, and starry flounder in the Sacramento-San Joaquin Delta with observations on food of sturgeon. pp. 115-129, In: J.L. Turner and D.W. Kelly, (eds.). Ecological studies of the Sacramento-San Joaquin Delta (Part 11, Fishes of the Delta). Calif. Dept. Fish Game Fish. Bull. 136
Rosen, R.A., and Hales, D.C. 1981. Feeding in paddlefish, *Polyodon spathula.* Copeia 1981: 441-455.
Ruelle, R., and Hudson, P. 1977. Paddlefish (*Polyodon spathula*): growth and food of young of the year and a suggested technique for measuring length. Trans. Am. Fish. Soc. 106: 609–613.
Russell, D.F., Wilkens, L.A., and Moss, F. 1999. Use of behavioural stochastic resonance by paddlefish for feeding. Nature 402: 291–294.
Sanderson, S.L., Cech, J.J., and Cheer, A.Y. 1994. Paddlefish buccal flow velocity during ram suspension-feeding and ram-ventilation. J. Exp. Biol. 186: 145-156.
Sbikin, Y. N. 1974. Age related changes in the role of vision in the feeding of various fish. J. Ichthyol. 14: 133-138.
Schreiber, M.R. 1962. Observations of the food habits of juvenile white sturgeon. Calif. Fish Game 48: 79-80.
Semakula, S.N., and Larkin, P.A.. 1968. Age, growth, food and yield of the white sturgeon, *Acipenser transmontanus*, of the Fraser River, British Columbia. J. Fish Res. Bd. Can. 25: 2589-2602.
Sillman, A.J., O'Leary, C.J., Tarantino, C.D., and Lowe, E.R.. 1999. The photoreceptors and visual pigments of two species of Acipenseriformes, the shovelnose sturgeon (*Scaphirhynchus platorynchus*) and the paddlefish (*Polyodon spathula*). J. Comp. Physiol. A 184: 37–47.
Smith, T.I.J., and Clugston, J.P. 1997. Status and management of Atlantic sturgeon, *Acipenser oxyrinchus*, in North America. Env. Biol. Fish. 48: 335-346.
Teeter, J.H., Szamier, R.B., and Bennett, M.V.L. 1980. Ampullary electroreceptors in the sturgeon *Scaphyrhynchus platorynchus* (Rafinesque). J. Comp. Physiol. 138: 213-223.
Traquair, R.H., 1887. The ganoid fishes of the British Carboniferous Formations. 31: 1-60.
Van Eenennaam, J.P., Webb, M.A.H., Deng, X., Doroshov, S.I., Mayfield, R.B., Cech, J.J, Hillemeier, D.C., and Willson, T.E. 2001. Artificial spawning and larval rearing of Klamath River green sturgeon. Trans. Amer. Fish. Soc. 130: 159–165.
Vecsei, P., Sucui, R., and Peterson, D. 2002. Threatened fishes of the world: *Huso huso* (Linnaeus, 1758) (Acipenseridae). Env. Biol. Fish. 65: 363-365.
Wilkens, L.A., Russell, D.F., Pei, X., and Gurgens, C. 1997. The paddlefish rostrum functions as an electrosensory antenna in plankton feeding. Proc. R. Soc. Lond. B. Biol. Sci. 264:1723–1729.
Wilkens, L.A., Wettring, B., Wagner, E., Wojtenek, W., and Russell, D. 2001. Prey detection in selective plankton feeding by the paddlefish: Is the electric sense sufficient? J. Exp. Biol. 204: 1381-1389.
Wojtenek, W. Pei, X., and Wilkens, L.A. 2001. Paddlefish strike at artificial dipoles simulating the weak electric fields of planktonic prey. J. Exp. Biol. 204: 1391-1399.

CHAPTER 5

STURGEON ECOMORPHOLOGY: A DESCRIPTIVE APPROACH

P. VECSEI AND D. PETERSON
Warnell School of Forest Resources, Fisheries Division, University of Georgia, Athens GA 30602 USA.

5.1 Introduction

The interactions of the environment and the physical features of an organism provide insights into how that organism survives and reproduces. The study of these associations, termed ecomorphology (ecological morphology), examines the relationship between an organism's morphology and its environment (Van der Klaauw 1948). Among sturgeons, the study of ecomorphology is critical in understanding how specific morphological characters function in meeting the life-history requirements. Historically, sturgeons have received much attention because of their large size, secretive natures, and prehistoric appearance. The morphology of sturgeons was first described by early naturalists during the 19th Century. However, only recently have ichthyologists considered form-function relationships. During the last decade, sturgeon have become a focal organism for many evolutionary biologists because of their ancient lineages and because extant forms have retained many "primitive" characters.

5.2 Evolution and Brief Overview of Family Acipenseridae

The family Acipenseridae consists of 25 extant sturgeon species (Birstein and Bemis 1997) and, along with the family Polyodontidae, comprises the order Acipenseriformes. Members of Acipenseridae are holarctic in distribution and well adapted to a wide range of aquatic habitats, possessing a variety of physical, chemical, and biological attributes (Choudhury and Dick 1998). The distribution of extant species within the group is indicative of the ancient relationships among fish faunas of North America, Europe and Asia (Berg 1968, Grande and Bemis 1991, Birstein and DeSalle 1998).

Acipenseriform probably evolved from a paleonisciform ancestor through paedomorphic reduction of endoskeleton ossification. The branchiostegal bone of the teleosts is absent in sturgeons and the specialized hyostylic jaw suspension and flexible palatoquadrate creates a protrusible mouth, permitting an outward projection that creates focused suction during feeding (Bemis & Grande 1997, Findeis 1997, Moyle &

G.T.O. LeBreton et al. (eds.), Sturgeons and Paddlefish of North America, 103–133.
© 2004 *Kluwer Academic Publishers. Printed in the Netherlands.*

Cech 2000). This unique arrangement, thought to have arisen during the Triassic in northern Asia (E. Artyukhin pers. comm), was probably paedomorphic in origin, however, because it facilitated a highly efficient and unique feeding capability, it ultimately emerged as a key morphological character shared by all sturgeon species (Gregory 1933).

Early works on acipenserid evolution theorized that the first sturgeon-like fishes were strictly benthic. Such an assumption would suggest that shovelnose sturgeons (Scaphirhynchini) are the most "primitive" sturgeon genera because they possess the most benthic morphologies and life history traits (see ecomorphology of benthic sturgeon, this paper). However, recent osteological and phylogenetic studies (Grande and Bemis 1996; Bemis et al. 1997, Findeis 1997) suggest that Huso is the most primitive genus, with Pseudoscaphirhynchus and Scaphirhynchus possessing the most highly evolved benthic morphologies within the Acipenseridae (Findeis 1997). Thus, within this phylogenetic arrangement, the acipenserines are phylogenetically intermediate to Huso and Scaphirhyincus in the degree of their benthic specialization. While this theory identifies Huso as the most primitive of sturgeons, it is based on the assumption that the extant Huso species have lost many of the earliest sturgeon characters as they adapted to a pelagic existence. Despite the widespread acceptance of this phylogenetic arrangement, Artyukhin (1995) has argued that Huso may actually be more closely related to more recently evolved species within the genus Acipenser.

5.3 Form, Function, and Movement

Just as the unique mouth arrangement has facilitated the development of a specialized feeding ecology, sturgeons utilize several other unique physiological characters that are closely linked to their basic life-history strategies. Among the contemporary members of the Acipenseriformes, the genera Huso and Acipenser are capable of osmoregulation in both fresh and marine or, at least, brackish water. Even the purely freshwater members of Acipenser have the ability to efficiently regulate blood ion concentrations in water with salinity as high as 15ppt (LeBreton and Beamish 1998).

Anadromous sturgeons moving between fresh and salt water, require precise control of serum water and ion concentration to maintain osmotic balance (Potts & Rudy 1972, Krayushkina et al. 1995). Sturgeons possess osmoregulatory mechanisms similar to those of other teleosts - when salinity levels are high, water lost across body surfaces is replaced by the consumption of sea-water and chloride cells of the gill epithelium remove excess sodium and chloride (Krayushkina et al. 1995). The larger ions, such as calcium and potassium, are excreted through the kidneys (LeBreton and Beamish 1998, Cataldi et al. 1995). This process ensures stable osmotic and ionic balances within the serum of anadromous sturgeons (McEnroe and Cech 1985), enabling them to move freely between fresh and marine systems as needed.

The ecological flexibility afforded by their osmoregulatory abilities is well suited to a migratory life history that allows sturgeons to use different habitats throughout their salutatory development. It also enables them to take advantage of temporal and spatial changes in benthic prey abundances within both fresh and saltwater portions of large estuarine systems. Freshwater species such as the lake sturgeon, *A. fulvescens*, often inhabit saline regions of large rivers, such as the lower St. Lawrence River and Hudson

Bay Tributaries. LeBreton and Beamish (1998) have shown that even freshwater species like the lake sturgeon, are capable of some degree of osmoregulation in brackish water environments. Sturgeons are highly migratory but the distance of the spawning run varies among species and even among stocks (Berg 1968, Scott and Crossman 1973, Bemis and Kynard 1997). Within the Acipenseridae, evolutionary processes have created many divergent life histories – some potamodromous (e.g. *A. fulvescens*), some amphidromous (*A. brevirostrum*), and some anadromous (*A. oxyrinchus*). Because all larval sturgeon are intolerant of salt water, freshwater spawning habits are critical for all species. The varying life history patterns account for the present biogeography of sturgeons worldwide (Bemis and Kynard 1997).

Respiration in sturgeons is facilitated by a unique arrangement of the mouth, buccal cavity, gill arches, and sub-operculum. Burggren (1978) found that the postero-dorsal margin of the operculum is emarginated – an adaptation that allows water to flow into the dorsal region of the opercular chamber, through the posterior gill arches, and back out the ventral opening of the sub-operculum (Figure 5-1). Although most sturgeon are continuous cruisers, the use of ram ventilation is precluded by the subterminal placement of the mouth and the ability for continuous feeding. Members of Huso (beluga and kaluga) are notable exceptions - the mouth of these piscivorous species projects forward (Vecsei et al. 2002).

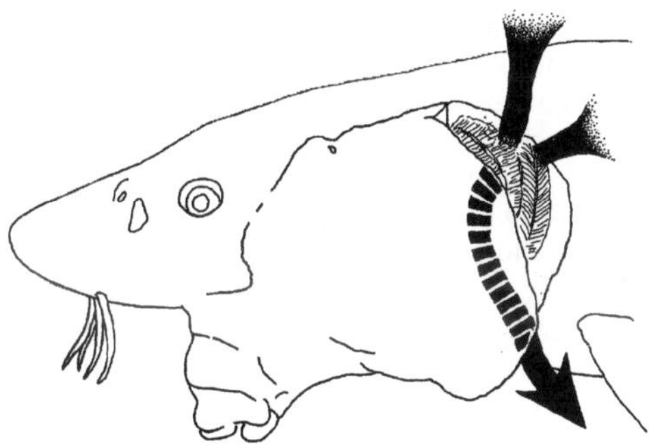

Fig. 5-1. Ventilation of sturgeon gills during feeding.

Although all sturgeon share a similar body design with many specializations common to all species, substantial anatomical differences exist among the four genera depending on phylogenetic relationships (Billard and LeCointre 2001, Billard 2002). In Acipenser, for example, the rostral bones (rostralia) are widened, forming interlocking plates that essentially sheath the snout, protecting the rostrum both dorsally and ventrally. Members of Huso, however, possess poorly defined dermal elements within the rostrum, skull roof, orbit and opercular areas (Findeis 1997).

5.4 Descriptive Morphology

Like most other fishes, sturgeons have a body plan comprised of three regions: head, body, and fins. Each of these regions is characterized by a number of specialized adaptations – some typical of all sturgeons while others are species specific. These sometimes subtle morphological variations are used to identify different species or, in some cases, to identify different populations within a species (Figure 5-2).

A: Rostrum or snout (terminal portion of head) slightly down-curved but with an upturned tip. The snout length is measured from the anterior rostrum tip to the anterior edge (origin) of the eye. The rostrum used in ecomorphological measurements as a criteria relating to flow and/or feeding methods.

B: Spiracle: a small opening above the gills in Huso and Acipenser - absent in Scaphirhynchus and Pseudoscaphirhynchus

C: Interorbital width: distance between eyes (measured from a dorsal perspective). This is a useful character to distinguish different stocks of the same species.

D: Barbels: single row of four barbels situated approximately mid way between tip of snout and upper lip. Barbels are covered with chemoreceptor and act more like " tasters" rather than "feelers". The length of the 2 inner barbels relative to the outer pair is also an important criteria for distinguishing the different shovelnose sturgeons. Barbels of some species are fimbriated (see barbels of shovelnose sturgeons).

E: Mouth: subterminal, protractile, and lacking teeth. Thick lips surround the mouth and the upper lip is interrupted in all species except ship sturgeon (*A. nudiventris*). The sturgeon mouth is unique in lacking a jaw bone, and the mouth is capable of extensive downward and outward projection. Mouth dimensions in sturgeon are often used in ecomorphological analyses: mouth width or gape relates to diet (e.g.: prey size), and hence, ecology. In proportion to body size, the smallest mouth occurs in the Atlantic sturgeon that feeds primarily on bryozoans and small invertebrates. In contrast, white sturgeon have the largest mouth to accommodate a much more varied diet which, at times, may include adult salmon (both live and dead).

F: Eyes: small in proportion to head. Compared to other fishes that feed by sight, visual acuity of sturgeon is poor. Eye size and position are often important ecomorphological measures, as they may relate to water clarity and method of feeding.

G: Operculum and pre-operculum: absent in sturgeon, with sub-operculum acting as gill cover. Note the gap in the upper gill opening, a feature common to all sturgeons thought to increase ventilation efficiency during feeding.

H: Dorsal fin: posteriorly located on dorsum of trunk. Like other sturgeon fins, the dorsal fin is comprised of numerous segmented rays incapable of independent movement and lacking inter-ray membranes.

I: Caudal fin: heterocercal in form, with notochord extending into the upper portion of fin (small number of ganoid scales present). The upper lobe is longer than the lower lobe but in large or old individuals, upper and lower lobes may appear equal in length (epibatic). Lower lobe is typically reduced in the more benthic species (e.g. Scaphirhynchus sp.)

J: Caudal peduncle: appears narrow and round in cross-section, more flattened in the shovelnose sturgeons.

K: Spines at snout tip: occur only in two genera of shovelnose sturgeons and not all species within the genera. While these spines may be pronounced in juveniles, they typically disappear with age.

L: Preorbital spines: situated directly anterior to the eyes in shovelnose sturgeons.

M: Parietal spines: situated on the dorsum of the head in shovelnose sturgeon.

N: Frontal spines: situated on the dorsum of the head, posterior to the parietal spines in shovelnose sturgeon.

O: Ventral squamation: refers to the presence and type of scalation (dermal denticles of varying shape and size) on the ventral surface. This character is useful in distinguishing various Eurasian and North American species of shovelnose sturgeons.

P: Anal fin: situated on the midline of the ventral surface, below or slightly posterior to origin of dorsal fin. The shape and relative size of the anal fin is similar among all sturgeon species.

Q: Pelvic fins: sub-abdominal, paired fins occurring on either side of the body midline.

R: Pectoral fins: paired fins positioned behind and slightly below the sub-opercular opening. Sturgeon pectoral fins are large, with an ossified leading edge.

S: Scutes: prominent bony plates occurring in longitudinal rows extending from the base of skull to tail. All sturgeon possess five rows of scutes; a single dorsal row and two lateral and ventral rows. The number of scutes within a row is a key meristic character used to distinguish species. Scutes are also variable in shape and relative size but most possess a rearward projecting, "apical hook". The prominence of this hook varies with age and species.

T: Scutelets: similar to scutes but much smaller and not part of the 5 longitudinal rows of principal scutes. Scutelets are usually confined to laterodorsal portions of trunk except in the green sturgeon.

U: Denticles: tiny star-like ossification present over entire trunk. Density and shape varies with species or age of individual.

V: Plates: similar to scutes except they lack the apical hook. The post-dorsal and anal plates (post or pre-anal) are a key diagnostic character in the identification of Acipenserids. Tightly sutured plates also cover the head and dorsum of the snout.

W: Posterior edge of branchiostegal membranes: sub horizontal in shovelnoses but v-shaped in Acipenser.

X: Anterior edge of clavicle: part of clavicle closest to the head.

Y: Clavicle keel: plates of bone situated ventrally, posterior to head and below pectoral fin insertion.

Z: Life colours: not a useful diagnostic in adult sturgeons, however, may be useful in distinguishing juveniles of different species.

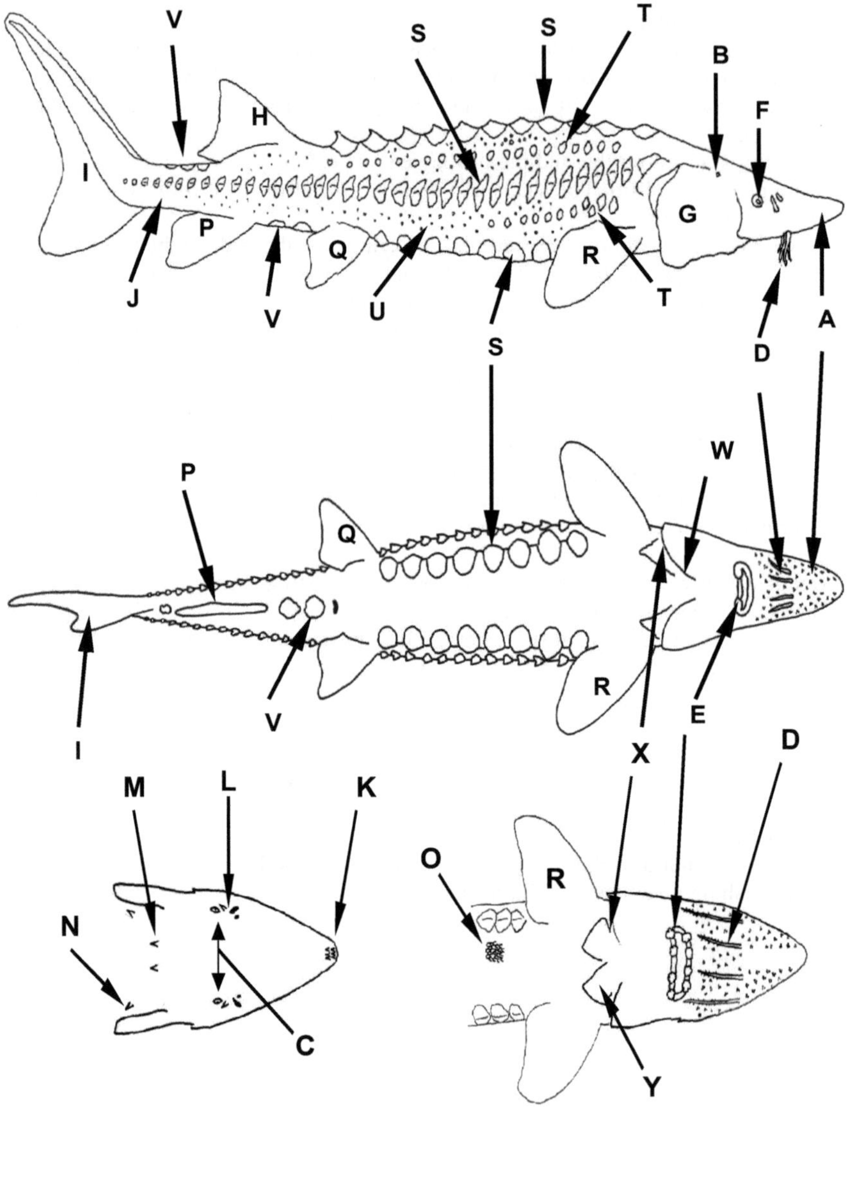

Fig. 5-2. Gross morphology of sturgeon

5.5 Ecomorphology

"Ecological morphology (ecomorphology) is the study of the relationship between the morphology of organisms and their environment"
- Van der Klaauw, 1948

Ecomorphology is the study of how specific morphological characters function in meeting the life-history requirements of the species being studied. While the term is still used in a general sense to refer to the ecological functions of various anatomical structures, modern scientists have subdivided this field into 3 distinct sub disciplines: functional, ecological, and descriptive. Functional morphology tests, through quantitative means, the ability of various structures to perform functions. A functional ecomorphologist, for example, would be interested in determining the drag coefficients from pectoral fins of Acipenser versus those from Huso. Studies in functional morphology help quantify the unique adaptations of different sturgeons. Ecological morphology examines how differing morphologies relate to particular environments. The digging-feeding method of foraging used by *S. platorhinchus*, with its wide spatulate rostrum illustrates such a relationship. Ecological analysis of sturgeon morphology has helped fisheries biologist understand the species-specific habitat requirements of sturgeons. Descriptive morphology relies on the combined use of meristics, morphometrics, and gross morphology for taxonomic identification. While this approach has obvious uses in the identification of sturgeon species, it also has lead to the discovery of subtle morphological differences among different populations – a discovery that has important implications for future management of many depleted stocks. In the following discussion of sturgeon ecomorphology, we will draw upon all three sub disciplines of ecomorphology in emphasizing the associations of physical form and ecological function.

Descriptive morphology of sturgeons has been well documented by Berg (1968) Scott & Crossman (1978) Holcik (1989) and Billard (2002). Although these authors provide thorough species-specific descriptions of morphology, most do not describe the gross morphological differences typical of juveniles, sub-adults, and adults within species. Useful summaries of North American sturgeon ecology are discussed in Moyle & Cech (1999) and Bemis et al (1997), Findeis (1997). Some ecomorphological studies summarize numerous morphological traits through multivariate analysis. The various morphological features are then linked to specific ecological features of the species in question.

5.6 A New Ecological Guild: The Supra-Benthos Cruisers

Early attempts at sturgeon classification placed the Acipenserids with many species of sedentary benthic species such as suckers, catfishes, darters, and others. While trends in Acipenserids have resulted benthic life strategies, these fish have not developed benthic specializations seen in other groups.

Benthic fish	*Sturgeons*	*Pelagic fish*
↑	↑	↑
Flattened trunk	cylindrical, tapering trunk	oval trunk
Stationary lifestyle	cruising while foraging	highly active swimmers
Highly camouflaged	basic one colour	little camouflage
|	|	|
Scaphirinchinae	Acipenser	Huso

A hyostylic jaw suspension and downward jaw projection is typical of North American Acipenser. This differs markedly from the forward projecting jaws of Huso (Figure 5-3). When not feeding, the Acipenser jaw remains entirely within the ventral surface of the head. Once food has been "sucked" into the mouth, it is ground by the palatal complex (flat cartilages) on the roof of the mouth that moves across the tongue pad (Moyle and Cech 2000, Findeis 1997).

Sturgeons display varying degrees of flattening of the rostrum (Plate 1). *Huso* possess a conical shaped snout while Scaphirhynchus and Pseudoscaphirhynchus possess flat, broad snouts that are used to stir-up and expose buried prey (Moyle and Cech 2000, Mayden and Kuhajda 1997). Acipenser is nested somewhere in the middle, having a rostrum sensory ampullae, some of which are present on the sub-operculum and posterior to the eye orbit. This allows them to detect faint electrical fields emitted by prey, buried in the substrate. The ventral surface of the rostrum is adorned with two sets of barbels covered by chemoreceptive organs. This allows sturgeons cruising along the bottom substrate to taste the items their barbels come in contact with.

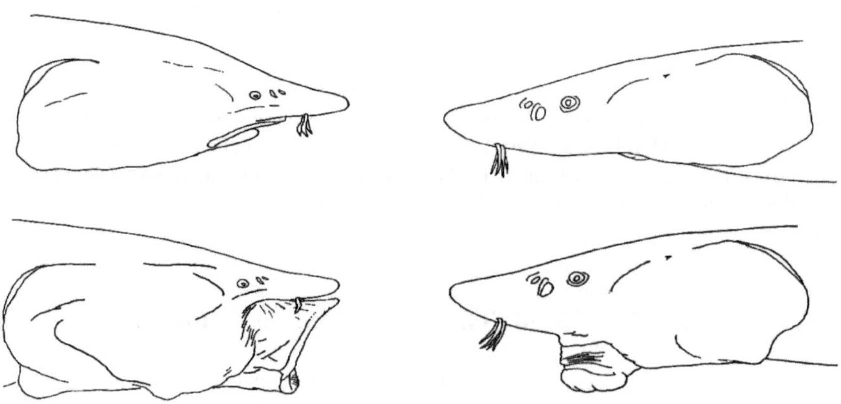

Fig. 5-3. Comparison of mouth projection in (a) Huso and (b) Acipenser

Acipensers are strong swimmers, cruising above the bottom substrate in search of food. The shorter lower lobe of the caudal fin (hypochordal lobe) allows tail movement while swimming near the bottom substrate. Pectoral fin spines remain continually erected during swimming. Sturgeons have developed some ingenious solutions to the problems of moving through a highly viscous and dense medium. Similar to most fishes, sturgeons share a common design; fins, acting as control surfaces for moving through the water medium. The sturgeon's heterocercal tail produces both lift and thrust. This differs markedly from the homocercal tail of more advanced fishes that generates thrust without lift. Earlier works suggested the pectoral fins generated lift to counter the lift and movements created by the heterocercal tail (Videler 1993). The classic model states that the beating of the sturgeon tail tends to pitch the head downward. An upward inclination of the pectoral fins would generate a counteracting lift force. This net upward lift is balanced by the weight of the fish. In this scenario, the vertical forces of downward and upward lift, along with thrust, are balanced. Recent experiments by Wilga & Lauder (1999) using Digital Particle Image Velocimetry to describe and quantify the three-dimensional flow patterns around sturgeon, suggest that the lift generated by the pectoral fins is negligible. Rather, they propose four vertical components of force acting on a sturgeon during horizontal swimming (Figure 5-4):

1. Downward force at centre of mass.
2. Upward force due to dorsally pitched ventral surface anterior to centre of mass.
3. Upward force due to dorsally pitched ventral surface posterior to centre of mass.
4. Longitudinal force from caudal fin oscillation.

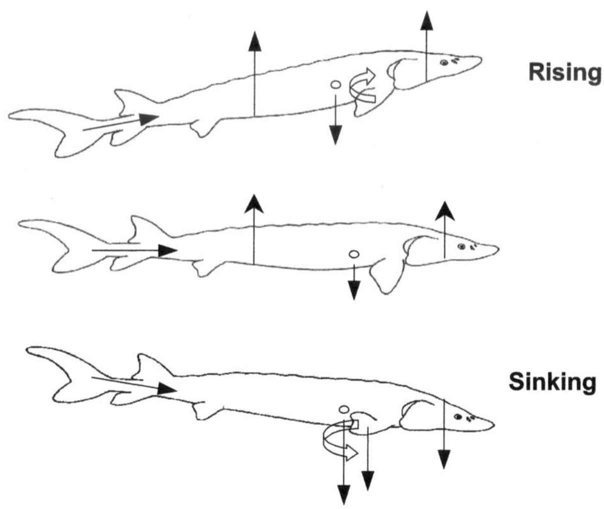

Fig. 5-4. Vertical components of force during swimming

This new hypothesis demonstrates the minimal contribution of the pectoral fins towards creating lift during horizontal benthic cruising. Electromyographic and kinematic data shows that the posterior flap of the pectoral fin is adjusted to reorient the head and body. This pitching, in turn, induces rising and sinking of the sturgeon (Wilga & Lauder 1999).

An interesting example of extreme benthic specialization is illustrated by the Amu-Darya shovelnose (*Pseudoscaphirhynchus hermanni*). The outer margin of the enlarged pectoral fin is curved upward to facilitate a side-to-side "shuffling" motion when the fish is foraging along the benthos (Figure 5-5) (Berg 1968, Findeis 1997). The flattened body shape of Scaphirinchinae allows them to maintain position in the large swift rivers they inhabit.

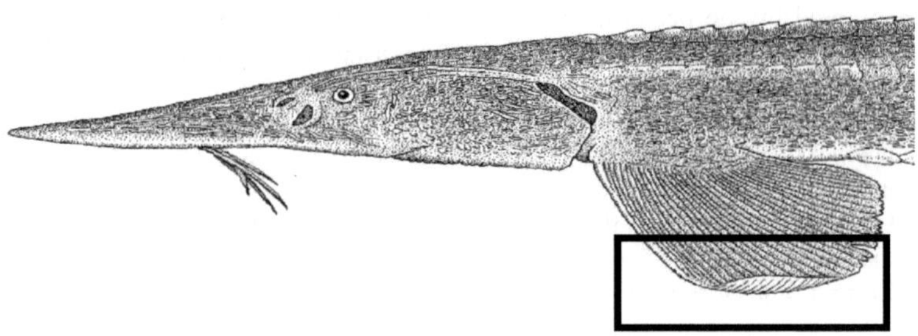

Fig. 5-5. Pectoral fin adaptation of Pseudoscaphirynchus

Many benthic fishes possess extensive scalation as a means of defence against predators, or as protection against substrate. The role of scutes in sturgeon is primarily for defence. Marine species of Acipenser maintain their scutes throughout their lives, living in an environment with more potential predators. This is in stark contrast to freshwater species, which undergo scute absorption as adults (Artyukhin 1995, Holchik 1989). The close proximity of scutes and denticles become reduced with increase in size in all Acipenserids. Body armouring in the form of scutes, plates, scutelets and denticles is an adaptation for juveniles vulnerable to predation and associated with benthic living (Figure 5-6).

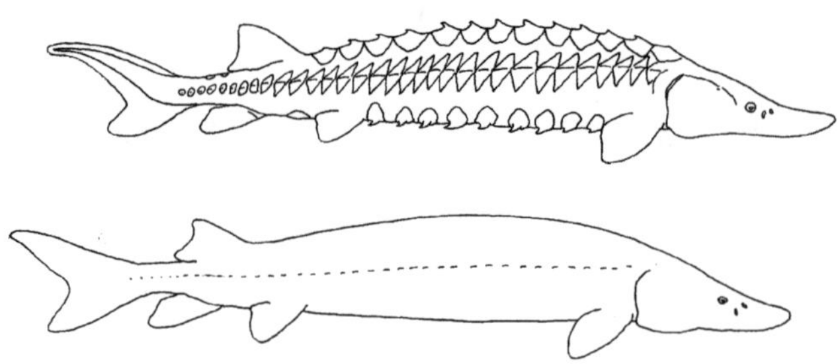

Fig. 5-6. Armouring of juvenile (top) versus adult (bottom) lake sturgeon

The shovelnose sturgeons represent the extreme in body armouring. Almost no part of these sturgeons is left exposed. The head of the shovelnose sturgeon (*S. platorhinchus*), besides being covered by closely joined plates, is protected by numerous sharp backward curving (retrose) spines on the dorsum of the snout (Figure 5-7).

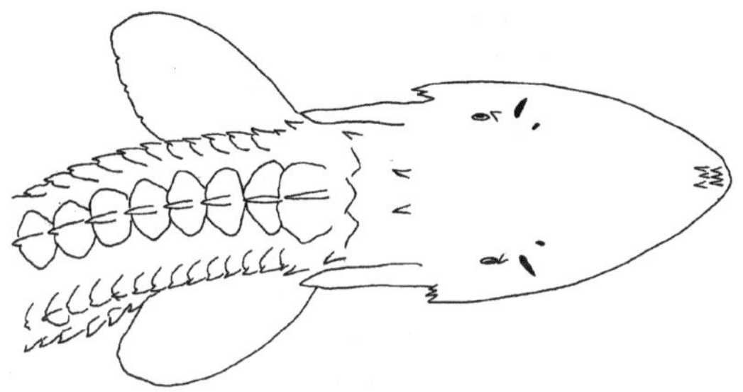

Fig. 5-7. Retrose spines on head of shovelnose sturgeon.

The retrose orientation of these spines maximizes protection, but also permits forward movement while in direct contact with the benthos during foraging. Single spines are present on the post temporal, tabular, and parietal bones (Mayden and Kuhajda 1996). Unique to the genus Scaphirhynchus, is a single retrose spine located anteriordorsal to the orbit (Berg 1968, Mayden & Kuhajda 1996, 1997). The entire caudal peduncle is covered with scutes and plates (Figure 5-8) and the skin surface is similar to chain armour in the form of ventral squamation or scutelets and denticles on the lateral and upper surface. These morphological adaptations permit sturgeons to be successful bottom feeders, preying on molluscs, worms, and insects and to a lesser degree, small fishes.

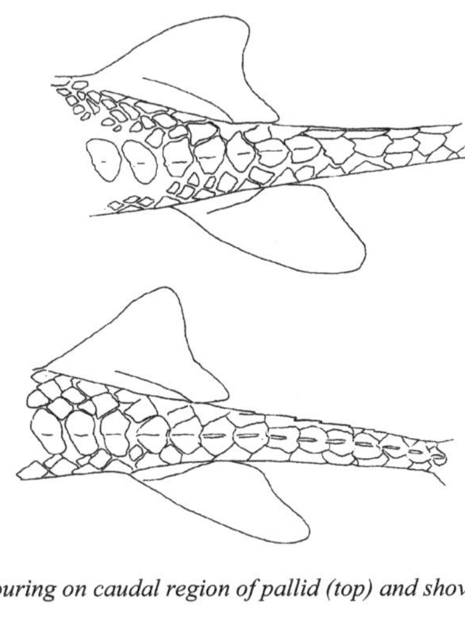

Fig. 5-8. Armouring on caudal region of pallid (top) and shovelnose sturgeon (bottom).

5.7 Descriptive Morphology of North American Species

5.7.1 PHYLOGENETIC AND ONTOGENETIC MORPHOLOGICAL VARIATION

Phylogenetic variation, as defined by Hilton and Bemis (In press), is simply the morphological differences among closely related taxa. In sturgeon, phylogenetic differences among species provide the basis for species recognition, although, these difference may be minimal in closely related species. For example, the morphologies of the Eurasian kaluga (*H. dauricus*) and the Ponto-Caspian beluga (*H. huso*) differ only in the size of their first dorsal scutes (Berg 1968). Similarly, the Persian sturgeon (*A. persicus*) is nearly indistinguishable from the Russian sturgeon (*A. gueldenstaedtii*) except by experienced ichthyologists using a combined meristic and morphometric approach.

Unlike phylogenetic differences in morphology, ontogenetic variation refers to the morphological changes that occur within the lifespan of an individual. While ontogenetic variation may occur as "punctuated" events associated with discrete life stages, it also may occur gradually and continuously throughout the entire lifespan. For a detailed species description, both of these morphological variations should be included. Sturgeon species vary in their degree of ontogenetic morphological variation. The lake sturgeon, however, (*Acipenser fulvescens*) is a good example of a species where juvenile morphology differs significantly from that of sub-adults and adults. Hence, competency in sturgeon identification requires not only an understanding of the morphological differences among species, but also a recognition of the complex differences in morphology attributable to ontogenetic variation (e.g.: egg, larva, juvenile, adult).

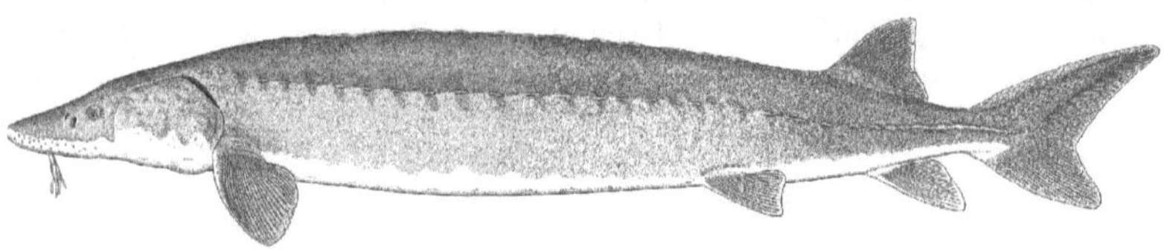

Fig. 5-9. Lake Sturgeon - Dorsal scutes 9-17, Lateral scutes 29-42, Ventral scutes 7-12, Dorsal fin rays 35-45 Anal fin rays 25-30

5.7.1.1 Lake Sturgeon (Acipenser fulvescens)

The lake sturgeon has an elongated body, pentagonal in cross section in juveniles but becoming rounded in adults (Scott & Crossman 1973). Rostrum length shows little variability in adults but is proportionally larger in juveniles. The tip of the rostrum is slightly upturned. Mouth of lake sturgeon is transverse and approximately 70% of interorbital width. The top lip is continuous and the bottom lip is interrupted. Mouth shape and size in relation to head is most similar to that of shortnose sturgeon (Plate 2). There are two sets of barbels situated closer to tip of snout than to the origin of mouth. The barbels are proportionally longer on large adults than smaller fish. Body armouring is extensive on juveniles but becomes progressively reduced in adults (Plate 3). Laterodorsal and lateroventral surface tightly peppered with denticles in individuals up to 100cm in length. All five principal rows of scutes are gradually reabsorbed after maturation and disappear completely in very large individuals. The numerous bony plates of the head are evident in juveniles, some young adults, but eventually become covered by skin. The parietal bony plates rarely reach as far forward as the midpoint between the eyes (Plate 4). The number of post dorsal plates may vary in number, size and grouping (Plate 5). Typically, they are 1-2 in number and unpaired (Hochleithner and Gessner 2001). Some individuals have two large plates followed by a pair of smaller ones. The pre-anal plates are in single file, relatively large and 1-2 in number (Plate 6) (Hochleithner and Gessner 2001). On very large individuals, both post-dorsal and post-anal plates become completely absorbed. The anal fin originates behind the origin of the dorsal fin. The caudal peduncle is longer than that of the shortnose sturgeon. The tip of the anal fin rarely surpasses the caudal fulcrate plate, unlike the shortnose sturgeon where it reaches origin of anal fin. Most adult lake sturgeon range in size from 15-70kg. A specimen weighing 310lb (145kg) and 238cm was captured from Lake Superior in 1922. Numerous individuals over 100kg have been captured throughout the Great Lakes basin. For detailed review of maximum size for this species, see Harkness & Dymond (1961).

Body colouration is variable among species within a stock or between populations but is most typically dark brown, pale brown, slate grey, and rarely black. Colouration along lateroventral surface lightens gradually towards ventrum. Entire ventral surface is white but some individuals have grey or black pigmentation on the ventrum of the head on the lips and barbels. Some rare adult individuals have white blotches or spots on the lateral surface of body. Rarely, an adult may have a few remaining black spots from its juvenile stage. In juveniles and sub-adults, scutes and other ossifications are of the same colour as surrounding body. Lateral scutes may sometimes be lighter than body. Juveniles have overall coppery-metallic sheen, especially evident on the lateroventral surface of the body. Juveniles have large black saddle marks and peppered with smaller black spots that disappear prior to maturation. Some juveniles and sub-adults are overall light grey with a yellowish tinge apparent in all principal rows of scutes. Juveniles have dark, irregular blotches and/or spotting. Post-yolk sac larvae have a characteristic dark lateral band.

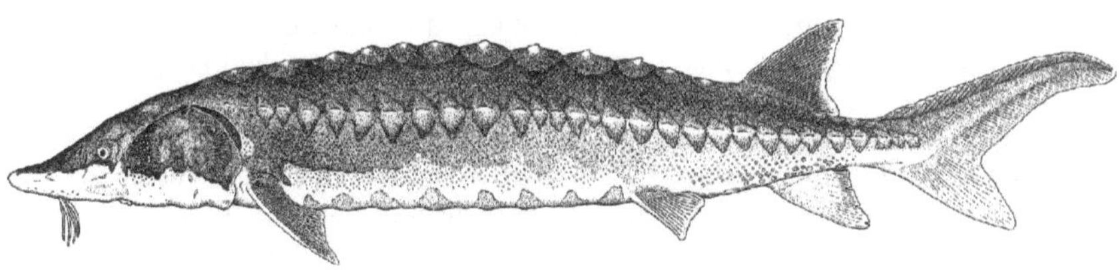

Fig. 5-10. Atlantic Sturgeon - Dorsal fin rays 38-46, Anal fin rays 26-28, Dorsal scutes 7-16, lateral scutes 24-35, ventral scutes 6-9.

5.7.1.2 Atlantic sturgeon (Acipenser oxyrinchus)

The Atlantic sturgeon has an elongated body, "pentagonal" in cross section, with each of the five rows of hooked scutes constituting the "apex" of the pentagon shape. The rostrum is the most elongated of North American Acipensers comprising 30-50% of total head length. Among North American sturgeons, the fontanelle on the dorsum of the head is present only in *A. oxyrinchus* (Plate 4). The two pairs of barbels originate anterior to the midline between tip of snout and mouth origin. The football shaped mouth is distinctly different (Plate 2) from the straight or slightly crescent-shaped oval mouth typical of other Acipensers (Vecsei 2002). Eyes of the adults are positioned within a prominent brow ridge not found in other members of the genus.

The major rows of scutes are the largest of the Acipenser, the apex of which forms a sharp posteriorly curved hook. The laterodorsal and lateroventral surface of the body is covered by rhombic platelets – distinguishable from abdominal plates by their smaller size and from scutes by the absence of an apical hook. All scutes, plates and platelets are retained in adulthood. All principal rows of scutes are enormous and have a radiating alveolar-tubercular pattern originating at the scute apex. This is particularly pronounced on the lateral row of scutes but may fade somewhat in very large or old adults. The characteristic of complex adult armouring also is typical of Pacific marine Acipenser species. Unique to the Atlantic sturgeon are the large rhombic plates found along the lateral base of the anal fin (Plate 6). The laterodorsal and ventrolateral surface of the body is densely armoured with rhombic plates arranged in oblique rows (Plate 3) (Berg 1968, Artyukhin & Vecsei 1999). Unlike other sturgeon species in North America, the distance on the laterodorsal and lateroventral surface, between ossifications (rhombic plates) is less than the size of the plates. In other species, the distance between ossifications (denticles) on the body is greater than the size of the denticles. The Atlantic sturgeon is almost entirely covered by bony armour. Pre-anal plates are enormous, paired and often followed by a second pair of plates and another, larger single plate. The presence of these paired pre-anal plates is unique to this species (Plate 6). The post dorsal plates are also paired and most individuals possess two or three sets, both of which are much larger than those found in other sturgeons (Plate 5). In juveniles, these post dorsal plates form a carapace by being in contact with one another (Artyukhin and Vecsei 1999). The dermal plates of the head possess some unique features; the parietals are short, rarely reaching the midpoint of the eyes. The frontal dermal plates are larger than the parietals and the rostralia are set close to one another (Plate 4). The rostralia also lack the strong indentations of *A. transmontanus*, *A. medirostris* and, to a lesser degree *A. brevirostrum*. In North America, the Atlantic sturgeon is second only to the white sturgeon in maximum size. Adults returning to spawn in fresh water range in size from 30-150kg. Most individuals caught on the St. John River, New Brunswick are less than 100kg. A 150kg individual is approximately 3m in length and 60 years of age.

The colouration of Atlantic sturgeon is consistent throughout the species' range. The dorsum of head, body and dorsal scutes dark are brown in colour. The laterodorsal surface is brown with slight metallic, copper sheen in juveniles and sub-adults. Generally, the lateral scutes are of the same colour or slightly lighter than surrounding body. The lateroventral portion of trunk is similar to laterodorsal colouration but somewhat lighter. The central portions of lateral and dorsal scutes are lighter than margin of scutes or surrounding body, a characteristic most prominent in juveniles. The central portion of head plates is also lighter in juveniles. The ventral scutes are ivory or pale beige in colour and entire ventral surface is white. Adult members of the genus are easily identified by a prominent black patch located on the ventral tip of the rostrum (Plate 2). The size and shape of this marking may vary and is not present in juveniles (Vecsei & Peterson 2000). A distinct band formed at the lateral "edge" of the rostrum may readily identify juvenile Atlantic sturgeon. This colour character is lost shortly after maturity. The fins of Atlantic sturgeon are of the same colour as the lateral surface of trunk but the leading edge of all fins is white. A white tip is often visible on the apex of all scutes but this fades with advanced age or great size. The rhombic platelets are of same colouration as surrounding body while the lateral scutes are usually lighter than surrounding body.

Fig. 5-11. Shortnose sturgeon - Dorsal fin rays 38-42, anal fin 19-22, dorsal scutes 8-13, lateral scutes 22-33, ventral scutes 7-11.

5.7.1.3 Shortnose sturgeon (Acipenser brevirostrum)

Shortnose sturgeon have an elongated body with greatest depth occurring at the 2nd -4th dorsal scute posterior to the head. The head is 22-28% of fork length (Scott & Crossman 1973). The parietal dermal plates of the head are short but reach well past the midpoint of the eyes. The frontals are wide and large relative to the parietals (Plate 4). The rostrum length is very variable, consisting approximately 25-40% of total head length (Walsh et al. 2001). The two pairs of barbels originate closer to the tip of the snout than upper lip. The mouth is wide with the upper lip indented in the middle and the lower lip interrupted. Numerous small denticles are distributed over the entire laterodorsal and lateroventral surface. The scutes remain prominent in adults but the hook appendage loses its sharp edge. There are 1-3 pre-anal plates posterior to the rectum, never paired (Plate 6). There are 1-2 small, paired post-dorsal plates (Plate 5)) (Vecsei & Peterson 2000, Hochleithner 2001). The caudal fin is heterocercal but in old or senescent individuals, the upper and bottom lobe are of the same length. The eye is considerably larger in adults than other species. The average size of shortnose sturgeon varies between populations. On the Hudson River, adults are usually 600-850mm in length and weigh between 3-5kg. Similar sizes are reported elsewhere except maximum size for this species is greatest on the St. John River, New Brunswick, Canada. Numerous individuals over 15kg are captured and individuals of 23kg are caught almost yearly (M. Litvhak pers. comm.)

The dorsum of the head, body and lateral scutes is dark brown, peanut brown or black-brown in colour. A dark contour enhances the periphery of dorsal and lateral scutes. Adults often have a faded black spot on tip of the snout dorsum. The laterodorsal surface gradates from dark to light brown. The region surrounding the lateral scutes may have an iridescent, coppery sheen. The ventrolateral surface is considerably lighter than upper surface with colouration often ending abruptly before reaching the ventral scutes or ventrum. The entire ventral surface is white or off-white. All paired fins are pigmented and have a white outline along the margins.

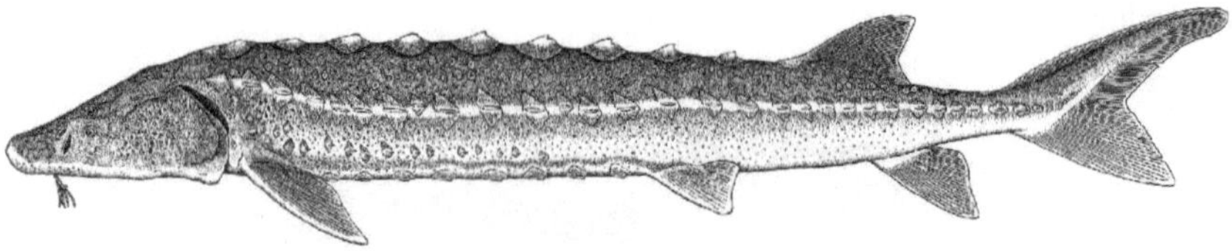

Fig, 5-12. Green sturgeon - Dorsal fin rays 12-36, Anal fin rays 11-19, dorsal scutes 7-12 lateral scutes 22-33, ventral scutes 7-12.

5.7.1.4 Green sturgeon (Acipenser medirostris)

The green sturgeon has a slender, tapering body with maximum depth occurring between the first and third dorsal scute, ranging from 9-13% of total length. The overall morphometric proportions and meristic counts are extremely variable (North et al 2002). The rostrum is elongated and somewhat flattened with a width 10.9-18.2% of head length. The rostrum shape is variable and ranges from 8-14% of head length (Figure 5-12) (North et al. 2002). The head length is 16.2-21.6% of total length. The parietal bones are roughly twice the size of the squamosums. The frontal bones are prominent and the medials sometimes consist of 2 or even 3 pieces. Several small bony plates are often present at the supratemporal-nuchal-parietal junction. The rostralia are usually numerous and elongated but some rare specimens have a very large single rostralia covering the anterior portion of the rostrum surface (Plate 4). Unlike most other sturgeons, the tip of the rostrum is not upturned in *A. medirostris* (Plate 1). Unusual among Acipenser is the two pairs of barbels which are situated equidistant between the upper lip and tip of snout or closer to the mouth (North et al. 2002). The mouth is large for an Acipenser and the bottom lip is thick and interrupted in middle (Plate 2). Unlike the white sturgeon, the position of the anus is anterior to the dorsal fin insertion.

The green sturgeon exhibits a high degree of ossification. Principal rows of scutes are maintained throughout life and the trunk is covered by smaller scutelets, platelets and denticles. A row of lateroventral scutelets is often present (see lateral view) (Vecsei 2002). The pectoral fins are more pointed than *A. transmontanus* with a moderately developed bony w-spine (Findeis 1997).

All five rows of principal scutes remain prominent throughout life and harbour sharp hook-like appendages. Hooks on dorsal and lateral scutes become progressively more prominent posteriorly. Green sturgeon have one to two postanal and postdorsal plates. The postdorsal plates are elongated and sometimes paired (Plate 5). There are also one to three preanal plates (Plate 6) (Hochleithner 2001). Average size of adults in Eastern Pacific waters is 120-200cm and 30-50kg. Commercial fisherman report maximum weight for this species is somewhere between 200-300lb.

The colouration on the dorsum of head, body and dorsal scutes is olive to bright green. The dorsal scutes are lighter with a slight yellowish tinge. The dorsal head plates are light green to yellowish green. The contour of each dermal head plate is pronounced due to the dark suturing between the various plates. The laterodorsal surface is the same or lighter than the dorsum with denticles and scutelets noticeably lighter. Dorsal scutes are lighter than dorsum. The lateral scutes are lighter than the surrounding body with a white stripe passing laterally through the entire row of scutes. The colouration on the lateroventral surface ends abruptly midway between the lateral and ventral scutes. Characteristic to this species is the lateroventral dark band running from the anterior of the body to the ventral fin origin. Most individuals also have a distinct, dark band on the midline originating between pectoral fins to midway between pectoral fin and pelvic fin origin. The apex of scutes and scutelets is white, creating a brilliant contrast against the green overall body colouration. Denticles covering the body lack any white colouration and appear the same colour or slightly darker than surrounding body. *A. medirostris* also has an olive coloured longitudinal band on the ventral portion of the anterior trunk region. The ventral surface is bright white or cream in colour.

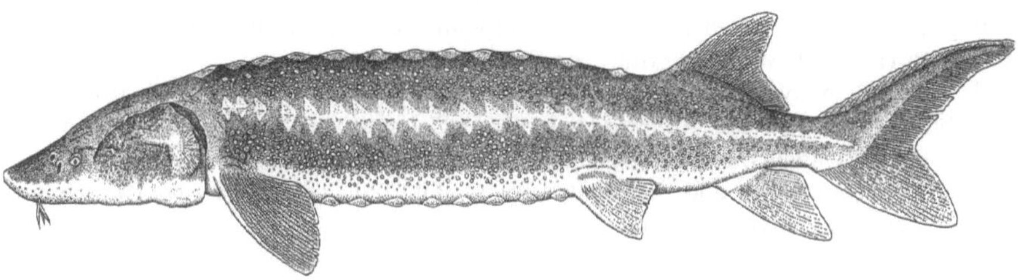

Fig. 5-13. White sturgeon - Dorsal fin rays 42-53, anal fin rays 27-32, dorsal scutes 11-14, lateral scutes 36-48, ventral scutes 9-12

5.7.1.5 White sturgeon (Acipenser transmontanus)

The body of white sturgeon is more robust than the sympatrically occurring green sturgeon. Adults vary from slender to rotund with maximum body depth occurring between the 3rd and 6th dorsal plate. The head is proportionally large and broad, approximately 25% of total length. The parietal dermal plates of the head are long, with the anterior portion often passing the midpoint of the eyes. There seems to be no discernable medial dermal plate, normally situated in between the frontals. The parietals come in direct contact with the rostralia which are numerous, small and serrated (Plate 4). The rostrum of white sturgeon is more prominent in juveniles (longer than postorbital length) and becomes proportionally smaller in adults (shorter than postorbital length). The rostrum shape and size is somewhat variable but less than in *A. medirostris*. Large, adult white sturgeon have a more blunted snout than younger individuals. The mouth of this species is the largest among North American Acipenser (Plate 2). The barbels are situated approximately equidistant between the mouth and tip of snout but closer to tip of snout. The pectoral fins are rounded with a moderately developed bony w-spine.

The white sturgeon exhibits a high degree of ossification. Principal rows of scutes are maintained throughout life and the trunk is covered by smaller scutelets, platelets and denticles. The five principal rows of scutes are never absorbed but lose the sharp hooks in very large individuals. A laterodorsal row of scutelets is sometimes present, restricted to the region posterior to the head and anterior to the origin of the dorsal fin.

The denticles are complex in shape and often snowflake-like (Plate 3). The density of denticles covering the entire body is high and apparent because of their white colouration in stark contrast to the body. The white sturgeon has 6-9 tiny, paired postdorsal plates, often missing outright. This is a major morphological character used to distinguish it from the sympatrically occurring green sturgeon. The anus is positioned posterior to the dorsal fin insertion, unlike *A. medirostris*, where it is anteriorly situated (North et al. 2002). There are two rows of four to eight post anal plates (Plate 6) (North et al. 2002, Hochleithner and Gessner 2001). Juveniles are similar to adults in overall appearance except for the proportion of the rostrum to total head length and the fin size in relation to body length.

The white sturgeon is the largest North American sturgeon, reaching a maximum size of over 1000 lb. There is a report of a white sturgeon weighing over 1800 lb (850kg) taken from the Fraser River near Mission, British Columbia. Another giant from the same river was caught in 1897 and was authenticated to weigh 1387 lb (600kg) (Scott & Crossman 1973). A 1506 lb white sturgeon from the Fraser measured 167 inches (Fraser River White Sturgeon Study Program).

The overall body colouration of the dorsum and lateral surface varies from slate grey, blue-grey to brown. All principal rows of scutes and denticles are white and contrast strongly against surrounding body. The numerous denticles covering the entire body are light in colour and surrounded by a white halo (Figure 5-13). The leading edge of pectoral, pelvic and anal fin is lighter than the rest of fin. Approximately midway down the lateroventral surface, colouration abruptly ends. The rest of the lateroventral surface and ventrum is white. Fins are lightly coloured and the ventral fins and lower caudal fin are white or off-white. Individuals in captivity take on a dark overall appearance, including scutes and scutelets. Juveniles are similarly coloured.

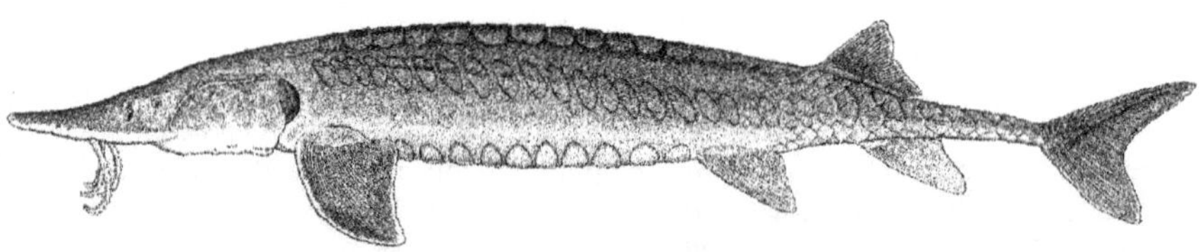

Fig. 5-14. Shovelnose sturgeon – Dorsal scutes 13-16 lateral scutes 38-50, ventral scutes 9-14

5.7.1.6 Shovelnose sturgeon (Scaphirhynchus platorynchus)

The shovelnose sturgeon has an elongated, tapering body. The caudal peduncle is narrow and long. The upper lobe of the caudal fin extends into a long filament. The body is more flattened in cross-section than Acipenser or Huso. The rostrum is wide, flat and shovel shaped. Bemis et al (1997) show allometric lengthening of the rostrum of shovelnose sturgeon. The two sets of barbels are extensively fimbriated with inner barbels being more than half of the length of the outer barbels (Plate 2) (Robison & Buchanan 1984). The lower lips have four papillose lobes (Plate 2). The eyes are tiny and spiracles are absent. The 5 rows of scutes remain prominent throughout the individuals' life. All scutes have a prominent retrose hook appendage. The narrow, elongated caudal peduncle is entirely covered with scutes and plates (Figure 5-8). The filament of the upper lobe of the heterocercal tail is present only in small, immature individuals. Adult shovelnose sturgeon are usually 1.5-4kg. Carlander (1969) reported a maximum weight of 4.5kg with most individuals weighing less than 2.5kg. However, Keenlyne (1997) reported individuals of over 7kg in the upper Missouri River. Keenlyne (1997) suggests that the larger shovelnose sturgeon of the upper Missouri represent a different strain or "stock" from those lower downstream.

Shovelnose sturgeon colouration is consistent throughout this species' range. The dorsum is dark brown, becoming progressively lighter on the lateroventral surface. The principal rows of scutes are slightly darker than the surrounding body. The ventral surface of the head and body is white or cream coloured. The fins are pigmented and similar in colour to the body region of origin.

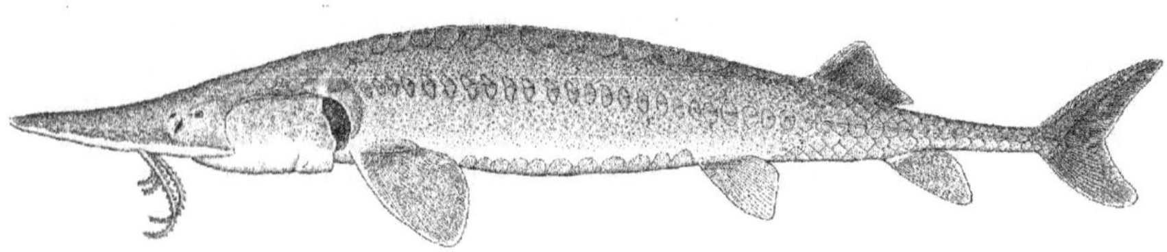

Fig. 5-15. Pallid sturgeon - dorsal scutes 14-18, lateral scutes 40-48, ventral scutes 9-13

5.7.1.7 Pallid sturgeon (Scaphirhynchus albus)

The pallid sturgeon has a robust body tapering posteriorly into a long and narrow caudal peduncle. The snout is broad, flat and spatulate in shape. Viewed dorsally, the pallid sturgeon's rostrum is more elongated than that of other species of Scaphirhynchus (Figure 5-15). The inner barbels are situated slightly anterior to the outer barbels and are heavily fimbriated. The Pallid sturgeon's mouth is transverse with four papillose lobes on the continuous lower lip (Figure 5-15). The branchiostegal connection is narrow and the posterior edge of the branchiostegal membranes is sub horizontal (Mayden and Kuhajda 1996). The pectoral fins are large and rounded. Most specimens, especially adults, lack retrose spines on the tip of the snout, parietal spines and preorbital spines typical of *S. suttkusi* and *S. platorynchus*. The frontal spines are always absent. The squamation on the ventral surface is weak and reduced (Mayden and Kuhajda 1996, 1997). Similar to other shovelnose sturgeons, the spiracle is absent in *S. albus*. The anterior edge of the clavicle keel is smooth. The flattened caudal peduncle is entirely covered with scutes and the upper lobe of the caudal fin has a distinct filament, but absent in large individuals. The post anal plates range in number from 7-8. There are 9 post dorsal plates. Both these counts overlap meristically with *S. suttkusi* and *S. platorynchus*. The average size of adult pallid sturgeon is 8-15kg with a historic maximum size of 40kg being noted in the literature.

The colouration of *S. albus* is lighter than other species of Scaphirhynchus, with the entire fish having an overall greyish-white appearance (Trautman 1981; Robison and Buchanan 1984). The principal rows of scutes are slightly darker than the surrounding body. The transition from the pale dorsum colouration to the white ventrum is subtle compared to other shovelnose sturgeons.

Fig. 5-16. Alabama sturgeon - dorsal scutes 15-21, Lateral scutes 40-49, Ventral scutes 11-14

5.7.1.8 Alabama sturgeon (Scaphirhynchus suttkusi)

The Alabama sturgeon has an elongated and slightly flattened body with a long, flattened caudal peduncle. In comparing S Scaphirhynchus with *S. platorynchus*, Mayden and Kuhajda (1996) found the former to have a wider caudal peduncle. The snout is broad, flat and shovel shaped. There are two pairs of large and heavily fimbriated barbels situated midway between tip of snout and upper lip. The inner and outer barbels are of the same length. Large chemoreceptor cells cover the entire ventral surface of the rostrum. These cells also present throughout the preorbital region. Some chemoreceptors are present on the dorsum of the anterior portion of the snout. The mouth is transverse and the lower lip is continuous with four papillose lobes. Spines located on the head of *S. suttkusi* are much reduced in comparison to *S. platorynchus*, and *S. suttkusi* lacks left and right preorbital and snout spines. The lack of spines at the tip of the snout is a useful diagnostic in distinguishing *S. suttkusi* from *S. platorynchus* except when identifying large individuals of *S. platorynchus* on which rostral spines have disappeared. Mayden and Kuhajda (1996) found the anterior lateral plates of *S. platorynchus* to be more vertical than those of S. suttkusi. The squamation on the ventral surface is weak and reduced with some small plates sometimes present in the interpelvic region (Mayden and Kuhajda 1996). There are 7-9 post dorsal plates, 4-6 postanal plates and 7-8 preanal plates. The Alabama sturgeon is similar in size to S. suttkusi with individuals averaging 2-3kg.

Colouration of the Alabama sturgeon is a useful diagnostic feature since it differs considerably form the much paler S. albus and less colourful *S. platorynchus*. The dorsal surface of the body and head is brown to brownish-orange. Dorsal plates are slightly darker than the surrounding body. The laterodorsal surface is brownish orange and the lateral plates slightly are darker than the surrounding body. The lateroventral surface is light brown or yellowish brown and the ventral plates are brassy or cream coloured. Entire ventral surface is white or cream coloured. All fins are pigmented and similar in colour to the body region from which they originate. The orange component in the overall colouration is more distinct in males.

5.8 Intraspecific Morphological Variation

[Intraspecific variation] is perhaps the most complex form of variation to study or recognize, but is often the most interesting"
- Hilton & Bemis, in press.

Morphological variation among different populations is well documented for many vertebrate species, however only a few such detailed analyses or descriptions of sturgeons exist (Walsh et al. 2001, Hilton and Bemis, in press). The two aspects of intraspecific morphological variation typical to sturgeon are: 1) variation among allopatric stocks and 2) variation within a stock.

5.8.1 MORPHOLOGICAL VARIATION AMONG ALLOPATRIC STOCKS

Morphologies unique to particular stocks of Ponto-Caspian acipenserid species are poorly documented (see: Antipa 1909, Berg 1968, Holcik 1989). No such attempt has been made for similar examination concerning North American species of Acipenserids. Slight morphometric and meristic differences are documented among regional stocks of Quebec lake sturgeon throughout the Saint Lawrence watershed. L. Hildebrand (pers.com) reported white sturgeon of the upper Fraser River have a more elongated rostrum than those of the lower Fraser River. The Green sturgeon is rare throughout its range and comparative morphological work between stocks is lacking. Morphological variation of shortnose sturgeon among eastern coastal rivers has been described by Walsh et al. (2001) who observed morphometric and meristic differences among stocks. Such differences are expected in reproductively isolated organisms within species and are the basis for speciation. North American sturgeons show little of the distinct stock morphologies typical of Ponto-Caspian species such as Beluga (*Huso huso*), Russian (*Acipenser gueldenstaedtii*), Persian (*Acipenser persicus*) and stellate sturgeon (*Acipenser stellatus*). However, species such as the lake sturgeon (*A. fulvescens*) has only recently invaded post glacial environments and it is likely that in time, various stocks will develop particular morphologies.

5.8.2 MORPHOLOGICAL VARIATION WITHIN AND AMONG STOCKS

Individual morphological variation occurs "between individuals of the same species at the same ontogenetic stage" (Hilton and Bemis, in press) and can be further subdivided into:

a) Correlated:

- temporal (associated with particular life history stages)
- ecomorphological (variation among morphs)
- geographic (variation between individuals of the same species from different locations)

b) Uncorrelated:

- natural (no particular cause)

The morphological variation of North American species has not been well studied. Early taxonomists were primarily concerned with creating workable diagnostic keys or, at least, descriptions that could be used to distinguish one species from another. Unfortunately, this task was complicated by the surprising degree of morphological variation found within many wild stocks. The green sturgeon (*A. medirostris*) from the Columbia River, Pacific Northwest show high degrees of variability in rostrum shape, head width, mouth size, meristic counts, and body armouring. The species offers a good example of variation within a stock.

On the Hudson River, for example, a fall census of adult shortnose sturgeon (*A. brevirostrum*), yielded a wide range of adult morphology among individuals of similar sizes (Vecsei 1999). A high degree of variation exists in all external morphological characters, particularly rostrum length and width, mouth width, meristic counts, caudal fin shape and colouration. These characters vary interspecifically, ontogenetically and geographically (Vladykov & Greeley 1963). A traditional fish taxonomist might have easily concluded that several species were present. This example illustrates how expertise in sturgeon taxonomy can only be attained through a comprehensive knowledge of the morphological variations that occur both ontogenetically and phylogenetically.

We examine the possible causes behind morphological variation present within sturgeon species. Are these ecologically driven variations or can they be explained by genetic drift and/or founder effect? In founder effect, a new sturgeon population would have been initially formed by a small number of individuals having broken off from a larger population. Such a limitation, likewise limits the potential for future morphological variations because of the limited genetic variation within the stock.

Founder effects in sturgeon have not been well studied, however the biogeographical history of sturgeon speciation may support the notion that many intraspecific variations in sturgeon morphology have arisen from founder effects. A comparison of shortnose and Atlantic sturgeon illustrates the different mechanisms by which genetic processes may influence morphological variation within a population. In the case of shortnose sturgeon, the founder effect may explain much of the inter- and intra-morphological stock variability. In the case of the Hudson River shortnose sturgeon, the early colonizers of this stock probably consisted of many individuals from numerous adjacent systems. Founder effects, in this instance, were minimal because genetic diversity of the founding population was high. This would explain why the current population is comprised of many haplotypes (Waldman et al. 2002). In contrast, Wirgin et al. (1997) have shown that the St. John River population was probably not extirpated by the Pleistocene Glaciation, but rather, was likely repopulated by individuals surviving this period within a glacial refugium. Hence, haplotype numbers are even higher than if it had been re-colonized by a small number of individuals after the Pleistocene. High haplotypic diversity and unique haplotypes are more likely the result of surviving the Pleistocene via a refugium rather than high mutation rates (Waldman et al. 2002). The shortnose sturgeon of the St. John River were thus able to preserve some, or most of, their original genetic diversity and this is partly expressed by their morphologies being as variable as in the Hudson.

Another possible explanation for the varying degrees of intraspecific morphological variation observed among different sturgeon stocks can be traced to the forces of genetic drift as influenced by changes in population abundance through geological time. Effects of genetic drift are most pronounced in small populations, where random chance may gradually influence the distribution of allelic frequencies within the population. The accumulation of these random genetic changes - occurring purely by chance within a small population - are often accompanied by parallel changes in morphology. While these effects can lead to significant genetic and morphological divergence among stocks, the level of their influence is not static through time – that is, the importance of genetic drift within a population is largely determined by changes in population size. Because most sturgeon populations have survived dramatic declines at various times throughout their evolutionary history, many stocks have been genetically "bottlenecked" at some point in geological time. During the late Pleistocene, for example, many Atlantic coast populations of shortnose sturgeon may have been reduced to only a few individuals. During such periods of low abundance, remnant sturgeon populations become extremely susceptible to the forces of genetic drift and, consequently, morphological divergence.

In contrast to the varying morphologies among shortnose sturgeon stocks, Atlantic sturgeon exhibit very little morphological variation across their range. While the selective forces acting on each species are probably quite different, Atlantic sturgeon are truly anadromous, and hence, more prone to straying when returning to spawn in their natal rivers. Over time, this important life history difference between Atlantic and shortnose sturgeon has resulted in an imperfect isolating mechanism that has allowed at least some gene flow among the different stocks of Atlantic sturgeon of each river system (Wirgin et al. 2002). That is not to say that the various populations of Atlantic sturgeon are genetically identical – they are not – however, the differences are so slight that they are not expressed phenotypically.

Regardless of evolutionary history, it is clear that much of the morphological variation documented among contemporary stocks was likely established thousands of years ago. One interesting evolutionary question frequently posed by sturgeon researchers focuses on the relative importance of genetic drift versus natural selection in driving phenotypic change. While morphological divergence among sturgeon stocks may have been partially shaped by natural selection, contemporary ecomorphologists must consider the evolutionary history of the stock in question and consider the possibility (or in some cases, the likelihood) that morphological variations often result from random genetic changes that foster ecologically neutral changes in phenotype. These ecologically "neutral" morphological characters such as caudal fin shape, rostrum length/depth, fin shape, mouth size, etc., exhibit consistent variation among and within stocks of shortnose sturgeon. For ecomorphologists, this is sometimes difficult because we tend to look for ecological explanations for even the slightest variations in morphology.

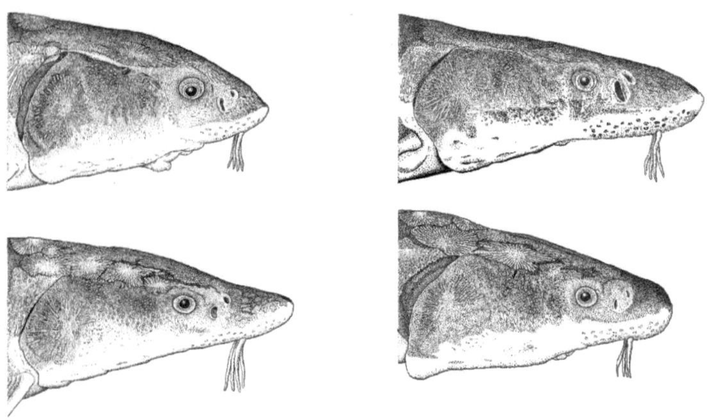

Fig. 5-17. Intraspecific variation among four adult shortnose sturgeon captured in the Hudson River by Peterson and Vecsei, 1997.

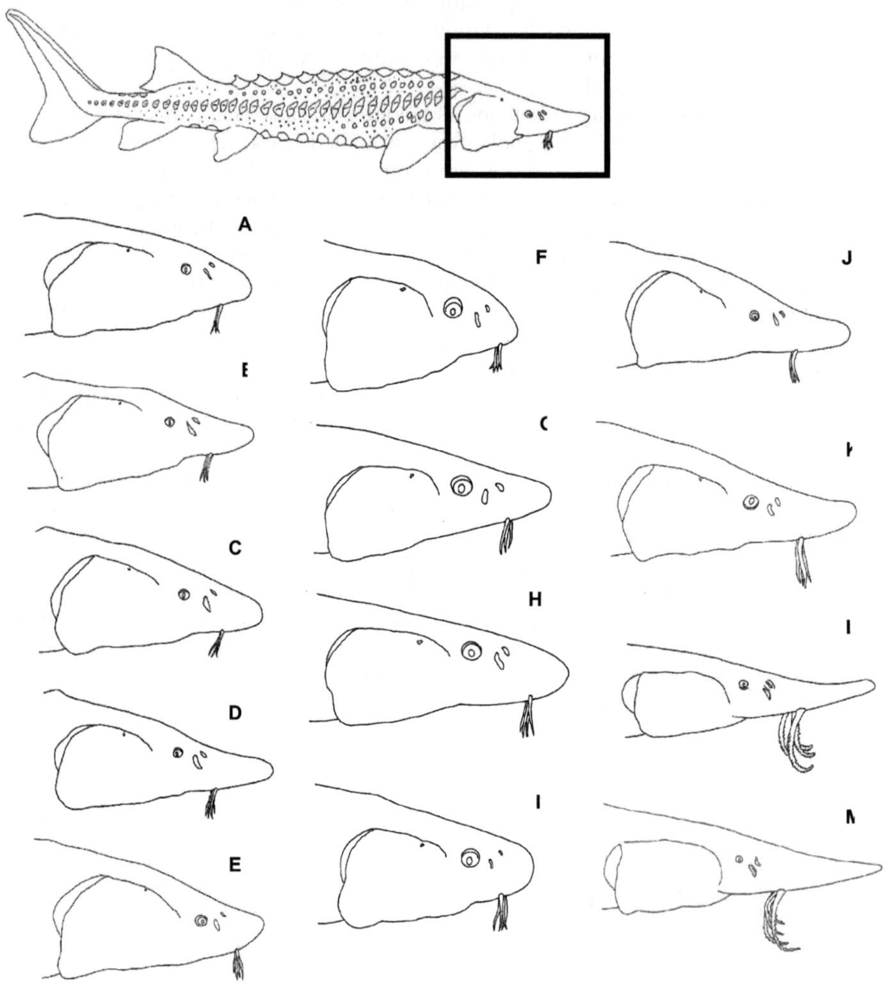

A: *A. medirostris*, *B:* *A. medirostris*, *C:* *A. medirostris*, *D:* *A. medirostris*, *E:* *A. transmontanus*, *F:* *A. brevirostrum*, *G:* *A. brevirostrum*, *H:* *A. brevirostrum*, *I:* *A. brevirostrum*, *J:* *A. oxyrinchus* *K:* *A. fulvescens*, *L:* *S. platorynchus*, *M:* *S. albus*

Plate 1. Comparison of sturgeon head shape.

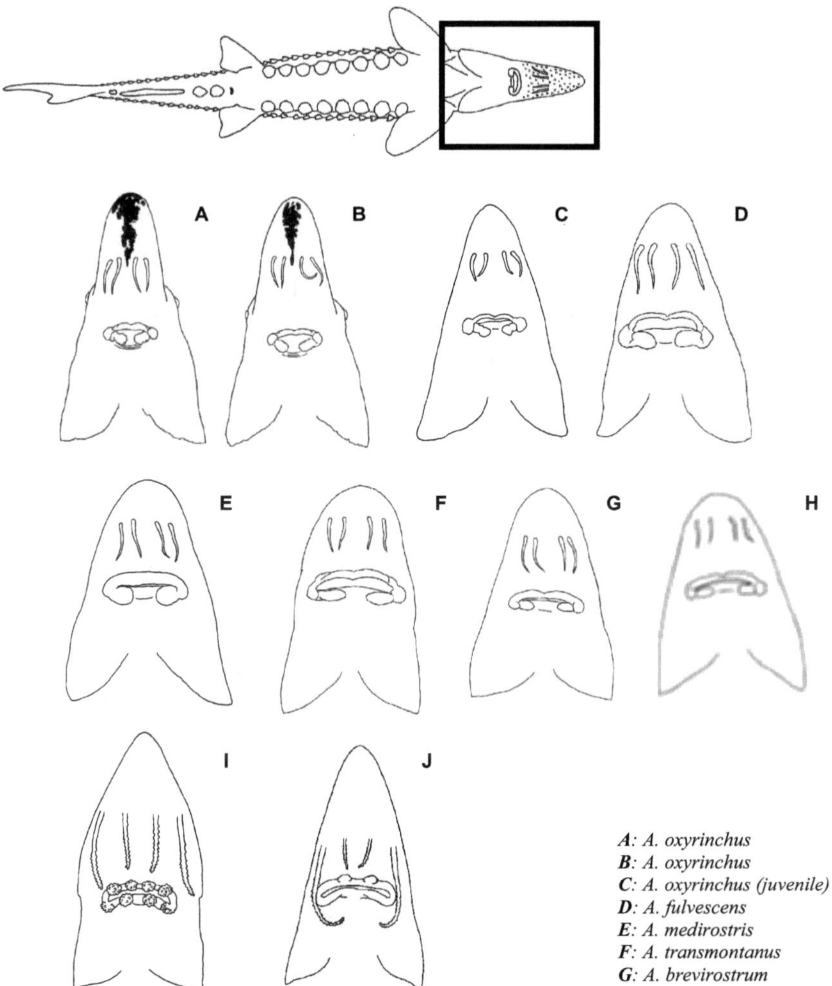

A: *A. oxyrinchus*
B: *A. oxyrinchus*
C: *A. oxyrinchus (juvenile)*
D: *A. fulvescens*
E: *A. medirostris*
F: *A. transmontanus*
G: *A. brevirostrum*
H: *A. brevirostrum (typical)*
I: *S. platorynchus*
J: *S. albus*

Plate 2. Head morphology of North American sturgeon (ventral view).

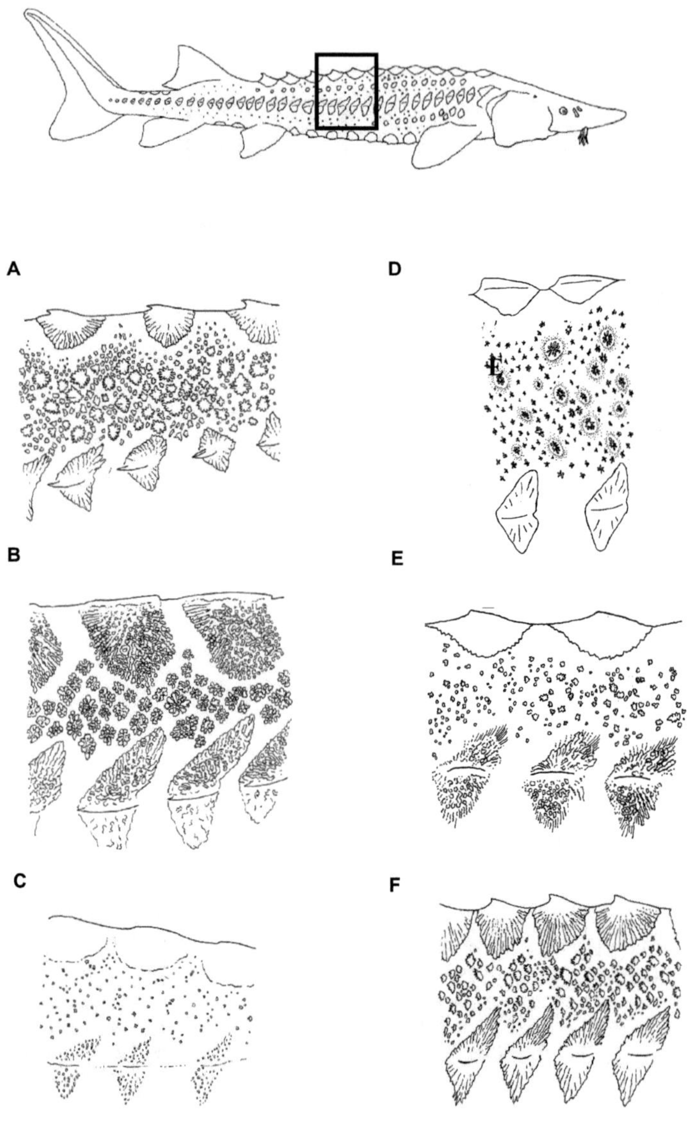

A: *A. medirostris*, ***B***: *A. oxyrinchus*, ***C***: *A. fulvescens (adult)*, ***D***: *A. transmontanus*, ***E***: *A. brevirostrum*, ***F***: *A. fulvescens (juvenile)* ***E***: *A. brevirostrum*, ***F***: *A. fulvescens (juvenile)*

Plate 3. Ossification of laterodorsal surface of sturgeons

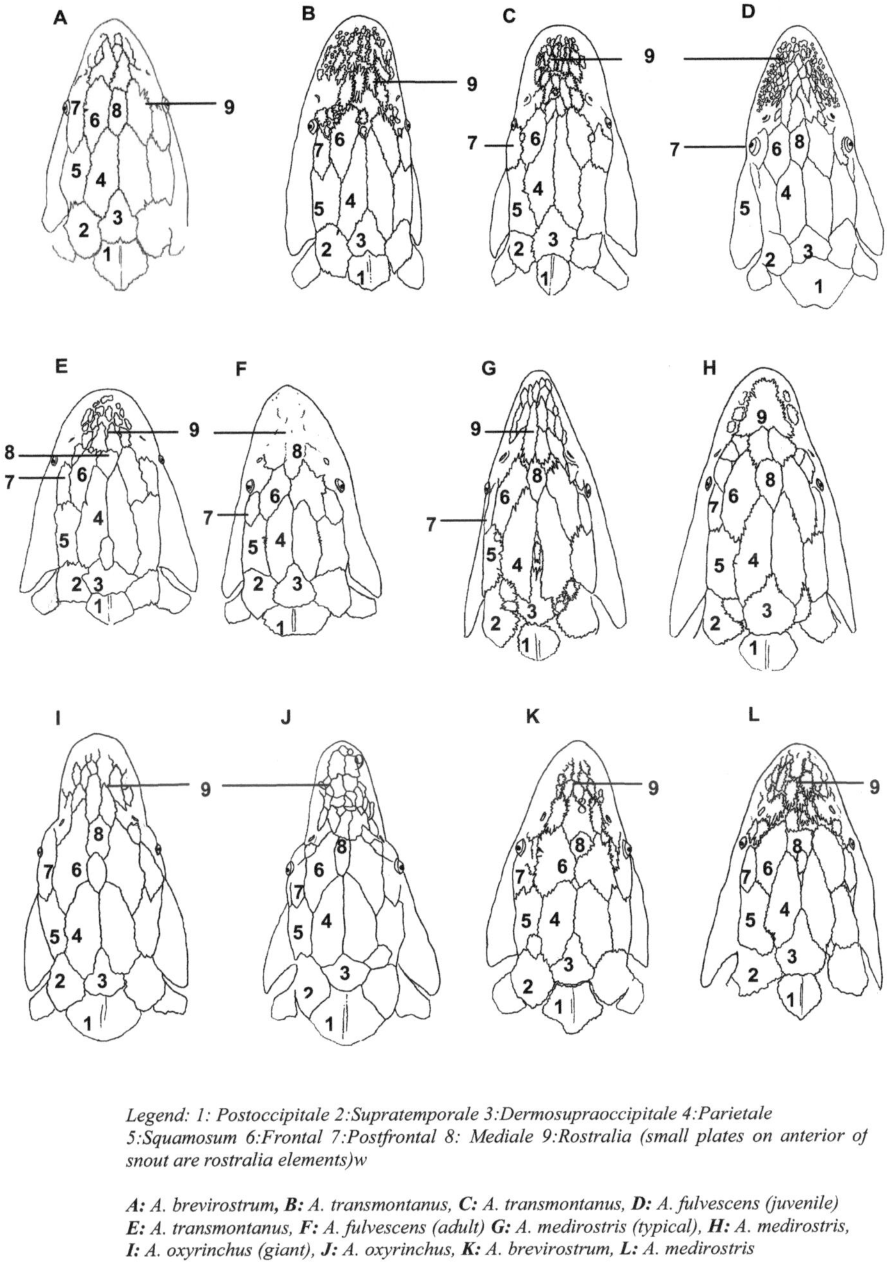

Legend: 1: Postoccipitale 2:Supratemporale 3:Dermosupraoccipitale 4:Parietale 5:Squamosum 6:Frontal 7:Postfrontal 8: Mediale 9:Rostralia (small plates on anterior of snout are rostralia elements)w

***A:** A. brevirostrum, **B:** A. transmontanus, **C:** A. transmontanus, **D:** A. fulvescens (juvenile) **E:** A. transmontanus, **F:** A. fulvescens (adult) **G:** A. medirostris (typical), **H:** A. medirostris, **I:** A. oxyrinchus (giant), **J:** A. oxyrinchus, **K:** A. brevirostrum, **L:** A. medirostris*

Plate 4. Head plates of the genus Acipenser

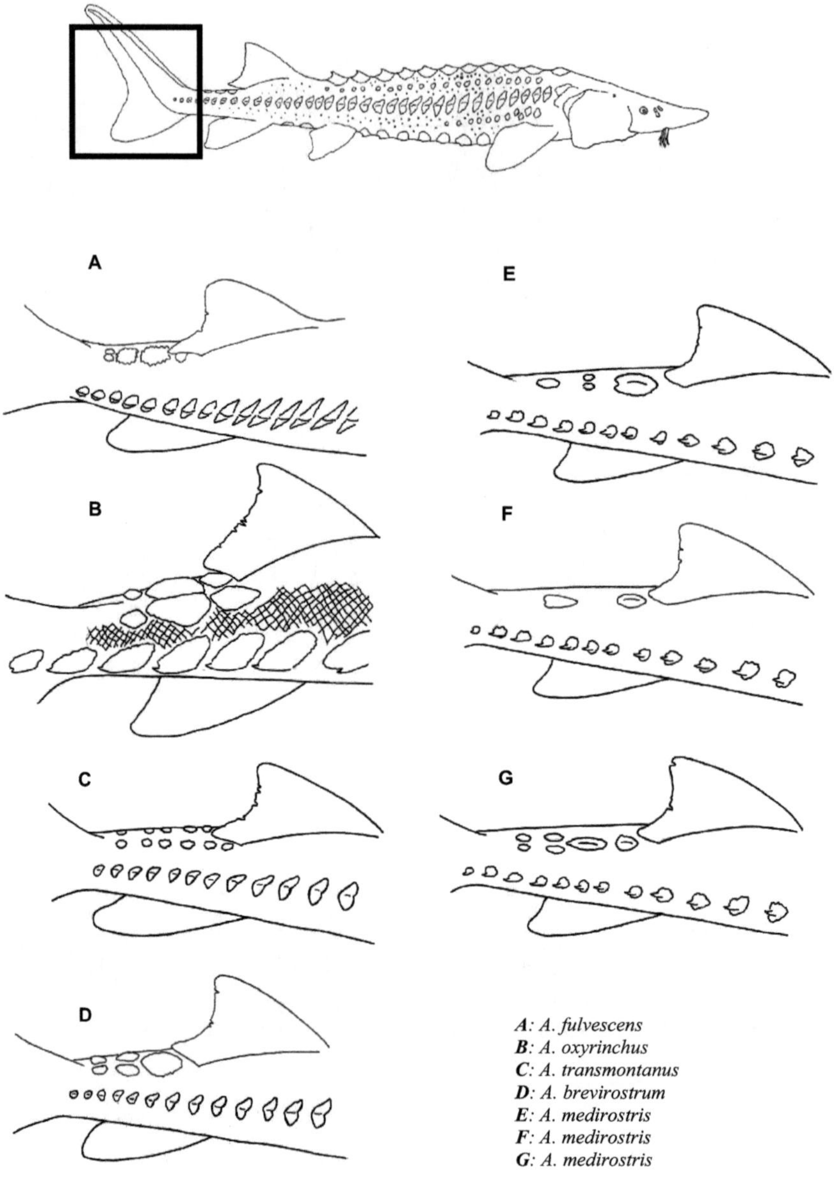

A: *A. fulvescens*
B: *A. oxyrinchus*
C: *A. transmontanus*
D: *A. brevirostrum*
E: *A. medirostris*
F: *A. medirostris*
G: *A. medirostris*

Plate 5. Post-dorsal plates of genus Acipenser.

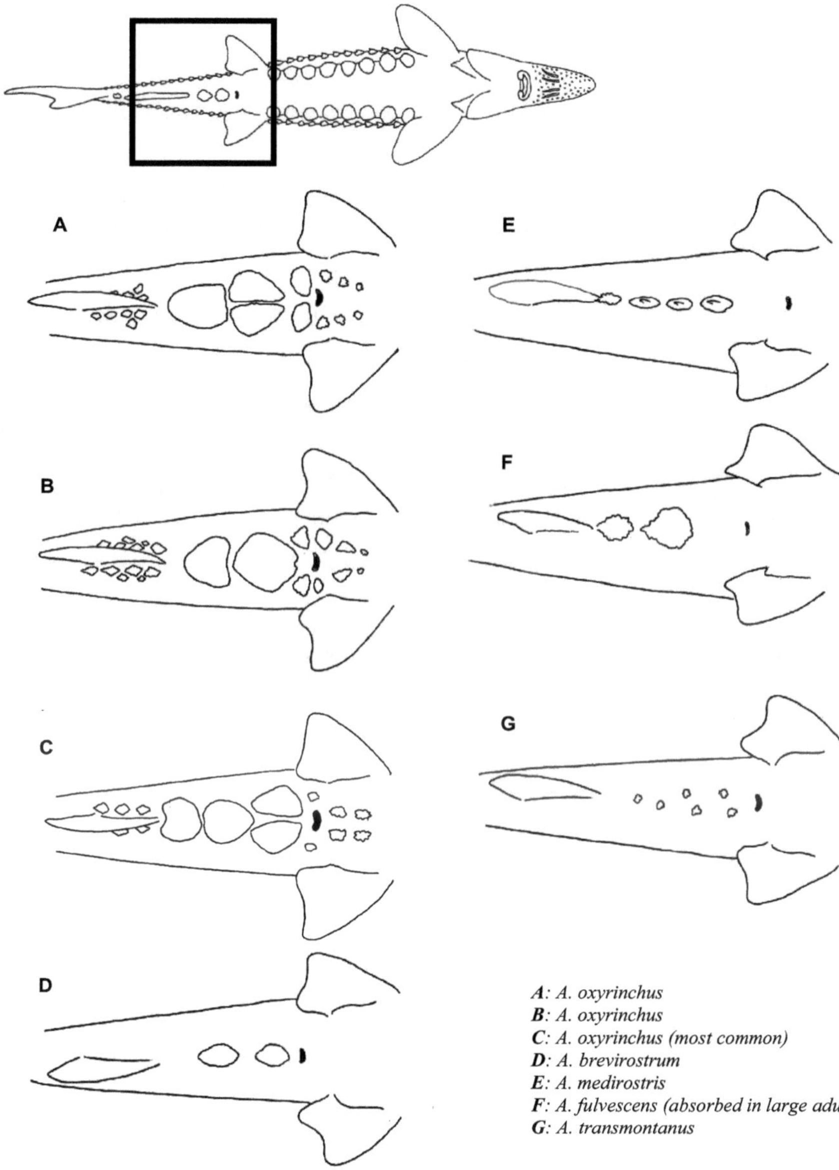

A: A. oxyrinchus
B: A. oxyrinchus
C: A. oxyrinchus (most common)
D: A. brevirostrum
E: A. medirostris
F: A. fulvescens (absorbed in large adults)
G: A. transmontanus

Plate 6. Anal plates of the genus Acipenser

5.9 References

Antipa, G. 1909. Fauna ihtiologica a Romaniei (Fish fauna of Romania). Publ. fond. Adamachi, Bucuresti. 294 pp.

Artyukhin, E.N. 1995: On Biogeography and Relationships within the Genus Acipenser. 3(2): 6-8.

Artyukhin, E.N. and P. Vecsei: On the status of Atlantic sturgeon: conspecificity of European *Acipenser sturio* and North American *Acipenser oxyrinchus*. J. Appl. Ichthyol. 15: 35-37.

Bain, M.B. 1997: Atlantic and shortnose sturgeon of the Hudson River: common and divergent life history attributes. Env. Biol. Fish. 48, 347-358.

Bemis, E.W., Findeis, E.K. and L. Grande. 1997: An overview of Acipenseriformes. Env. Biol. Fish. 48, 25-71.

Bemis, W.E., Kynard, B. 1997: Sturgeon rivers: an introduction to acipenseriforms biogeography and life history. Env. Biol. Fish. 48, 167-183.

Berg, L.S. 1968: The freshwater fishes of the USSR and adjacent countries, Vol. 1, Part 1. A. Israel Program for Scientific Translations, Jerusalem. 505 pp.

Billard, R. 2002. Esturgeons et Caviar. Editions TEC & DOC. 298 pp.

Billard, R. and G. Lecointre. 2001: Biology and conservation of sturgeon and paddlefish. Rev. Fish. Biol. Fisheries. 10: 355-392.

Birstein, V.J. and W.E. Bemis. 1997: How many species are there within the genus Acipenser? Env. Biol. Fish. 48: 157-163.

Birstein, V.J. and R. DeSalle: 1998: Molecular Phylogeny of Acipenserinae. In: Molecular Phylogenetics and Evolution. 9(1): 141-155.

Burggren, WW. 1978. Influence of intermittent breathing on ventricular depolarization patterns in chelonian reptiles. The Journal of Physiology, Vol 278, Issue 1 349-364.

Carlander, K.D. 1969: Handbook of freshwater fishery biology. Iowa State University Press, Ames. 752 pp.

Cataldi, E., P. Bronzi, E. Ciccotti, P. Di Marco, O. Di Santo, G. Monaco and S. Cataudella. 1995. Morphology of gills, digestive tract and kidney of Italian sturgeon, *Acipenser naccarii* in fresh and saline water: preliminary results. Pp. 52-61. *In*: A.D. Gershanovich and T.I. J. Smith (ed.) International Symposium on Sturgeons Proceedings, VNIRO Publ., Moscow.

Choudhury, A. and T.A. Dick. 1998: The historical biogeography of sturgeons (Osteichthyes: Acipenseridae): a synthesis of phylogenetics, paleontology and paleogeography. Journal of Biogeography 25: 623-640.

Findeis, E.K. 1997: Osteology and phylogenetic interrelationships of sturgeons (Acipenseridae). Env. Biol. Fish. 48, 73-126.

Grande, L. and W.E. Bemis. 1991. Osteology and Phylogenetic Relationships of Fossil and Recent Paddlefishes (Polyodontidae) with Comment on the Interrelationships of Acipenseriformes. J. Vertebrate Paleontology, Special Memoir Number 1 (supplement to Volume 11). Pages 1-121

Grande, L. and W.E. Bemis. 1996: Osteology and phylogenetic relationships of fossil and recent paddlefishes (Polyodontidae) with comments on the interrelationships of Acipenseriformes. J. Vert. Paleo. 11, supplement 1: 1-121.

Gregory, W.K. 1933. Fish skulls: A study of the evolution of the natural mechanism. Trans. Amer. Phil. Soc. 23: 75-481.

Harkness, W.J.K. and J.R. Dymond. 1961: The lake sturgeon. The history of its fishery and problems of conservation. Ont. Dep. Lands Forests, Fish Wildl. Br. 121 pp.

Hilton, E.J., Bemis, W.E., In press. External morphology of shortnose sturgeon, *Acipenser brevirostrum* (Acipenseriformes: Acipenseridae), from the Connecticut River, with notes on variation as a natural phenomenon.

Hochleithner, M. and J. Gessner. 2001: The sturgeons and paddlefishes of the world: Biology and aquaculture. AquaTech Publications, Kitzbuehl, Austria, 207 pp.

Holcik J. (ed.). 1989 The Freshwater Fishes of Europe. Wiesbaden: AULA- Verl., Vol.1, p.2. 469 p.

Keenlyne, K.D. Henry, C.J. Tews, A. and P. Clancey. 1994: Morphometric comparisons of Upper Missouri River sturgeons. Trans. Amer. fish. Soc. 123: 779-785.

Keenlyne, K.D. 1997: Life history and status of the shovelnose sturgeon, *Scaphirhynchus platorynchus*. Env. Biol. Fish. 48: 291-298.

Krayushkina, L.S., W.T.W. Potts, A.A. Gerasimov and A.A. Panov. 1995. Peculiarities of ionic regulation in young sturgeons (Acipenseridae) during adaptation to sea water. Pp. 43-51. *In*: A.D. Gershanovich and T.I. J. Smith (ed.) International Symposium on Sturgeons Proceedings, VNIRO Publ., Moscow.

LeBreton, G. and F.W.H. Beamish. 1998. Ionic and osmotic regulation in juvenile, *Acipenser fulvescens*. Env.Biol.Fish. 52:477-482.

Ludwig, A. Debus, L. Lieckfeldt, Wirgin, I. Benecke, N. Jenneckens, I. Williot, P. and J. R. Waldman. 2003. When the American sea sturgeon swam east. Nature, 3 October, 447.

Matthews, W.J. 1992: Patterns in Freshwater Fish Ecology. Chapman & Hall. 510 pp.

Mayden, R.L. and B.R. Kuhajda. 1996. Systematics, taxonomy, and conservation status of the endangered Alabama sturgeon, *Scaphirhynchus suttkusi* Williams and Clemer (Actinopterygii, Acipenseridae). Copeia 1996: 241-273.

Mayden, R.L. and B.R. Kuhajda. 1997. Threatened fishes of the world: *Scaphirhynchus suttkusi* Williams & Clemmer, 1991 (Acipenseridae) Env. Biol. Fish. 48: 418-419.

Mayden, R.L. and B.R. Kuhajda. 1997. Threatened fishes of the world: *Scaphirhynchus albus* (Forbes & Richardson, 1905) (Acipenseridae) Env. Biol. Fish. 48: 420-421.

McEnroe, M. and J.J. Cech Jr. 1985. Osmoregulation in juvenile and adult white sturgeon, *Acipenser transmontanus*. Env. Biol. Fish. 14: 1: 23-30.

Moyle, P.B. and J.J. Cech. 2000. Fishes, An Introduction to Ichthyology, 4th Edition. Prentice Hall. 612 pp.

North, J.A., Farr, R.A. and P. Vecsei. 2002: A comparison of meristic and morphometric characters of green sturgeon *Acipenser medirostris*. J. Appl. Ichthyol. 18. 234-239.

Potts, W.T.W. and P.P. Rudy. 1972. Aspects of osmotic and ionic regulation in the sturgeon. J. Exp. Biol. 56: 703-715.

Robison, H.W. and T.M. Buchanan. 1984: Fishes of Arkansas. Univ. of Arkansas Press, Fayetville. 536 pp.

Scott, W.B and E.J. Crossman. 1973: Freshwater fishes of Canada. Bull. 184, Fisheries Res. Board of Canada, Ottawa. 966 pp.

Trautman, M.B. 1981: The fishes of Ohio. 2nd ed. Ohio State Univ. Press, Columbus.

Van Der Klaauw, C. J. 1948. Ecological studies and reviews. IV. Ecological morphology. Bibl. Biotheor. 4:27-111.

Vecsei, P. and D. Peterson. 1999: Threatened species of the world: *Acipenser oxyrinchus*, Mitchel 1889, Acipenseridae. Env. Biol. Fish. pp

Vecsei, P. and D. Peterson. 2000: Threatened fishes of the world: *Acipenser brevirostrum* Lesueur, 1818 (Acipenseridae). Env. Biol. Fish. 59: 270

Vecsei, P. 2001. Cites Identification Guide-Sturgeons and Paddlefish: Guide to the Identification of Sturgeon and Paddlefish Species Controlled under the Convention on International Trade in Endangered Species of Wild Fauna and Flora. Ed: Yvan Lafleur and Richard Charette, Environment Canada

Vecsei, P., Suciu, R. and D. Peterson. 2002: Threatened fishes of the world: *Huso huso* (Linnaeus, 1758) (Acipenseridae) Env. Biol. Fish. 65: 363-365.

Vecsei, P., Artyukhin, E. and D. Peterson. 2002: Threatened species of the world: *Acipenser nudiventris* Lovetsky, 1828 (Acipenseridae). Env. Biol. Fish. 65: 455-456.

Vecsei, P. 1999: Illustrating sturgeon species: The art in science in highlighting morphological figures. J. Ichthyol. 15: 7-8.

Videler, J.J. 1993: Fish Swimming. New York, Chapman & Hall. ??pp.

Vladykov, V.D. and J.R. Greenley. 1963: Order Acipenseroidei. Fishes of the western North Atlantic. Part III. Sears Found. Mar. Res., Yale University, New Haven, CT. 1(3): 24-60.

Waldman, J.R., Grunwald, J.S., Stabile, J. and I. Wirgin. 2002: Impacts of life history and biogeography on the genetic stock structure of Atlantic sturgeon *Acipenser oxyrinchus oxyrinchus*, Gulf sturgeon *A. oxyrinchus desotoi*, and shortnose sturgeon A. *brevirostrum*. J. Appl. Ichthyol. 18 509-518.

Walsh, M.G. Bain, M.B. Squiers, T., Waldman, J.R. and I. Wirgin. 2001: Morphological and genetic variation among shortnose sturgeon *Acipenser brevirostrum* from adjacent and distant rivers. Estuaries 24: 41-48.

Wilga, C.D. and G.V. Lauder. 1999: Locomotion in sturgeon: function of the pectoral fins. The Journal of Experimental Biology. 202: 2413-2432.

Wirgin, I., Waldman, J.R., Rosko, J., Gross, R., Rogers, S.G. and J. Stabile. 2000:Genetic structure of Atlantic sturgeon populations based on mitochondrial DNA control region sequences. Trans. Am. Fish. Soc. 129: 476-486.

Wirgin, I. Stabile, J.E. and J.R. Waldman. 1997: Molecular analysis in the conservation of sturgeons and paddlefish. Env. Biol. Fish. 48: 385-398.

CHAPTER 6

EMBRYOLOGY

JESSICA A. BOLKER
University of New Hampshire, Department of Zoology, 46 College Road, Durham NH 03824 USA

6.1 Introduction

This chapter reviews the early embryological and larval development of acipenserids, as well as some of the developmental abnormalities that occur in hatchery populations of sturgeons. The majority of acipenserid developmental research has been carried out within the context of aquaculture needs and studies, though these fishes can also serve as informative models for more general developmental questions.

6.2 History of Developmental Research in Sturgeons

Sturgeons have long attracted the attention of embryologists and developmental biologists because of their status as a "primitive" fish, and the obvious peculiarities of their skeletons. Early studies often had an explicitly comparative and evolutionary focus: it was hoped that information about development in sturgeons might resolve questions about the evolution of different patterns of early development, such as those seen in Amphioxus, Lepisosteus, sharks, and teleosts, and about the relationships among chordate taxa.

The first major study of sturgeon embryology was Salensky's work with *Acipenser ruthenus* (Salensky 1881) in which he described development from the oocyte through the development of larval features. Ryder (1888) undertook an extensive study of the development and aquaculture potential of Atlantic sturgeon in the Delaware River, and published the first illustrations of this species (mostly of larvae and adults). Dean (1895) presented original descriptions of early development in Atlantic sturgeon and Lepisosteus, and a detailed comparison between the two.

After the demise of the commercial sturgeon fisheries in the U.S. around 1900, research on sturgeon development was limited to studies using preserved material, generally of larval and later stages, and to researchers in the U.S.S.R. who had access to the hatchery systems there. A great deal of research, including experimental studies, was carried out in the U.S.S.R. following the establishment of extensive aquaculture programs for several sturgeon species in the 1950s (e.g. Dettlaff 1962; Dettlaff and Ginsburg 1954; Ginsburg 1961; Ignatieva 1960, 1961, 1963, 1965; see Schmalhausen 1991 and Ginsburg and Dettlaff 1991 for reviews).

G.T.O. LeBreton et al. (eds.), Sturgeons and Paddlefish of North America, 134–146.
© 2004 *Kluwer Academic Publishers. Printed in the Netherlands.*

Studies of sturgeon development in the U.S. became impossible as natural populations were depleted and initially optimistic aquaculture efforts abandoned (Beer 1981); virtually no new information on the development of North American species was available until 1981. Beer (1981) presented the first description of development in white sturgeon, including a staging table suitable for aquaculture work. His results are summarized by Conte et al. (1988).

6.3 Scope of Modern Research

Most knowledge of acipenserid development is based on hatchery- or laboratory-reared fish, particularly the most important cultured species: the European sturgeons *Huso huso*, *A. gueldenstaedti* Brandt, *A. stellatus*, and *A. nudiventris*, and the North American white sturgeon and paddlefish (Dettlaff et al. 1993; Doroshov 1985; Conte et al. 1988). The most detailed descriptions of development in Russian sturgeon species are those of Dettlaff, Ginsburg and Schmalhausen, including staging tables for *A. stellatus* and *A. gueldenstaedti* (Dettlaff and Ginsburg 1954; Ginsburg and Dettlaff 1991; Schmalhausen 1991). Ballard and Ginsburg's study of several species includes a fate map of the late blastula based on vital-dye marking experiments (Ballard and Ginsburg 1980).

Developmental studies of acipenserid fishes include general descriptions of embryonic and larval development, largely though not exclusively for aquacultural purposes (e.g. Schmalhausen 1991; Beer 1981; Conte et al. 1988; Dettlaff et al. 1993; Deng et al. 2002; Van Eenennaam et al. 2001); reproductive physiology, which is extremely important for control of gamete maturation and spawning (Doroshov et al 1997); and studies of larval growth and nutrition. Bridging the gap between applied and basic research is a body of Russian work on developmental timing, beginning with the seminal paper of Dettlaff (1953), which is valuable to aquaculture and also addresses basic developmental questions. A small body of research has been carried out on acipenserid embryos for phylogenetic or comparative reasons (Poole 1988; Bemis and Grande 1992; Leung and Wourms 1991; Ballard and Needham 1964; Ballard and Ginsburg 1980; Bolker 1993b, c). Toxicological studies (Dettlaff et al. 1993; Shagaeva et al. 1993; Akimova and Ruban 1993, 1996) have been motivated largely by the knowledge that natural habitats of most sturgeon species are heavily polluted, and long-lived bottom-feeding fish such as sturgeons are particularly vulnerable to the effects of environmental pollutants. In fact, reproductive disturbances due to pollution, combined with physical disruption or inaccessibility of spawning grounds due to dams, have brought many central European sturgeon species to the brink of extinction in the wild.

The account of early development given here is based primarily on the white sturgeon. Early development in different species, and even genera, of sturgeons is generally very similar. Indeed, paddlefish and sturgeon embryos are almost indistinguishable through early organogenesis; thus, an account based on one species is applicable to others, and much information can be generalized (Schmalhausen 1991).

6.4 Developmental Stages and Timing

Several different embryological and larval staging systems, or normal tables, have been developed for sturgeons. The primary one is that of Dettlaff and Ginsburg (Dettlaff and Ginsburg 1954; Dettlaff et al. 1993), which comprises 45 stages from unfertilized egg through prelarva. Originally based on *A. stellatus*, *A. gueldenstaedti*, and *H. huso*, this description is also broadly applicable to other acipenserid species (Bemis and Grande 1992; Ballard and Needham 1964; Bolker 1993b; Deng et al. 2002). A simpler table with fewer stages was developed by Beer in 1981 specifically for use in the context of white sturgeon aquaculture. Bolker (1993b) provides a table that cross-references these two staging schemes.

The timing of development is strongly dependent on temperature in sturgeons (Dettlaff 1953; Dettlaff et al. 1993; Wang et al. 1985). The viable range extends from 8.3°C for northern species such as lake sturgeon (Kempinger 1988) to 15-20°C for the Gulf sturgeon (Chapman and Carr 1995); within the viable range for a given species, early development goes faster at warmer temperatures, and may be considerably slowed at lower ones. In the white sturgeon, development can be temporarily suspended when embryos are shipped at low temperature during mid- to late blastula stages. Normal development resumes when the temperature rises (Bolker, unpub. obs.). Both for aquaculture purposes, where hatchery water temperatures may not be closely controllable, and for basic research, where it is useful to be able to compare stages between species that normally develop at different temperatures, it is desirable to have a temperature-independent measure of developmental rate. Dettlaff and others (Dettlaff 1953; Dettlaff and Dettlaff 1961; Dettlaff et al. 1993) developed such a metric for some European species, using as the base unit (referred to as $\tau 0$) the time from fertilization to first cleavage. Their scheme offers a baseline against which any species, developing at any temperature, can be compared (Dettlaff et al. 1993), and has been particularly useful for timing and standardizing chromosome manipulation (Shelton et al. 1997; Van Eenennaam et al. 1996).

6.5 Major Stages of Embryonic Development

6.5.1 FERTILIZATION

Embryological development begins with the union of the male and female gametes. The unfertilized, mature egg (Dettlaff and Ginsburg stage 1) is already distinctly polarized, with more yolk at the vegetal end. Concentric rings of differential pigmentation (generally lighter rings on a dark background) are centred on the animal pole of the slightly ovoid egg. Pigmentation can vary significantly between individual females of the same species, though it is generally uniform throughout a single batch of eggs. The unfertilized egg rests with its animal-vegetal axis horizontal. Fertilization is accomplished by the entry of a sperm through the micropyles present in the chorion above the animal pole. Fifteen to 20 minutes after fertilization, the egg rotates 90° so that its animal pole is upward, and then retains this orientation. Time-lapse movies of newly-fertilized white sturgeon eggs reveal a rapid series of cortical contractions accompanied by shifts in the pigmentation of the concentric rings around the animal pole within a few minutes of fertilization. The shifts in pigmentation result in the formation of a single light spot at the animal pole, centred within a dark area that is in turn surrounded by a broad zone of reduced pigment.

The first externally noticeable event after the rotation is formation of the light crescent, though this may not be very visible in lightly-pigmented eggs. Rotation of the darkly pigmented cortical (peripheral) layer of the egg relative to the lighter internal cytoplasm leads to the appearance of a light crescent that extends one-third to one-half of the way around the egg, roughly 50° above the equator; the exact latitude varies in different species. The formation of the crescent indicates that cortical rotation has occurred. This active rotation of the egg cytoplasm is a key step in determining the dorsoventral axis in other vertebrate groups, particularly amphibians, where the grey crescent has long been recognized as the first morphological marker of the future dorsal side of the embryo (Gerhart et al. 1989). Much work has been done in amphibians to characterize the underlying mechanism and molecular correlates of cortical rotation; few such studies have been done in sturgeons. Since the rotation itself is a conserved, ancestral feature of vertebrate development, it is reasonable to assume that many of the results from amphibian studies probably hold for sturgeons as well.

6.5.2 CLEAVAGE AND BLASTULA FORMATION

Sturgeons have holoblastic cleavage: each cleavage plane completely divides the egg cytoplasm. However, the large size of the egg, significant quantity of yolk, and concentration of the yolk in the vegetal cytoplasm mean that the progress of cleavage furrows through the egg to the vegetal pole is relatively slow. Each furrow begins at or near the animal pole and gradually progresses vegetally, but because cleavage of vegetal cytoplasm is delayed, by the time the first furrow is complete several more have already formed at the animal pole. By the time the first cleavage is completed, the second is well begun, and sixteen animal pole blastomeres (the result of three subsequent cleavage furrows) may be visible by the time the second cleavage nears the vegetal pole (stage 7). The yolk concentration and resulting slowing of cleavage in the vegetal hemisphere give rise to a graded distribution of blastomere sizes at the end of the cleavage phase: those in the animal hemisphere are much smaller than the large, yolky vegetal cells, which can often be distinguished even without the aid of a microscope. Animal pole cells are also more uniform in shape and size than those in the vegetal hemisphere.

The first cleavage crosses the animal pole, dividing the egg cytoplasm in half (stage 4); it normally bisects the light crescent formed by cortical rotation. The second furrow is perpendicular to the first, creating four equal-sized blastomeres (stage 5). The third cleavage, forming eight cells at the animal pole (stage 6), consists of a pair of furrows roughly perpendicular to the second division furrow and located to either side of the first cleavage plane: the result is two rows of four cells, with the central quartet somewhat smaller than the peripheral blastomeres. By this stage only the first cleavage plane has completely surrounded the egg; the blastomeres are therefore still continuous with the uncleaved cytoplasm at their periphery. Blastomeres vary in colour, reflecting the initial distribution of pigmentation in the egg cortex; often, the cells nearest the animal pole appear distinctly darker than the larger cells around them.

At the end of cleavage, the animal pole cytoplasm is subdivided into cells too small to be distinguished under a dissecting microscope. At this point (stage 11) the animal hemisphere appears uniformly light in colour, while the larger vegetal cells retain some variation in pigmentation, though they are generally darker than the animal cells. A transitional zone of intermediate-sized cells, known as the marginal zone, encircles the embryo roughly at the equator. The location of the marginal zone is constant within a species, though may differ between species. Though not visible from the outside, the fluid-filled blastocoel cavity has formed inside the embryo, which is now termed a blastula. The size and location of the blastocoel vary somewhat between species, but it is generally located in the centre of the animal hemisphere, with its floor at about the level of the marginal zone. The blastocoel in sturgeon embryos is proportionally smaller and thicker-walled than that of most amphibians, a morphological difference that has significant consequences for the mechanisms of gastrulation (Bolker 1993c).

The blastula is covered by a superficial layer of flattened epithelial cells, surrounding blockier internal cells that increase in size from the animal to the vegetal pole, reflecting the gradient in yolk distribution. At the end of cleavage, the blastocoel roof is about five cells thick at the animal pole in the white sturgeon; the blastocoel walls thicken as they approach the blastocoel floor. The roof thins significantly as gastrulation begins; eventually becoming a translucent sheet consisting of only one to two layers of cells (Bolker 1993c).

Ballard and Ginsburg (1980) prepared a fate map of the late blastula stage, indicating the prospective fates of different areas of the embryo by staining small groups of cells with vital dyes, allowing development to continue, and examining the distribution of dyed cells at later stages. Their map is broadly similar to those for many amphibians: most animal-hemisphere cells will form ectoderm, with the dorsalmost ectoderm fated to become neural plate and tube; the marginal zone forms primarily mesoderm, with prospective notochord cells located close to the dorsal midline, prospective somitic mesoderm in two symmetrical regions lateral to the notochord, and prospective lateral plate and ventral mesoderm farther around the embryo; and the large yolky cells of the vegetal hemisphere will contribute to endoderm and yolk, with the superficial layer of cells on the dorsal side becoming the lining of the archenteron or primary gut.

6.5.3 GASTRULATION

Gastrulation is the process by which the structurally simple, hollow blastula forms a multilayered embryo with an internal archenteron, a distinct anterior-posterior axis, and clearly delineated germ layers. It is an assemblage of dynamic processes rather than a definable morphological stage, involving active movement and rearrangement of cells within the embryo. Many of the component mechanisms of gastrulation have been studied in detail in the frog *Xenopus laevis*; experimental work in sturgeons indicates that while many of the cellular-level processes and mechanisms may be the same (and thus presumably retained from a common ancestor), their morphogenetic roles may differ as a function of their different structural and mechanical contexts (Bolker 1993c, 1994).

Gastrulation begins with the formation of a short, dark pigment line perpendicular to the dorsal midline (stage 13; Figure 6-1). This line represents the contracted apices of bottle cells, a population of cells in the superficial layer of the marginal zone that ingress into the interior of the embryo to begin the process of involution, or inrolling of cells from the surface. The lip forms at different distances above the "equator" of the egg in different species: in white sturgeon gastrulation begins in the equatorial marginal zone midway between the animal and vegetal poles, while in other species (e.g. Ignatieva 1965) the bottle cells form closer to the animal pole. The pigment line is the first indication of the formation of the dorsal lip of the blastopore.

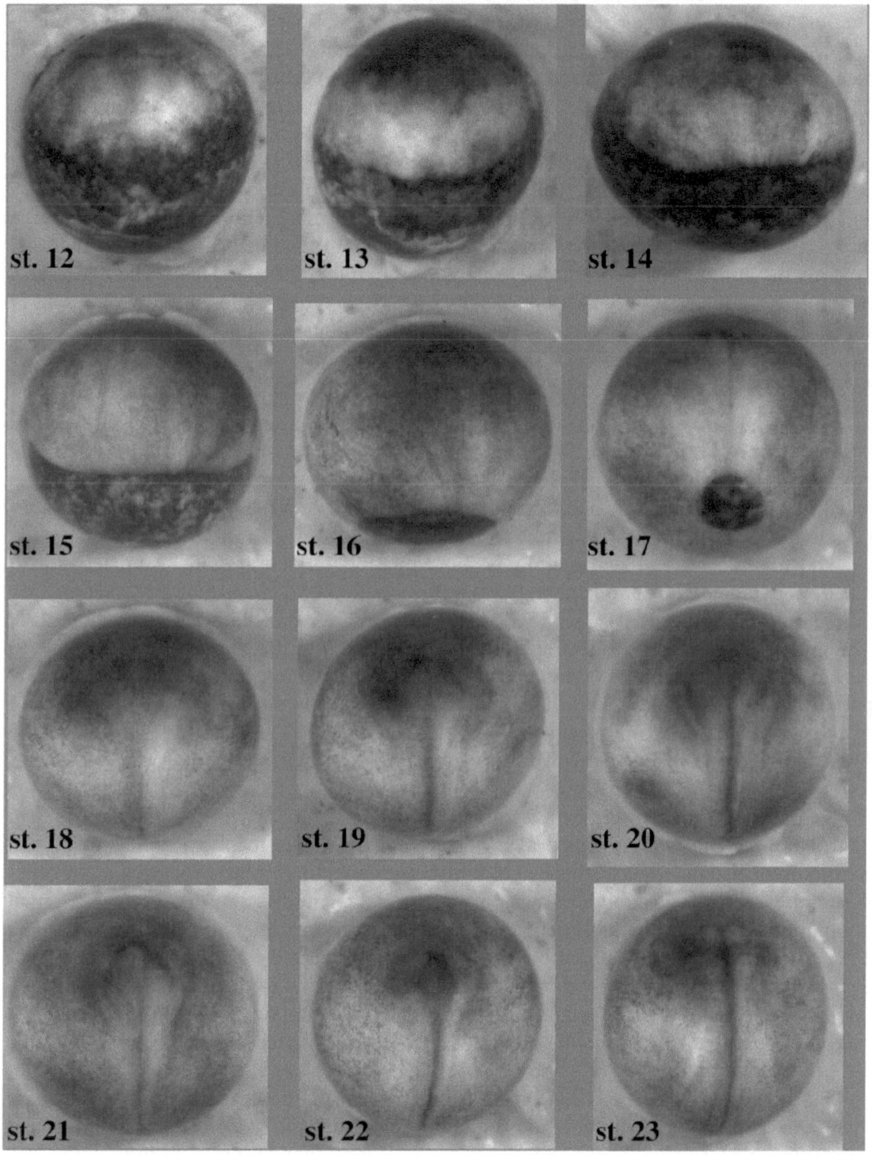

Fig. 6-1. Dorsal views of living white sturgeon embryos at stages from late blastula (stage 12) through neurulation (stage 23); staging follows Dettlaff and Ginsburg (1954). Embryos are ~2.3mm in diameter.

The main event of gastrulation is the involution (internalization) of surface layers, comprising prospective endoderm and mesoderm, through the blastopore. In the white sturgeon, as in Xenopus, this process is driven by active extension and convergence of the dorsal marginal zone. Epiboly (expansion) of animal-cap tissue, which occurs early in gastrulation, displaces the marginal zone vegetally so that its constriction is mechanically effective in producing involution. Experimental prevention of this displacement by removing the animal cap results in equatorial constriction of the embryo (Bolker 1993c).

As involution continues, the lighter and much smaller cells of the animal hemisphere override the dark, yolky vegetal cells. The blastopore lip extends laterally and ventrally as it migrates toward the vegetal pole, forming a complete circle around a large yolk plug at stage 16. At this point, the internal movement of involuted material towards the animal pole is frequently visible through the dorsal surface. The blastopore continues to constrict the yolk plug (stages 16-18), eventually closing down to a slit at stage 18, the transition between gastrulation and neurulation.

6.5.4 NEURULATION

During neurulation (stages 18-23; Figure 6-1) the rudiments of the central nervous system (CNS) form from surface ectoderm on the dorsal side of the embryo as the embryonic axis continues to extend. At the same time, neural crest migration occurs and the early morphological differentiation of the mesodermal somites and excretory system become visible alongside the forming neural tube.

The major event of neurulation is the elevation of neural folds from the dorsal ectoderm (stages 19-22) and subsequent approximation and closure of the folds to form the neural tube (stages 20-23), which later differentiates into the brain and spinal cord. The neural folds first appear around the anterior margin of the thickened neural plate (a region of columnarized ectodermal cells on the dorsal side), and progress posteriorly parallel to the midline, where a faint neural groove is often visible. Because the anterior of the neural plate (the prospective forebrain region) is its widest part, closure of the folds along the trunk is generally completed before final apposition of the anterior folds.

Neural crest cells migrate out from the dorsal surface of the neural tube late in neurulation. While sturgeon neural crest has not been fate-mapped, comparative and morphological data suggest that it probably gives rise to similar structures as in other vertebrates. Crest cells migrate away from the dorsal midline in dense mesenchymal populations. In the white sturgeon (Bolker 1993a), as in paddlefish (Bemis and Grande 1992), distinct regional populations of crest migrate around the forming placodes and evaginations of the CNS. Scanning electron microscopy (Bolker 1993a) also reveals individual cells apparently migrating along the proximal surface of somites, a crest migration pathway well-documented in a variety of vertebrate species (Hörstadius 1950; Le Douarin 1982).

As the neural folds converge dorsally, rudiments of the excretory system become visible as thickenings to either side, beginning at about the level of the hindbrain and extending posteriorly. These ridges enclose a lyre-shaped area within which the mesodermal somites begin to develop.

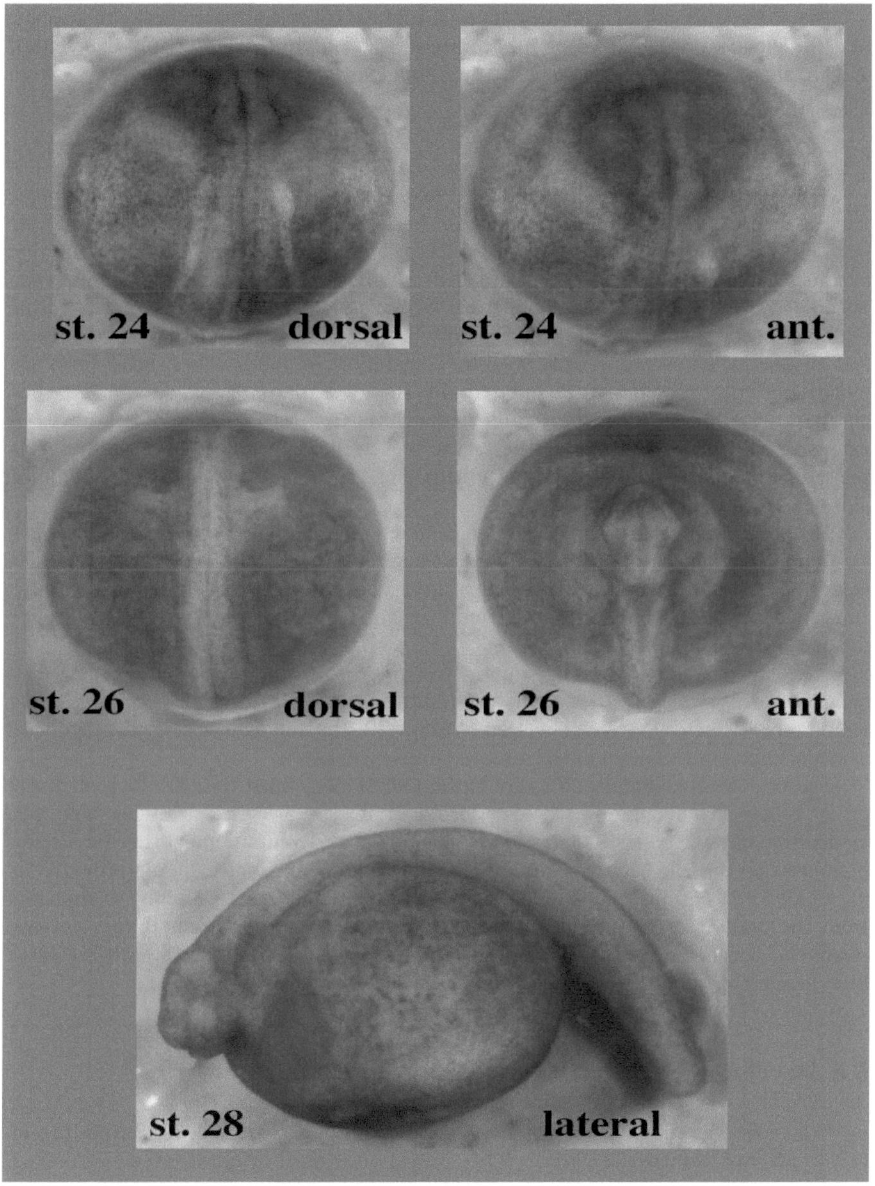

Fig. 6-2. White sturgeon embryos from late neurulation through early organogenesis (stages 24-28); staging follows Dettlaff and Ginsburg (1954). Embryos are ~2.3mm in diameter. Dorsal, anterior, or lateral views as indicated; stage 28 embryo has been dechorionated to show head and tail separation from yolk sac.

The relative timing of CNS and mesodermal development varies between species; in consequence, beyond stage 24 it becomes increasingly difficult to apply the Dettlaff and Ginsburg stage designations based on *A. gueldenstaedti*, *A. stellatus* and *H. huso* to other species.

6.5.5 ORGANOGENESIS

Between gastrulation and hatching, all major organ systems of the animal must be laid down, and most become functional. Early stages of this process are visible beginning in late neurulation (Figure 6-2), when regionalization of the brain first becomes apparent. As development proceeds, the body axis becomes well-defined and raised above the profile of the endodermal yolk mass. The tail begins to separate from the mass at about stage 26, by which point optic placodes, pronephric duct rudiments, and distinct somites are readily visible. Separation of the anterior head from the underlying yolk mass begins at stage 29-30. However, the yolk sac remains attached anteriorly nearly as far as the mouth in hatchlings, so head separation is never as extensive as that of the tail. Hatched larvae present the appearance of a bulbous yolk sac with a long, active tail; the proportionally small head is distinctly visible only upon close examination.

The heart rudiment appears as a straight tube in the midline just anterior to the head at stage 27, lengthens during stage 29, and begins to form a C-curve at stage 29. By stage 31 the heart is S-shaped, and its active contraction drives circulation through superficial vessels on the dorsal and lateral surfaces of the yolk sac, as well as through major vessels within the body.

The placodes and CNS evaginations that give rise to the major sense organs first appear as early as stage 24, when the eye rudiments are visible as bulges of the anterior neural tube. Olfactory sacs can be seen by stage 29, and auditory vesicles distinguished at stage 32 on either side of the thin-roofed myelencephalon.

The primary locomotory structure is the tail. Well-defined muscle blocks segment progressively from anterior to posterior, starting out as vertical ridges and then elongating and extending posteriorly at their dorsal and ventral ends to form anterior-pointing chevrons. In embryos and larvae, as in adults, axial stiffening is provided by the notochord. The fin fold running along the dorsal and ventral midline extends out from the body axis to provide increased propulsive surface. Pectoral fins develop relatively late, being barely visible in hatchlings of many species, and probably serve little locomotory function until the yolk sac has been largely resorbed.

6.6 Developmental Bottlenecks and Abnormalities

Common abnormalities in hatchery populations are reviewed by Dettlaff et al. (1993) and Schmalhausen (1991).

6.6.1 FERTILIZATION AND ONSET OF CLEAVAGE

The first readily visible indicator of successful fertilization is the timely onset of the first cleavage (stage 4), and its symmetrical division of the egg into two equal-sized blastomeres. Significantly delayed cleavage may indicate poor gamete quality, while the formation of extra cleavage furrows is evidence of polyspermy (Ginsburg 1961).

Shifts in pigmentation, particularly formation of the light crescent, suggest that the egg has been fertilized and development activated; however, since these changes are relatively subtle and may not be visible in lightly-pigmented eggs, they are less useful as diagnostic signs. Cleavage can occur in eggs that are not developing normally, but it is commonly delayed, less regular than in healthy eggs, and results in a poorly-organized group of blastomeres of variable size. Often the vegetal pole cleaves little or not at all in such eggs. Normally-developing eggs fertilized at the same time proceed through cleavage with little variation in timing. One can therefore make a good initial assessment of fertilization success and egg quality within a batch once first cleavage is well underway (the exact timing depends on temperature as well as species).

6.6.2 GASTRULATION

There is relatively little mortality during cleavage and blastula stages among embryos that have successfully begun development. The next point at which widespread problems occur is gastrulation, which requires extensive, well-coordinated morphogenetic movements and inductive signals throughout the embryo. Failure to commence or to complete gastrulation occurs even among embryos that develop apparently normally through stage 12 (late blastula). In embryos that begin gastrulation but cease blastopore closure with a wide yolk plug, differentiation of dorsal and anterior structures may continue for some time, including the formation of the anterior CNS. However, posterior structures fail to form and the embryos usually die before hatching. If blastopore closure is incomplete but most of the yolky endoderm is internalized and the yolk plug is relatively small, development through hatching can continue without major disruption.

6.7 Formation of the Body Axis, CNS and Heart

Various malformations of the CNS and body axis have been documented in European species; many deformities are known or suspected to be the result of maternal or embryonic exposure to environmental toxins (Dettlaff et al. 1993). Similar defects also occur spontaneously in hatchery-reared white sturgeons, though at much lower frequency. In contrast to the embryos with arrested gastrulation described above, some specimens show complete posterior development but absence or severe underdevelopment of anterior structures, particularly the brain and heart. Such animals die before hatching. Occasional embryos show unilateral failures of development, though this is rare. In some cases the main body axis appears normal, but heart morphogenesis ceases at the straight-tube stage with an abnormally thin rudiment that degenerates rather than progressing through the C- and S-shaped stages and commencing normal function. Such embryos may be otherwise well-formed, but do not survive to hatching.

As in other fish species, scoliosis or spinal curvature (usually consisting of a downward curve around the yolk sac) is a relatively common axial defect, and is often associated with an abnormally short body axis. Strongly curved larvae may move vigorously and hatch spontaneously, but are unable to swim effectively and do not survive long-term. Occasional individuals show relatively minor, localized kinks in the axis, which may not have significant effects on locomotion.

6.8 Conclusion

A central component of any successful aquaculture venture is a thorough understanding of the basic biology of the species of interest. Such knowledge allows both optimization of production conditions, and the design of effective solutions to common problems. Knowing the sequence of normal embryogenesis allows early screening of hatchery production runs, which can help establish the value of broodstock individuals, and inform decisions about the allocation of resources such as rearing space to different batches of embryos. At this point, too little is known of sturgeon developmental mechanisms and the defects induced by different environmental factors to permit use of patterns of developmental abnormalities for diagnosing specific problems. Nevertheless, since early embryos are often more sensitive than later stages to environmental conditions, they can serve as general bioindicators in assessing hatchery conditions or water quality (particularly as observations are not confounded by effects of larval diets or feeding success). Understanding the necessary conditions for successful early development is also essential to conservation efforts, whether these involve habitat preservation or restoration, or hatchery rearing of larvae for release (McCabe and Tracy 1994).

Finally, now that embryonic stages are accessible through aquaculture programs, sturgeons and paddlefishes have great potential importance for comparative and evolutionary studies of development. Their phylogenetic position as an outgroup to teleost fishes, combined with their amphibian-like early development, makes them a uniquely valuable point of comparison for both teleost and amphibian studies. The recent emphasis on the zebrafish *Danio rerio* (a highly derived teleost) as a fundamental "model species" for vertebrate development (Roush 1996; Haffter et al. 1996; Metscher and Ahlberg 1999) increases the importance of comparative analyses that include species rooted deeper in the vertebrate phylogeny, and retaining more ancestral developmental characteristics.

Acknowledgments

I am grateful to Dr. Sergei Doroshov for his helpful suggestions on the manuscript, and for his ongoing encouragement and support.

6.9 References

Akimova, N.V. and Ruban, G.I. 1993. The condition of the reproductive system of the Siberian sturgeon, *Acipenser baeri*, as a bioindicator. J. Ichthyology. 33:15-24.

Akimova, N.V. and Ruban, G.I. 1996. A classification of reproductive disturbances in sturgeons (Acipenseridae) caused by an anthropogenic impact. J. Ichthyology. 36:65-80.

Ballard, W.W. and Ginsburg, A.S. 1980. Morphogenetic movements in acipenserid embryos. J. Exper. Zool. 213:69-103.

Ballard, W.W. and Needham, R.G. 1964. Normal embryonic stages of *Polyodon spathula* (Walbaum). J. Morphol. 114:465-478.

Beer, K.E. 1981. Embryonic and larval development of the white sturgeon (*Acipenser transmontanus*). Master's thesis, University of California, Davis.

Bemis, W.E. and Grande, L. 1992. Early development of the Actinopterygian head. I. External development and staging of the paddlefish *Polyodon spathula*. J. Morphol. 213:47-83.
Bolker, J.A. 1993a. Embryonic development in the white sturgeon, *Acipenser transmontanus*. Ph.D. dissertation, University of California, Berkeley.
Bolker, J.A. 1993b. Gastrulation and mesoderm morphogenesis in the white sturgeon. J. Exper. Zool. 266:116-131.
Bolker, J.A. 1993c. The mechanism of gastrulation in the white sturgeon. J. Exper. Zool. 266:132-145.
Bolker, J.A. 1994. Comparison of gastrulation in frogs and fish. Amer. Zool. 34:313-322.
Chapman, F.A. and Carr, S.H. 1995. Implications of early life stages in the natural history of the Gulf of Mexico sturgeon, *Acipenser oxyrhyncus desotoi*. Env. Biol. Fishes. 43:407-413.
Conte, F., Doroshov, S.I., Lutes, P.B. and Strange, E.M. 1988. Hatchery manual for the white sturgeon, *Acipenser transmontanus* Richardson, with application to other North American Acipenseridae. U.C. Regents, Oakland, CA.
Dean, B. 1895. The early development of gar-pike and sturgeon. J. Morphol. 11:1-62
Deng, X., Van Eenennaam, J.P. and Doroshov, S.I. 2002. Comparison of early life stages and growth of green and white sturgeon. In Biology, management, and protection of North American sturgeon. Edited by W. Van Winkle, P.J. Anders, D.H. Secor, and D.A. Dixon. American Fisheries Society Symposium 28. Bethesda, Maryland. pp. 237-248.
Dettlaff, T.A. 1953. Dependence of the rate of cleavage of sturgeon eggs on temperature. Dokl. Akad. Nauk. SSSR. 91:695-698.
Dettlaff, T.A. 1962. Cortical changes in acipenserid eggs during fertilization and artificial activation. J. Embryol. Exper. Morph. 10:1-26.
Dettlaff, T.A. and Dettlaff, A.A. 1961. On relative dimensionless characteristics of the development duration in embryology. Arch. Biol. 72:1-16.
Dettlaff, T.A. and Ginsburg, A.S. 1954. [Embryonic development in sturgeon (Russian sturgeon, sevruga, and beluga) in connection with problems of artificial propagation.] (In Russian.) USSR Acad. Sci. Press, Moscow.
Dettlaff, T.A., Ginsburg, A.S. and Schmalhausen, O.I. 1993. Sturgeon fishes: developmental biology and aquaculture. Springer-Verlag, New York.
Doroshov, S.I. 1985. Biology and culture of sturgeon Acipenseriformes (Ch. 7). In Recent advances in aquaculture. Edited by J. Muir and R. Roberts. Croon Helm Publishers, London. pp. 251-274.
Doroshov, S.I., Moberg, G.P. and Van Eenennaam, J.P. 1997. Observations on the reproductive cycle of cultured white sturgeon, *Acipenser transmontanus*. Environ. Biol. Fishes 48:2265-278.
Gerhart, J., Danilchik, M., Doniach, T., Roberts, S., Rowning, B. and Stewart, R. 1989. Cortical rotation of the Xenopus egg: consequences for the anteroposterior pattern of embryonic development. Development 107 Supplement.
Ginsburg, A.S. 1961. The block to polyspermy in sturgeon and trout with special reference to the role of cortical granules (alveoli). J. Embryol. Exp. Morph. 9:173-190.
Ginsburg, A.S. and Dettlaff, T.A. 1991. The Russian sturgeon *Acipenser güldenstädti*. I. Gametes and early development up to time of hatching. In Animal species for developmental studies. Vol. 2. Vertebrates. Edited by T.A. Dettlaff and S.G. Vassetzky. Plenum, NY. pp. 15-65.
Haffter, P., Granato, M., Brand, M., Mullins, M.C., Hammerschmidt, M., Kane, D.A., Odenthal, J., van Eeden, F.J.M., Jiang, Y.-J., Heisenberg, C.-P., Kelsh, R.N., Furutani-Seiki, M., Vogelsang, E., Beuchle, D., Shach, U., Fabian, C. and Nüsslein-Volhard, C. 1996. The identification of genes with unique and essential functions in the development of the zebrafish, Danio rerio. Development. 123:1-36.
Hörstadius, S. 1950. The Neural Crest. Oxford University Press, London.
Ignatieva, G.M. 1960. [The regional nature of the inductive action of chordamesoderm in embryos of acipenserid fish.] (In Russian.) Dokl. Acad. Nauk SSSR 134:233-236.
Ignatieva, G.M. 1961. [Inductive properties of the chordamesodermal rudiment prior to invagination and the regulation of its defects in sturgeon embryos.] (In Russian.) Dokl. Akad. Nauk SSSR 139:503-505.
Ignatieva, G.M. 1963. [A comparison of the dynamics of invagination of chordamesoderm material in embryos of *Acipenser stellatus*, *A. güldenstädti*, and axolotl.] (In Russian.) Dokl. Akad. Nauk SSSR 151:1466-1469.
Ignatieva, G.M. 1965. Relationship between epibole and invagination in sturgeon embryos during the period of gastrulation. Proc. Acad. Sci. USSR - Biol. Sect. 165:970-973.
Kempinger, J.J. 1988. Spawning and early life history of lake sturgeon in the Lake Winnebago system, Wisconsin. American Fisheries Society Symposium 5. pp. 110-122.
Le Douarin, N. 1982. The neural crest. Academic Press, London.
Leung, T.C. and Wourms, J.P. 1991. Gastrulation of the paddlefish, Polyodon. Amer. Zool. 31(5):81A.

McCabe, G.T. Jr. and Tracy, C.A. 1994. Spawning and early life history of white sturgeon, *Acipenser transmontanus*, in the lower Columbia River. Fishery Bulletin 92:760-772.
Poole, T.J. 1988. Cell rearrangement and directional migration in pronephric duct development. Scanning Microscopy. 2:411-415.
Roush, W. 1996. Zebrafish embryology builds better model vertebrate. Science 272:1103.
Ryder, J.A. 1888. The sturgeons and sturgeon industries of the eastern coast of the United States, with an account of experiments bearing upon sturgeon culture. Bull. U.S. Fish Comm. 8:231-324.
Salensky, A. 1881. Recherches sur le développement du sterlet (*Acipenser ruthenus*). Arch. Biol. 2:233-341.
Schmalhausen, O.I. 1991. The Russian sturgeon *Acipenser güldenstädti*. Part II. Later prelarval development. Ch. 3. In Animal species for developmental studies. Vol. 2. Vertebrates. Edited by T.A. Dettlaff and S.G. Vassetzky. Consultants Bureau, New York and London. pp. 67-88.
Shagaeva, V.G., Nikol'skaya, M.P., Akimova, N.P., Markov, K.P. and Nikol'skaya, N.G. 1993. A study of the early ontogeny of Volga sturgeon (Acipenseridae) subjected to human activity. J. Ichthyol. 33:23-41.
Shelton, W.L., Mims, S.D., Clark, J.A., Hiott, A.E., and Wang, C. 1997. A temperature-dependent index of mitotic interval ($\tau 0$) for chromosome manipulation in paddlefish and shovelnose sturgeon. Progressive Fish-Culturist 59:229-234.
Van Eenennaam, A.L., Van Eenennaam, J.P., Medrano, J.F., and Doroshov, S.I. 1996. Rapid verification of meiotic gynogenesis and polyploidy in white sturgeon (*Acipenser transmontanus* Richardson). Aquaculture 147:177-189.
Van Eenennaam, J.P., Webb, M.A.H., Deng, X., Doroshov, S.I., Mayfield, R.B., Cech, J.J. Jr., Hillemeier, D.C. and Wilson, T.E. 2001. Artificial spawning and larval rearing of Klamath River green sturgeon. Transactions of the American Fisheries Society. 130:159-165.
Wang, Y.L., Binkowski, F.P. and Doroshov, S.I. 1985. Effect of temperature on early development of white and lake sturgeon, *Acipenser transmontanus* and *A. fulvescens*. In North American Sturgeons. Edited by F. P. Binkowski and S. I. Doroshov. Dr. W. Junk, Dordrecht. pp. 43-50.

CHAPTER 7

SWIMMING AND RESPIRATION

STEPHAN J. PEAKE
University of New Brunswick, Canadian Rivers Institute and Department of Biology, Bag Service 45111, Fredericton, NB Canada E3B 1J6

7.1 Introduction

Swimming capacity and respiratory characteristics of fishes can be modulated, affected and constrained by many biological and environmental variables. In a comprehensive review of the subject, Beamish (1978) identified several biological constraints on swimming capacity and the environmental aspects that independently affect them. As fish show great morphological and physiological diversity, it is not surprising that Beamish (1978) found swimming and respiratory performance among species to be highly variable.

Sturgeon and paddlefish are of particular interest to evolutionary biologists because their morphology and physiology are quite unique and, in many ways, intermediate between sharks (selachians) and the more recently evolved teleosts. Studies conducted on the relationship between morphology, physiology, and locomotion of North American sturgeon species can be roughly divided into those that have (1) examined kinematics, behaviour, respiration, and performance of individuals forced to swim in water tunnels and respirometers, and (2) those that have measured behavioural and respiratory variables of free-swimming fish under less constrained conditions. This chapter will attempt to summarize and discuss existing knowledge on swimming capacity, behaviour, and respiration of sturgeon and paddlefish generated from both lines of research in relation to their unique anatomical features, and place this knowledge into an evolutionary context with that of teleosts and selachians.

7.2 Behaviour, Performance, and Respiration in Forced Swimming Tests

7.2.1 BACKGROUND

The range of swimming speeds over which a fish can operate has traditionally been divided into three distinct categories: sustained, prolonged, and burst (Beamish 1978). Sustained swimming speeds are relatively low, and are maintained primarily through the contraction of red, slow-twitch muscle fibres located in a narrow band along the lateral line. Metabolism in red muscle is predominantly aerobic in nature. Thus, sustained swimming speeds do not normally result in fatigue, and are characterized by

G.T.O. LeBreton et al. (eds.), Sturgeons and Paddlefish of North America, 147–166.
© 2004 *Kluwer Academic Publishers. Printed in the Netherlands.*

a lack of anaerobic metabolites in the blood and muscle during and post-exercise, a sub-maximal rate of oxygen consumption, little or no post-exercise hyperventilation, and no relationship between endurance and swimming velocity. The highest swimming speed that can be maintained aerobically is called maximum sustained speed, and is often estimated by measuring critical swimming speed, or U_{crit}, in a swim tunnel respirometer. Burst swimming speeds are relatively rapid and are maintained primarily by white, fast-twitch muscle fibres that make up a large portion of the axial trunk musculature. White muscle is predominantly anaerobic and glycolytic in nature. Burst speeds are, therefore, characterized by a build up of anaerobic metabolites in the plasma and muscle during and post-exercise, a maximal rate of oxygen consumption that is maintained for a considerable period of time post-exercise, fatigue within 15 to 20s, and a weak or absent relationship between endurance and swimming velocity. Those speeds that fall between maximum sustained and minimum burst represent prolonged activity, the characteristics and metabolic consequences of which are intermediate between those of sustained and burst swimming.

7.2.2 BEHAVIOUR

Most swimming performance tests have been conducted on freshwater teleost species and the majority of these have involved the ubiquitous rainbow trout (*Oncorhynchus mykiss*). Relatively few studies have provided details on swimming behaviour during these trials; however, it has been reported that rainbow trout often maintain low speeds through frictional contact with the base of the swim tunnel, small oscillations of the caudal fin, and slow pectoral fin paddling (Hudson 1973). Rainbow trout generally swim steadily at higher speeds using slow, rhythmic caudal tail beats; however, changes in swimming speed imposed by the researcher has resulted in periods of restless movement (Webb 1971) and occasional bursts of violent struggling (Hudson 1973).

In contrast, sturgeon have been characterized as relatively docile in swim tunnels (Burggren and Randall 1978), with bouts of restlessness and violent outbursts occurring rarely, if ever. Bottom holding, however, is a behaviour that many sturgeon species employ to hold station in a current, and individuals appear to use this strategy more readily than rainbow trout. Adams et al. (1999) reported that juvenile pallid sturgeon often held their position in a swim tunnel by pressing their bodies and pectoral fins against the bottom. Juvenile salmonids often exhibit a similar behaviour (Peake et al. 1997a), and are able to hold in currents 6 to 7 times faster than could be achieved by active swimming (Arnold et al. 1991). Interestingly, Adams et al. (1999) found that most bottom holding occurred at intermediate speeds, while free-swimming was most often observed at low (< 20cm/s) and high (> 50cm/s) water velocities. In a similar study, it was demonstrated that the frequency of bottom-holding behaviour by shovelnose sturgeon increased as water velocity exceeded 40cm/s (Adams et al. 1997). The authors hypothesized that the holding behaviour of pallid and shovelnose sturgeon may allow them to occupy fast-water microhabitats with a relatively small energetic cost, thereby reducing competition from more pelagic species. Although bottom holding is not specifically mentioned in studies conducted on white sturgeon (Burggren 1978; Counihan and Frost 1999) and Gulf sturgeon (Chan et al. 1997), the behaviour has been reported for lake sturgeon exposed to low to moderate water velocities (McKinley and Power 1992, Peake 1999). McKinley and Power (1992) suggested that lake sturgeon did not swim when faced with low water velocities because the flows were insufficient to lift the negatively buoyant fish off the bottom and into the water column.

Paddlefish appear to be more pelagic than sturgeon, as Burggren and Bemis (1992) reported that juveniles in still water were in constant motion, swimming at speeds of about 10cm/s (1.3bl/s). This is not surprising as paddlefish perfuse their gills with water by holding their mouths agape and swimming forward, an activity known as ram ventilation. This strategy essentially transfers the metabolic costs of ventilation from the buccal and opercular muscles to the axial trunk musculature, which operates with greater efficiency. Presumably, however, paddlefish exposed to moving water could hold to the bottom and simply allow the current to wash over their gills. Although this behaviour has not been documented experimentally, it would theoretically allow this species to breathe with little or no energy expenditure.

7.2.3 KINEMATICS

The locomotory movements of most teleost species exhibit the following characteristics: (1) swimming speed during steady locomotion is modulated primarily by the alteration of tail beat frequency (TBF), (2) tail beat amplitude (TBA) is usually constant over most of the swimming speed range, and (3) the length of the propulsive wave is independent of speed (Webb 1986). Locomotion in selachians tends to be kinematically similar to that of teleosts; however, tail beat frequency at a given swimming speed in the former tends to occur over a much greater range, and some speed-related variations in amplitude and propulsive wavelength have been reported (Webb and Keyes 1982).

Research has also indicated that sturgeon swimming has unique features but is kinematically more similar to teleosts than selachians (Webb 1986). The relationship between tail beat frequency and swimming speed (Table 7-1) is positive, linear, and similar in slope and intercept to that of rainbow trout (Webb 1986), yet unlike teleosts, tail beat amplitude in sturgeon does increase slightly with swimming speed (Table 7-1). Webb (1986) also reported that the propulsive wavelength of lake sturgeon was somewhat smaller than that of teleosts. These kinematic differences, although small, can affect swimming efficiency, as Webb (1986) found that overall thrust generated by lake sturgeon was 18% less than that of rainbow trout over much of its speed range. This situation was likely exacerbated by the further finding that lake sturgeon experienced 3.5 times more drag per unit area than the rainbow trout (Webb 1986).

In terms of locomotory gait, Wilga and Lauder (1999) reported that white sturgeon swam using continuous undulations of the body and tail at low to moderate speeds (0 to 70cm/s; 0 to 2.4bl/s), but switched to a more efficient (Weihs 1974) burst and glide type movement at higher speeds. When moving horizontally in the water column, white sturgeon maintained a positive body tilt relative to the flow that was inversely proportional to swimming speed (Wilga and Lauder 1999). At a low velocity of about 15cm/s (0.5bl/s), this angle was 20° or more; however, it became negligible at speeds approaching 90cm/s (3.0bl/s). When moving in the vertical plane, sturgeon adopted a positive body tilt when ascending and a negative tilt when descending (Wilga and Lauder 1999).

7.2.4 CRITICAL SWIMMING SPEED

Critical swimming speed has not been measured extensively for sturgeon and paddlefish; however, a compilation of available data suggests that U_{crit} for several species can be described as a linear function of fish length (Figure 7-1). Most studies that have measured swimming capacity of sturgeon have indicated that U_{crit} for a fish of given size is lower than that which would be expected for a salmonid (Adams et al. 1999, Peake et al. 1997b). However, it is noteworthy that sturgeon compare more favourably, in terms of swimming ability, with other warm water species such as walleye (*Stizostedion vitreum*; Figure 7-1). The relationship between U_{crit} and water temperature has not been examined in detail; however, data that exists for lake sturgeon over a range of 7 to 21°C (Figure 7-2) suggests that U_{crit} may follow the typically observed pattern, that is, a positive but decreasing slope that peaks and then becomes negative. The nature of the relationship at higher temperatures, however, is currently unknown.

Table 7-1. Intercepts, coefficients, and r^2 values for relationships between swimming speed, fork length, and water temperature and various dependent variables. TBF = tail beat frequency, TBA = tail beat amplitude, VO_2 = oxygen consumption rate) for lake sturgeon (LS), white sturgeon (WS), and pallid sturgeon (PS).

Dependent variable	Intercept	Speed (cm/s)	Length (cm)	Temp (°C)	r^2	Species	Reference
Movement (m/d)	167.51	-	-	75.54	0.89	LS	Diana et al. (1990)
TBF (Hz)	1.67	0.07	-	-	-	LS	Webb (1986)
TBA (cm)	3.2	0.02	-	-	-	LS	Webb (1986)
VO_2, 5 °C (mg/kg/hr)	12.32	1.28	-	-	0.93	LS	McKinley and Power (1992)
VO_2, 10 °C (mg/kg/hr)	43.00	1.10	-	-	0.81	LS	McKinley and Power (1992)
VO_2, 19 °C (mg/kg/hr)	54.67	1.15	-	-	0.76	LS	McKinley and Power (1992)
TBF (Hz)*	-0.036	7.25**	-	-	0.74	WS	Long (1995)
Log endurance (min)	- 1.58	0.051	0.24	-	0.71	PS	Adams et al. (1999)

7.2.5 ENDURANCE

Swimming endurance, or time to fatigue, over a wide range of speeds and temperatures has not been measured for many fish species because the protocol to collect this information is very time consuming. Despite this, endurance data provide a remarkably clear picture of the swimming capacity of fish, including the position of maximum sustained speed and the transition from prolonged to burst swimming. Traditionally, endurance data has been presented as a fatigue curve with time and speed on the y and x axes, respectively.

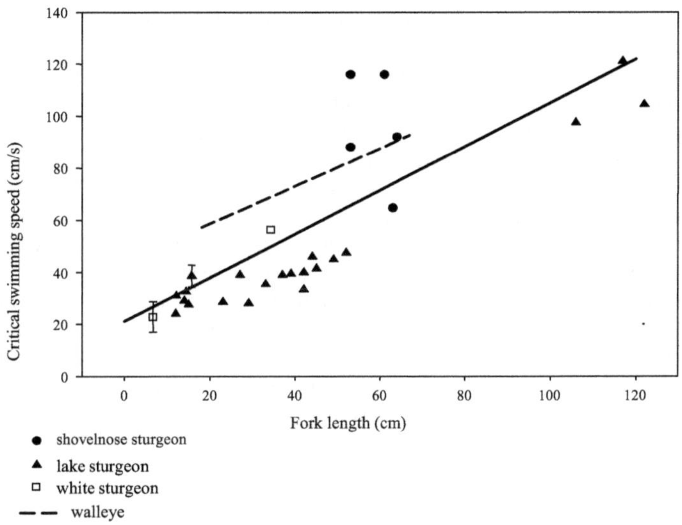

Fig. 7-1. The relationship between critical swimming speed and water temperature for lake sturgeon (44 to 52cm, Peake et al. 1995), coho salmon (4.6 to 9.0cm, Brett et al. 1958), and lake trout (27 to 82g, Gibson and Fry 1954).

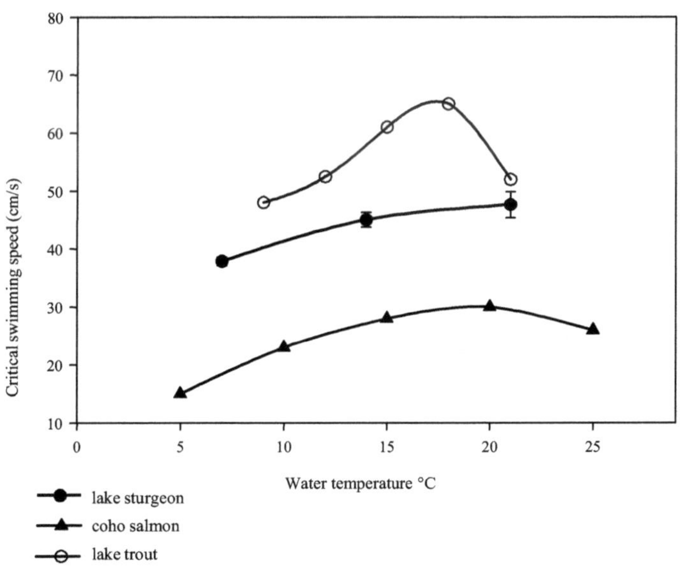

Fig. 7-2. Critical swimming speed data for shovelnose sturgeon (Adams et al. 1997), lake sturgeon (Webb 1986; Peake et al. 1995), white sturgeon (Bennett and Farrell 1998; Counihan and Frost 1999), and walleye (Peake et al. 2000). Error bars indicate standard error. The relationship between Ucrit (U, cm/s) and fork length (FL, cm) for these species is best described ($r^2 = 0.72$) by the linear model: $U = 21.05 + 0.84 \times FL$.

Fatigue curves typically consist of three distinct lines (Figure 7-3), the first being vertical and positioned at maximum sustained speed. Swimming velocities to the left of this line are considered sustained. The second line covers the spectrum of prolonged speeds, which are inversely related to endurance. The third line, representing burst swimming, typically begins at a speed that can only be maintained for a few seconds and ends at the maximum attainable velocity of the fish. The slope of the line in the burst performance envelope is generally negative or close to zero. Water temperature exerts little influence on time to fatigue at high prolonged and burst speeds; however, temperature tends to affect sustained and lower prolonged swimming capacity in a manner similar to that described for critical speed (Beamish 1978).

Although endurance data for sturgeon and paddlefish are scarce, fatigue curves generated for lake sturgeon follow the general pattern just described, except that a slope change between prolonged and burst swimming is not evident in this species (Figure 7-3). This suggests that lake sturgeon possess a relatively weak or inefficient capacity for high-speed, anaerobically-driven swimming.

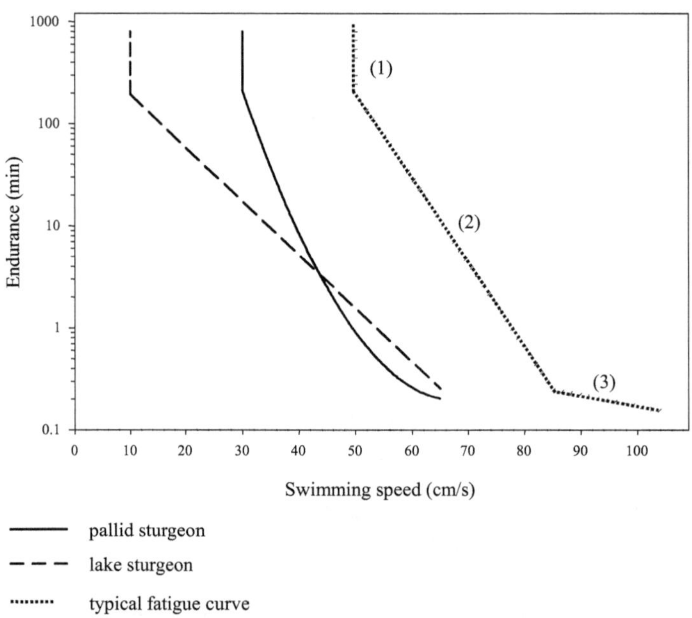

Fig. 7-3. Fatigue curves for pallid sturgeon (17 to 20cm fork length, 17 to 20 °C, Adams et al. 1999), lake sturgeon (17cm fork length, 18 °C, Peake et al. 1997a), and a typical curve showing hypothetical (1) sustained, (2) prolonged, and (3) burst performance envelopes.

On the other hand, an increasing number of studies have failed to detect a slope change at high swimming speeds in active species such as Atlantic salmon (*Salmo salar*), brook trout (*Salvelinus fontinalis*), brown trout (*Salmo trutta*; Peake et al. 1997a), lake whitefish (*Coregonus clupeaformis*) and lake cisco (*Coregonus artedii*; Bernatchez and Dodson 1985). Adams et al. (1999), however, were able to demonstrate a slope change at high speeds for pallid sturgeon (Figure 7-3: Table 7-1). Endurance at prolonged speeds appears to be quite similar for lake and pallid sturgeon; however, the latter possesses a greater maximum sustained speed (Figure 7-3).

Although no endurance data for paddlefish are available, Burggren and Bemis (1992) reported that maximum sustained speed for juvenile paddlefish was about 12 to 13cm/s (1.6 to 1.8bl/s). This value for paddlefish is similar to that for much larger lake and pallid sturgeon, indicating that paddlefish may possess a better developed capacity for aerobic swimming.

7.2.6 RESPIRATION

Energy for maintaining sustained swimming is derived primarily through oxidative metabolic processes. Sustained activity is, therefore, limited by the ability of the cardiovascular system to provide oxygen to the working musculature. Numerous studies have investigated the relationship between swimming performance and oxygen consumption in fish. As with most organisms, there is a basic oxygen requirement for operation and maintenance of essential organ systems. This is called the resting, standard, or basal metabolic rate. When a fish begins to swim, oxygen consumption increases beyond the resting level and continues to do so as speed increases up to maximum sustained, after which it levels off at the maximum, or active, metabolic rate. The relationship between oxygen consumption and swimming velocity is not typically linear as the drag that must be overcome increases with the square of velocity. Most often, the relationship is depicted as log-linear. The range of oxygen consumption values between the resting and active rates is defined as the aerobic scope for activity. The active rate for teleosts is generally between 5 and 10 times the resting rate, while the magnitude of the aerobic scope for activity tends to be approximately 300 to 800 mg O_2/kg/hr (Beamish 1978). Further increases in swimming speed beyond maximum sustained require that additional energy be provided through anaerobic processes.

Studies on the swimming energetics of sturgeon species are scarce. McKinley and Power (1992) correlated swimming speed with oxygen consumption for lake sturgeon and found a simple linear relationship at each of three different temperatures. The slope of this relationship remained constant with temperature; however, the intercept increased proportionally with water temperature (Table 7-1). Oxygen consumption in lake sturgeon at U_{crit} was only 3 to 4 times the resting rate, and the aerobic scope for activity was just 105, 90, and 75mg O_2/kg/hr at 19, 10, and 5°C, respectively. Similarly, Kieffer et al. (2001) reported that resting oxygen consumption rates in shortnose and Atlantic sturgeon only increased by a factor of about 2 following exhaustive exercise. Swimming energetics of white sturgeon have also been measured (Burggren 1978), and were found to be remarkably similar to those of lake sturgeon (Figure 7-4). While values associated with the aerobic scope of sturgeon were all low compared to more active species, they were comparable to those observed in other relatively sedentary species such as the flounder (*Platichthys flesus*), common dab (*Limanda limanda*) and lemon sole (*Microstomas kitt*; Duthie 1982). These species, unlike sturgeon however, demonstrated the expected log-linear relationship between oxygen consumption and swimming speed (Duthie 1982).

Paddlefish, unlike sturgeon, are ram ventilators (Burggren and Bemis 1992). As mentioned previously, this means that they perfuse their gills with water by swimming constantly rather than using the buccal and opercular musculature to irrigate the respiratory surface. Burggren and Bemis (1992) noted that, when fish were forced to remain stationary, regular buccal and opercular pumping did not occur. This, associated with the fact that stationary individuals became highly stressed, suggests that paddlefish are an obligate ram ventilator. Burggren and Bemis (1992) went on to find that swimming speeds lower than 6 cm/s (0.4bl/s) resulted in irregular buccal ventilation. At about 8cm/s (0.6bl/s) breathing became erratic, and at 10cm/s (0.8bl/s) ventilation was sporadic and shallow. A further increase to about 12cm/s (0.9bl/s) resulted in cessation of buccal breathing and a complete switch to ram ventilation. Sanderson et al. (1994) reported similar ventilatory dynamics for larger fish of the same species.

Although the resting metabolic rate of paddlefish seems to be similar to that of sturgeon, oxygen consumption in the former appears to increase faster with swimming speed (Figure 7-4). Burggren and Bemis (1992) reported that oxygen consumption at maximum sustained speed was approximately 5 to 6 times the standard rate and the magnitude of the aerobic scope for activity was about 230mg O_2/kg/hr at 22°C. In terms of swimming speeds, transition to ram ventilation was made at about 50% of the aerobic scope, while the preferred swimming speed (1.3bl/s) occurred at approximately 70% of the scope for activity. Therefore, it is clear that paddlefish, unlike many other active species (Briggs and Post 1997), routinely operate in the upper portions of their aerobic scope for activity.

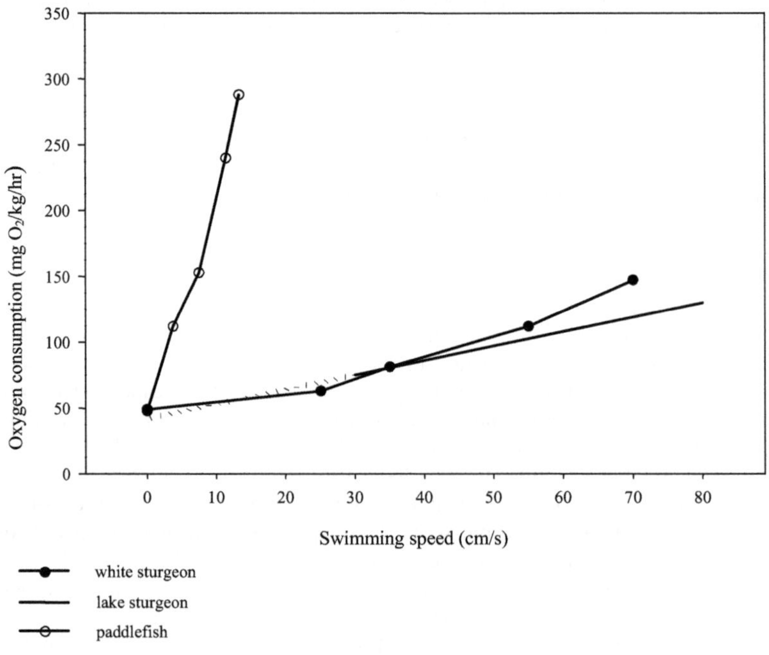

Fig. 7-4. The relationship between swimming speed and oxygen consumption for white sturgeon (62cm fork length, 10°C, Burggren 1978), lake sturgeon (80 to 95cm fork length, 10 °C, McKinley and Power 1992), and paddlefish (7.6cm fork length, 22°C, Burggren and Bemis 1992). The dashed line for lake sturgeon represents an extrapolation to zero swimming speed for estimation of resting metabolic rate.

7.3 Morphological Constraints on Swimming Activity and Respiration

7.3.1 BRANCHIAL CIRCULATION AND GILL MORPHOMETRICS

As the gills act as the primary respiratory structure in most fish, activity patterns and swimming capacity are often constrained by the morphology of these structures, and the relationship between the gill surface area and body mass. Burggren et al. (1979) examined gill morphometrics of adult white sturgeon and found that the surface area of the gill lamellar blood channels represented approximately 60% of the total surface area of the fish exposed to the environment. The authors also discovered that the weight-specific surface area was among the lowest reported for a wide variety of marine and freshwater fishes. While the number of lamellar filaments on the gill arches was typical of numerous teleost species, the density and size of the secondary lamellae on these filaments were comparatively small (Burggren et al. 1979). In addition, lamellae walls were somewhat thicker than those of other species suggesting that gas exchange through the membrane may be slow. If this is the case, however, white sturgeon must compensate for this through other mechanisms, as Burggren et al. (1979) found that dorsal aorta blood was fully oxygen saturated under normoxic conditions.

Ventilatory behaviour in white sturgeon is similar to that described for other teleosts. Water is drawn in through the mouth and passes over the gills by synchronous movements of the buccal cavity floor and the opercula. However, Burggren et al. (1979) discovered that white sturgeon can also ventilate through permanent openings in the upper regions of their opercular slits. It was further found that if water is prevented from entering through the mouth, a situation which would drown most fish, opercular slit ventilation can not only maintain the basal metabolic requirements of the animal, but can provide enough oxygen to the musculature to maintain swimming speeds up to two-thirds of maximum sustained. The authors speculated that this ability may be related to a feeding adaptation that allows fish to cruise along the bottom while their mouths are buried in the substrate.

The morphology of paddlefish mouths and gills reflect their classification as obligate ram ventilators. The anterior portions of the gill arches are elevated above the floor of the buccal cavity during ram ventilation, which exposes them maximally to the incoming flow and serves to direct water into the oral cavity (Burggren and Bemis 1992). This latter feature is also important for feeding in paddlefish, as they forage by filtering the incoming water for suspended zooplankton.

7.3.2 TAIL AND FINS

Most freshwater teleosts employ an undulatory body and caudal fin swimming mode and a subcarangiform class of locomotion. This type of swimming is intermediate between the anguilliform mode displayed by eels, where undulatory propulsive waves cover most of the body length, and the thunniform mode employed by tunas, where oscillatory propulsive waves are concentrated in the tail region (Webb 1998). To attain forward motion, any aquatic organism must generate enough thrust to overcome the drag forces exerted on its body by the surrounding water. In fish, the caudal fin is driven by the axial musculature, and it is the primary propulsive structure for subcarangiform swimmers. The caudal fin is generally deep and triangular with a low aspect ratio and a large trailing edge depth, characteristics that allow it to generate the considerable thrust required for high-speed swimming (Webb 1986).

Sturgeon, like teleosts, employ a subcarangiform swimming mode; however, their tails most resemble those of sharks. The lower lobe is smaller than the upper, a morphological arrangement referred to as heterocercal (Webb 1986). In addition, tail depth is substantially smaller in sturgeon than in teleosts, likely due to the limited supportive capacity of the cartilaginous skeleton. Webb (1986) reported that the net result of these differences was that the contribution of tail depth to thrust was one-third that of a rainbow trout of similar size over sustained and prolonged swimming speeds. Little information is available on the relationship between swimming capacity of paddlefish and their caudal fins; however, Webb (1986) suggested that the tail morphology of paddlefish is more advanced than that of sturgeon, and is associated with a deeper, more fusiform body that is better suited for active fish.

While the caudal fin is the primary locomotory propulsor in most fish species, pectoral fins can also contribute to low speed swimming and aid in manoeuverability (Lauder and Jayne 1996). The pectoral fins of teleosts tend to extend laterally in a transverse plane and are located on the sides of the body (Wilga and Lauder 1999). The basal and radial bones are internal and the portion of the fins exposed to fluid forces are composed of soft, flexible rays. In contrast, the pectoral fins of sturgeon extend laterally from a ventral insertion on the pectoral girdle (Wilga and Lauder 1999). The basal and radial bones extend into the fin and the rays are relatively firm and non-collapsible. This morphology essentially renders the pectoral fins of sturgeon immobile. Traditionally, it has been assumed that their function has been lift generation to compensate for unsteady forces created by the heterocercal tail, and to counteract negative buoyancy. Recent work, however, has indicated that pectoral fin movements promote vertical manoeuverability rather than stabilization. Wilga and Lauder (1999) found that, during steady horizontal swimming, a positive body tilt was maintained; however, it was not generated by the pectoral fins, which were held at an angle that essentially generated no lift. Rather, downward vertical movement was initiated when the posterior halves of the pectoral fins were deflected upwards and the leading edges were rotated downwards relative to the flow. This resulted in a negative body tilt and a vertical descent (Wilga and Lauder 1999). The opposite occurred when fish moved up in the water column.

7.3.3 THE NOTOCHORD

Bony fish possess a fully ossified vertebral column, which serves to transfer force generated by the musculature to the propulsors (Webb 1998). In contrast, the skeletons of sturgeon and paddlefish are made up of cartilage that is less resistant to compression and does not store as much energy as bone, making them a less efficient force transmitter (Webb 1998). In addition, adult sturgeon and paddlefish retain a notochord into adulthood, which is not jointed and contains a continuous internal lumen filled with vacuolated cells (Long 1995). This is in contrast to most other species that lose their notochord during embryonic development. As swimming capacity has been linked to the ability of a fish to bend or curve its body (Rome et al. 1992), and as this ability likely differs between bony and cartilaginous fish, it may be that swimming capacity of sturgeon and paddlefish is constrained by the retention of the notochord.

Long (1995) examined this hypothesis quantitatively and noticed that white sturgeon exhibited "slow" and "fast" locomotory modes, during which similar tail beat frequencies elicited different swimming speeds (Table 7-1). The author demonstrated that white sturgeon accomplished this feat by altering the axial curvature of their notochord depending on the swimming speed desired, and hypothesized that increasing the curvature of the notochord increased the excursion of the muscle sarcomeres, which increased the muscle-shortening velocity. This, in turn, resulted in a faster swimming speed by way of increased power output. The author pointed out, however, that the ability of white sturgeon to achieve greater speeds would eventually be limited when the extra power generated could not compensate for the increased drag induced by the greater body curvature (Long 1995).

Long (1995) also found that maximal midline notochord curvature of white sturgeon decreased near their pelvic fins and then increased towards the tail, whereas in teleosts, maximum curvature increases steadily from head to tail. Based on this finding, the author suggested that the sturgeon notochord has regional functionality, with the pre-caudal section acting as a resistor to bending, effectively limiting curvature and acting as a muscle antagonist. The caudal section appears to act like a cantilevered beam anchored at the pelvic fins, which resists bending moments generated by the tail.

7.3.4 THE MUSCULATURE AND VASCULAR SUPPLY

Muscle fibres in fish are categorized as red, pink, or white. Red, slow-twitch fibres have the smallest diameter, the greatest number of mitochondria and energy reserves, and are resistant to fatigue. White, fast-twitch fibres are relatively large, have few mitochondria, and fatigue rapidly. Pink fibres are intermediate in character between red and white muscle (see Bone 1978 for a review). The degree to which different muscle fibres are intermixed within a particular tissue depends on the type of fish in question. In selachians, different muscle types are completely separated (Kryvi 1977), while in teleosts some mixing of the fibres occurs (Bone 1978). Relatively little information exists on the muscular systems of sturgeon and paddlefish endemic to North America; however, there is evidence that gross muscle morphology is similar within the Acipenserids (Kiessling et al. 1993). As such, data from a non-North American species, the stellate sturgeon (*Acipenser stellatus*), will be discussed.

In terms of morphology, Kryvi et al. (1980) found that stellate sturgeon were similar to sharks in that the different fibre types were physically separated and sharply defined. However, sturgeon were dissimilar to sharks and teleosts in that the white fibres were relatively small and of variable diameter. In addition, red and pink muscle were less vascularized, while white fibres contained more blood vessels than in other species (Kryvi et al. 1980). The authors noted that the presence of blood capillaries in the sturgeon muscle was closely correlated with their mitochondrial concentrations, indicating that vascularization was greatest where intracellular demands for oxygen and metabolites were high. The authors also noted that the gross morphology of the muscle types and their patterns of vascularization suggested that sustained swimming capacity, normally associated with red muscle recruitment, is poorly developed in sturgeon. However, it was also suggested that sustained swimming might be augmented by partial recruitment of white muscle, which appears to have a greater capacity for aerobic metabolism than occurs in other species.

7.3.5 FISH SIZE

The effect of body size on forced swimming capacity of sturgeon and paddlefish has been discussed previously; however, additional information is available with respect to spontaneous activity and routine respiration. During experiments designed to elucidate substrate preference, Peake (1999) noted that small (14cm) lake sturgeon were much more active than larger fish (24cm). Diana et al. (1990) also found a negative correlation between body size and activity of free-swimming lake sturgeon in a biotelemetry study (Table 7-1). Finally, Crocker and Cech (1998a) reported that swimming activity of small white sturgeon (4 to 5cm) was almost 3 times that of larger (11 to 14cm) individuals. The authors also reported that resting oxygen consumption rate decreased with fish mass, a phenomenon that is typical for fish and clearly demonstrated by all sturgeon species studied to date (Figure 7-5).

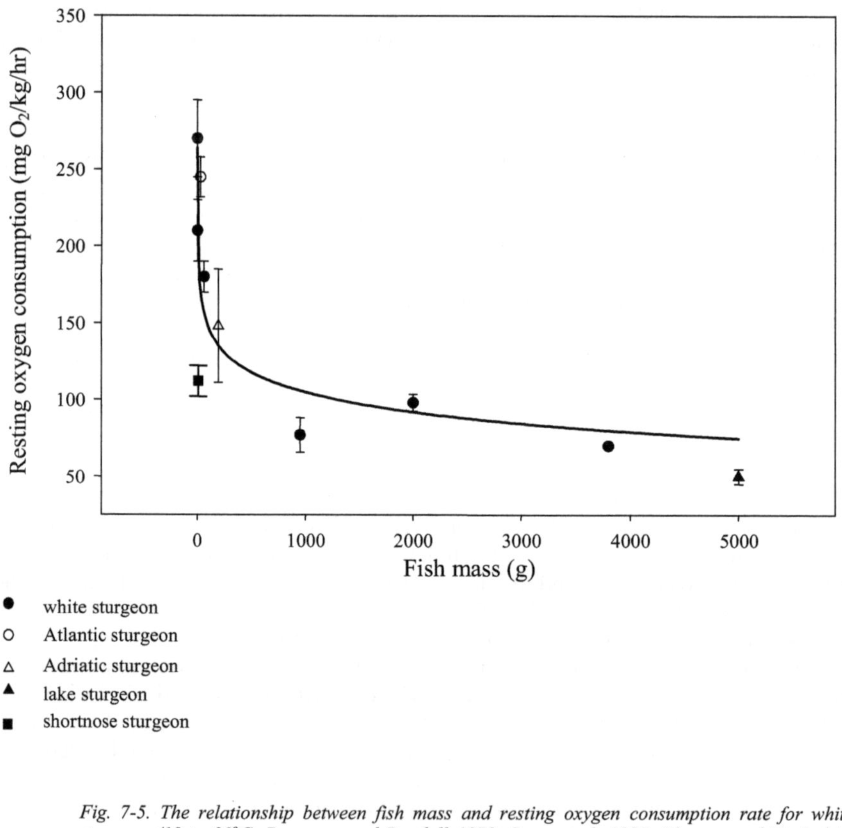

Fig. 7-5. The relationship between fish mass and resting oxygen consumption rate for white sturgeon (10 to 20°C, Burggren and Randall 1978; Ruer et al. 1987; Thomas and Piedrahita 1997; Crocker and Cech 1998a), Atlantic sturgeon (19°C, Secor and Gunderson 1998), Adriatic sturgeon (23°C, McKenzie et al. 1997), lake sturgeon (19°C, McKinley and Power 1992) and shortnose sturgeon (12°C, Kieffer et al. 2001). Error bars indicate standard error. The relationship between oxygen consumption (VO_2, mg/kg/hr) and fish mass (M, g) for these species is best described ($r^2 = 0.71$) by the model: $VO_2 = 233.9 - 18.7 \times \ln(M)$.

7.4 Environmental Constraints on Swimming Activity and Respiration

7.4.1 OXYGEN

As dissolved oxygen levels can have significant effects on swimming capacity (Beamish 1978) and general well-being, most fish tend to prefer water that contains an adequate supply of this essential gas. These conditions are referred to as normoxic. A considerable amount of effort has been made to examine the respiratory responses of fish to varying conditions of oxygen tension. In general, species are categorized as oxygen regulators if they increase their rate of oxygen consumption when exposed to oxygen-poor, or hypoxic, water. Those that maintain or decrease their respiratory rates under hypoxic stress, while actively reducing their metabolic demands, are referred to as oxygen conformers.

Several studies have examined the relationship between activity, respiration and ambient oxygen levels for sturgeon and paddlefish. In perhaps the most detailed of these, Burggren and Randall (1978) found that, under normoxic conditions, white sturgeon respiration was very similar to that displayed by teleosts. Mean branchial water flow was measured at 350ml/kg/min, oxygen utilization, or the relative amount of oxygen taken up by the gills, was 30 to 40%, and oxygen consumption at 15°C was about 80mg O_2/kg/hr. Burggren and Randall (1978), like Burggren et al. (1979), also found that white sturgeon blood was fully oxygenated by the gills during normoxia; however, it was noted by the former authors that the oxygen carrying capacity of the blood was low compared to that of more active species.

Burggren and Randall (1978) reported that white sturgeon under hypoxic conditions were oxygen conformers. As oxygen tension decreased, gill ventilation frequency and heart rate remained steady, although the former dropped slightly under the most severe hypoxic stress. Branchial stroke volume declined with hypoxia, a strategy that may have served to reduce water flow through the gills, which would have increased residence time and, therefore, oxygen extraction. Finally, routine oxygen consumption rate fell sharply with increasing hypoxia, reaching near negligible levels in very oxygen-poor water.

Burggren and Randall (1978) also described two characteristics of white sturgeon respiration under hypoxia that are very distinctive. The first was the absence of bradycardia, or increased heart rate, in response to low levels of dissolved oxygen. The second, and perhaps most interesting, was the lack of a compensatory increase in ventilation and oxygen consumption after a return to normoxic conditions. This indicates that white sturgeon do not meet their metabolic demands when oxygen is scarce by utilizing energy derived from anaerobic processes. If they did, recovery from hypoxic stress would include a period of elevated oxygen consumption to repay the resulting oxygen debt. This hypothesis was further strengthened by the fact that blood pH in the sturgeon examined by Burggren and Randall (1978) did not change during hypoxia, despite the fact that anaerobic activity generally results in a decrease in plasma pH. The mechanism by which white sturgeon were able to reduce oxygen consumption to almost negligible levels and still meet their metabolic requirements without resorting to glycolytic processes is unclear; however, it is apparent that the animals were dramatically suppressing their metabolic requirements and/or selectively diverting blood to essential organ systems during hypoxia.

Interestingly, Ruer et al. (1987) were unable to replicate the drop in oxygen consumption with moderate hypoxic exposure demonstrated by Burggren and Randall (1978). They suggested that insufficient time was allowed for fish used by Burggren and Randall (1978) to acclimate to hypoxia. Ruer et al. (1987) also found that swimming activity in caudal strokes/min was not affected by hypoxic exposure, which indicated that energy requirements were not reduced by decreasing metabolic demands associated with locomotion. These findings also contrast with those of Crocker and Cech (1998a), who were able to replicate both the drop in oxygen consumption with moderate hypoxic exposure, and the lack of a post-exposure compensatory increase in respiration demonstrated by Burggren and Randall (1978). Unlike Ruer et al. (1987), these workers showed that the decrease in metabolic requirements was associated with a reduction in spontaneous swimming activity. Crocker and Cech (1998a) suggested that sturgeon used by Ruer et al. (1987) may have been stressed as they were tested in groups, a condition that can increase oxygen consumption and elevate spontaneous swimming activity in non-schooling species (Ruer et al. 1987).

Data on other North American sturgeon species are scarce; however, Secor and Gunderson (1998) found that juvenile Atlantic sturgeon reduced their oxygen consumption rates from 250 to 440mg O_2/kg/hr in normoxic water to 100 to 200mg O_2/kg/hr under hypoxic conditions, indicating that this species, like white sturgeon, are oxygen conformers. Data from non-North American sturgeons, however, indicate that broad generalizations across species cannot be made. Randall et al. (1992) reported that Adriatic sturgeon (*Acipenser naccarii*) were oxygen regulators, and that this species did switch to anaerobic energy production in hypoxic water as evidenced by elevated post-exposure oxygen consumption and plasma lactate levels. Increased respiratory rates after hypoxic stress have also been reported for Siberian sturgeon (*Acipenser baeri*; Nonnotte et al. 1993; Maxime et al. 1995).

Paddlefish appear to be oxygen regulators, as Burggren and Bemis (1992) found that individuals did not reduce their oxygen consumption rate when exposed to moderate hypoxic stress. Slightly deeper levels of hypoxia, however, quickly resulted in stress and eventually ended in death. Paddlefish that did survive the hypoxia recovered quickly when oxygen levels were normalized and, like white sturgeon, they did so without an elevation in oxygen consumption rate (Burggren and Bemis 1992). As paddlefish cannot reduce their metabolic demands by decreasing spontaneous swimming activity, it may be that this species has a very limited capacity for anaerobic metabolism.

7.4.2 CARBON DIOXIDE

Beamish (1978) pointed out that swimming performance of some species is impaired by hypercapnia, or high CO_2 levels, while that of others is unaffected. The primary consequence of high levels of dissolved carbon dioxide is respiratory acidosis with a concomitant reduction in the affinity of the blood for oxygen via the Root effect (Beamish 1978). Crocker and Cech (1998b) reported that white sturgeon employ typical compensatory mechanisms to deal with this problem, including an increase in ventilation rate and plasma bicarbonate ions, the latter of which serve to increase blood pH. Although these mechanisms in other species reach the peak of their effectiveness (~ 70% pH compensation) within 48h of initial exposure, Crocker and Cech (1998b) found that white sturgeon exposed to environmental hypercapnia required up to 72 hrs to accomplish pH restoration toward normal acid-base status, and even then were only able to compensate for about 35% of their initial pH depression.

7.4.3 LIGHT

The effect of light on swimming capacity and respiration of fish has received little attention; however, light, and particularly photoperiod, has been shown to affect performance in forced swimming tests conducted on white crappie (*Pomoxis annularis*; Smiley and Parsons 1997), and largemouth bass (*Micropterus salmoides*; Kolok 1991). Beamish (1978) summarized existing knowledge of the effects of light on swimming speeds and movements of free-ranging fish in the field. Madison et al. (1972) found that migratory swimming speeds of sockeye salmon (*Oncorhynchus nerka*) were highest at midday and lowest during the hours of darkness. This pattern was also observed in migrating chinook salmon (*Oncorhynchus tshawytscha*; Johnson 1960). In contrast, free-swimming flounder (*Paralichthys olivaceus*) were most active at night (Liu et al. 1997), as were walleye (*Stizostedion vitreum*; Kelso 1978). Diurnal patterns of oxygen consumption have also been reported for several species (Liu et al. 1997, Mehner and Wieser 1994).

While no research has directly examined the effects of light on swimming performance of sturgeon and paddlefish, some behavioural data are available. Loew and Sillman (1998) discovered that newly hatched white sturgeon larvae were negatively phototactic, as they moved away from a light source, and negatively photokinetic because they reduced their swimming activity in high light situations. When light intensity was low, the larvae moved into the water column and alternated between rapid upward swimming and quiescence, while in the dark they swam almost continuously. Loew and Sillman (1998) reported that this swimming activity could, however, be inhibited by a single flash of light, and suggested that theis behaviour might correspond to predator avoidance and dispersal strategies critical to the survival of young sturgeon in the wild.

An increase in swimming activity during low light levels was also reported by Peake (1999) for juvenile lake sturgeon; however, this phenomenon was not evident in larger fish. If nocturnal behaviour is associated with predator avoidance, it may be that large sturgeon abandon the strategy as predation becomes less of a concern. Both McKinley and Power (1992) and Diana et al. (1990) were unable to detect a diurnal activity pattern in adult lake sturgeon during biotelemetry studies. In contrast, Long (1995) reported that adult white sturgeon displayed a slow swimming mode and a fast swimming mode and noted that fast swimming was only observed at night and in the morning before sunrise. This activity was, however, only observed in a single specimen.

7.4.4 WATER TEMPERATURE

Water temperature is an important constraint on swimming capacity as indicated by numerous forced performance studies. Resting oxygen consumption of fish also tends to increase with temperature, a trend that has been documented for Atlantic sturgeon (Secor and Gunderson 1998) and lake sturgeon (McKinley and Power 1992). While swimming capacity tends to increase with water temperature (Beamish 1978), an elevation in basal metabolic rate can compress the aerobic scope for activity and reduce maximum sustained speed (Alsop and Wood 1997). Although there is little direct evidence to indicate that this occurs in sturgeon, McKinley and Power (1992) did report that free-ranging, adult lake sturgeon moved at considerably lower speeds during the summer months when the water temperature was 19°C, than in fall when the water was 10°C, or in winter when the water temperature was 5°C. In contrast, Diana et al. (1990) found that mean hourly displacement (i.e. activity) of lake sturgeon was much greater in summer than in winter, although they could not directly estimate instantaneous swimming speeds.

Effects of temperature on spontaneous movements also appear to be related to fish size. Peake (1999) reported that swimming activity for small lake sturgeon was positively correlated to water temperature between 6°C and 19°C, but that the relationship was lost on larger fish. Crocker and Cech (1998a) found a similar pattern for juvenile white sturgeon in that small fish in cold water were more active than large sturgeon in warmer water.

7.4.5 FOOD AVAILABILITY

It has long been known that feeding can affect swimming capacity. Fish that have recently fed divert energy to the digestive system, which tends to elevate resting oxygen consumption levels, compress the aerobic scope for activity and reduce swimming capacity (Alsop and Wood 1997). It is for this reason that forced performance trials are typically performed on post-absorptive fish. However, relatively little work has focused on the effects of food availability and diet on spontaneous swimming activity and respiration of fish. An exception to this is Bjornsson (1993), who reported that free-swimming Atlantic cod (*Gadus morhua*) regulated swimming speed according to food availability. Only indirect information regarding the effects of food availability on swimming activity and capacity exists for sturgeon. Kiessling et al. (1993) studied the effects of limited food availability on the musculature of white sturgeon. During periods when food was scarce, sturgeon primarily mobilized protein from white muscle deep in the body to meet their metabolic requirements, while protein in the red muscle and white fibres near the lateral line was conserved. The authors suggested that protein in the muscle required for sustained and prolonged swimming was protected when food was scarce. While a correlation between muscle protein and swimming capacity has not been reported for any sturgeon species, such a relationship has been established for juvenile lake trout (*Salvelinus namaycush*; Beamish et al. 1989).

Beamish et al. (1989) also examined the impact of dietary protein and lipid content on respiration. Although diet did not affect oxygen consumption of lake trout, research on a non-North American sturgeon species (Adriatic sturgeon) reported that diet did affect oxygen consumption, spontaneous swimming activity (McKenzie et al. 1995) and respiratory responses to hypoxia (Randall et al. 1992). In general, these studies showed that an increase in dietary fat created a more active animal. Additionally, fish fed polyunsaturated fats showed higher routine oxygen consumption rates than did those fed saturated fats, and that they were more tolerant to hypoxic stress.

7.5 Management Implications

7.5.1 FISHWAY AND CULVERT PASSAGE

Increasingly stringent regulatory requirements concerning fish passage over and through migratory obstacles have created a need for information on swimming capacity and behaviour of sturgeon species. Fishways and culverts, when present, tend to be designed around the size, swimming behaviour, and capacity of salmonids, for which there is a greater body of knowledge (Schwalme et al. 1985). This chapter has highlighted some of the unique morphological and behavioural characteristics of sturgeon that must be taken into consideration when designing fish passage structures for them. It is important to note that although their morphology tends to work against them in terms of swimming capacity, large spawning-sized fish should be able to pass culverts containing considerable water velocities (Figure 7-6), similar to those navigable by smaller, but more proficient, swimmers. However, this only applies if the size and behaviour (e.g. non-jumping) of sturgeon is taken into account in fish passage designs (see Kynard 1998). It is in these areas that additional research is required.

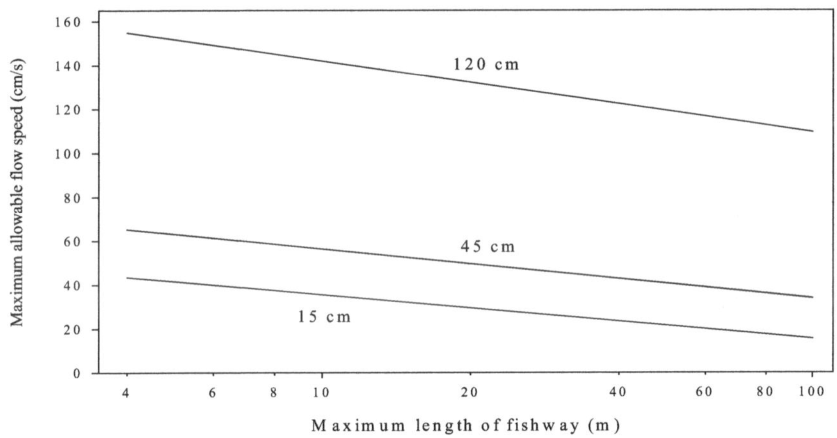

Fig. 7-6. Maximum allowable water velocities for given culvert lengths for lake sturgeon at 14°C based on swimming endurance data. Line positions were determined from models given in Peake et al. 1997b.

7.5.2 BIOTELEMETRY

Numerous studies have examined behaviour, habitat use, and migratory patterns of sturgeon using radio or acoustic biotelemetry equipment. These studies involve fitting a transmitter to the body of the fish, either internally in the body cavity, or externally. In the case of the latter, tags are held in place using wires passing through the dorsal musculature. A principal tenet of biotelemetry research is that the tagging procedure used does not influence the swimming capacity or behaviour of the fish being tracked. It is generally accepted that these requirements are met if the ratio of transmitter mass to fish mass does not exceed 2%. The large size of adult sturgeon has allowed researchers to attach relatively large transmitters to these animals; however, few studies have examined potential effects on behaviour and swimming capacity in light of their unique locomotory characteristics. A notable exception to this is recent work done by Counihan and Frost (1999). The authors externally attached dummy ultrasonic transmitters to white sturgeon (32 to 37cm fork length) and measured critical swimming speed. It was discovered that swimming performance was significantly impaired by the presence of the transmitters, even though the ratio of tag to fish mass was less than 1.3%. Clearly, more research is required on the effects of tagging on sturgeon swimming capacity, particularly for juveniles.

7.5.3 WATER QUALITY

With increased development and industrialization come problems with pollutants and general water quality. Based on experiments related to the effects of temperature and oxygen tension on Atlantic sturgeon, Secor and Gunderson (1998) hypothesized that increased levels of hypoxia in Chesapeake Bay, Maryland, USA were partly responsible for decimating the local population. Although some sturgeon species appear tolerant to hypoxic stress, others, including paddlefish, are not. Secor and Gunderson (1998) concluded their report by suggesting that restorative measures such as aquaculture are not justified until efforts to mitigate habitat loss due to water quality degradation can be addressed.

Research on the effect of pollutants on swimming capacity and survival of young white sturgeon has indicated that the species can be extremely sensitive to forest industry effluents, and specifically chlorinated phenolics, with significant acute effects (32 to 34% declines in U_{crit}) occurring at an exposure level of one part per million (Bennett and Farrell 1998). Information on the sensitivity of sturgeon to a wide range of common pollutants is clearly warranted.

7.6 References

Adams, S.R., J.J. Hoover and K.J. Killgore. 1999. Swimming endurance of juvenile pallid sturgeon, *Scaphirhynchus albus*. Copeia 3: 802-807.

Adams, S.R., G.R. Parsons, J.J. Hoover and K.J. Killgore. 1997. Observations of swimming ability in shovelnose sturgeon (*Scaphirhynchus platorynchus*). Journal of Freshwater Ecology 12: 631-633.

Alsop, D.H. and C.M. Wood. 1997. The interactive effects of feeding and exercise on oxygen consumption, swimming performance and protein usage in juvenile rainbow trout (*Oncorhynchus mykiss*). Journal of Experimental Biology 200: 2337-2346.

Arnold, G.P., P.W. Webb and B.H. Holford. 1991. The role of the pectoral fins in station-holding of atlantic salmon parr (*Salmo salar* L.). Journal of Experimental Biology 156: 625-629.

Beamish, F.W.H. 1978. Swimming Capacity. *In* Fish Physiology. Vol. 7. *Edited by* W. S. Hoar and D. J. Randall. Academic Press Inc., New York, pp. 101-187.

Beamish, F.W.H., J.C. Howlet and T.E. Medland. 1989. Impact of diet on metabolism and swimming performance in juvenile lake trout (*Salvelinus namaycush*). Canadian Journal of Fisheries and Aquatic Sciences 46: 384-388.

Bennett, W.R. and A.P. Farrell. 1998. Acute toxicity testing with juvenile white sturgeon (*Acipenser transmontanus*). Water Quality Research Journal of Canada 33: 95-110.

Bernatchez, L. and J.J. Dodson. 1985. Influence of temperature and current speed on the swimming capacity of lake whitefish (*Coregonus clupeaformis*) and cisco (*C. artedii*). Canadian Journal of Fisheries and Aquatic Sciences 42: 1522-1529.

Bjornsson, B. 1993. Swimming speed and swimming metabolism of Atlantic cod (*Gadus morhua*) in relation to available food: a laboratory study. Canadian Journal of Fisheries and Aquatic Sciences 50: 2542-2551.

Bone, Q. 1978. Locomotor muscle. *In* Fish Physiology. Vol. 7. *Edited by* W. S. Hoar and D. J. Randall. Academic Press Inc., New York, pp. 361-424.

Brett, J.R., Hollands, H. and Alderdice, D.F. 1958. The effect of temperature on the cruising speed of young sockeye and coho salmon. Journal of the Fisheries Research Board of Canada 15: 587-605.
Briggs, C.T. and J.R. Post. 1997. Field metabolic rates of rainbow trout estimated using electromyogram telemetry. Journal of Fish Biology 51: 807-823.
Burggren, W.W. 1978. Gill ventilation in the sturgeon, *Acipenser transmontanus*: unusual adaptations for bottom dwelling. Respiration Physiology 34: 153-170.
Burggren, W.W. and W.E. Bemis. 1992. Metabolism and ram gill ventilation in juvenile paddlefish, *Polyodon spathula* (Chondrostei: Polyodontidae). Physiological Zoology 65: 515-539.
Burggren, W.W. and D.J. Randall. 1978. Oxygen uptake and transport during hypoxic exposure in the sturgeon, *Acipenser transmontanus*. Respiration Physiology 34: 171-183.
Burggren, W., J. Dunn and K. Barnard. 1979. Branchial circulation and gill morphometrics in the sturgeon, *Acipenser transmontanus*, an ancient Chondrosteian fish. Canadian Journal of Zoology 57: 2160-2170.
Chan, M.D., E.D. Dibble and K.J. Kilgore. 1997. A laboratory examination of water velocity and substrate preference by age-0 gulf sturgeons. Transactions of the American Fisheries Society 126: 330-333.
Counihan, T.D. and C.N. Frost. 1999. Influence of externally attached transmitters on the swimming performance of juvenile white sturgeon. Transactions of the American Fisheries Society 128: 965-970.
Crocker, C.E. and J.J. Cech. 1998a. Effects of environmental hypoxia on oxygen consumption rate and swimming activity in juvenile white sturgeon, *Acipenser transmontanus*, in relation to temperature and life intervals. Environmental Biology of Fishes 50: 383-389.
Crocker, C.E. and J.J. Cech. 1998b. Effects of hypercapnia on blood-gas and acid-base status in the white sturgeon, *Acipenser transmontanus*. Journal of Comparative Physiology B. 168: 50-60.
Diana, J.S., D.F. Clapp and E.M. Hay-Chmielewski. 1990. Relative success of telemetry studies in Michigan. American Fisheries Society Symposium 7: 346-352.
Duthie, G.D. 1982. The respiratory metabolism of temperature-adapted flatfish at rest and during swimming activity and the use of anaerobic metabolism at moderate swimming speeds. Journal of Experimental Biology 97: 359-373.
Gibson, E.S. and F.E.J. Fry. 1954. The performance of lake trout, *Salvelinus namaycush*, at various levels of temperature and oxygen pressure. Canadian Journal of Zoology 132: 252-260.
Hammer, C.H. 1995. Fatigue and exercise tests with fish. Comparative Biochemistry and Physiology 112A: 1-20.
Hudson, R.C.L. 1973. On the function of the white muscle in teleosts at intermediate swimming speeds. Journal of Experimental Biology 58: 509-522.
Johnson, J. 1960. Sonic tracking of adult salmon at Bonneville Dam, 1957. U.S. Fish and Wildlife Service Fisheries Bulletin 176: 471-485.
Kelso, J.R.M. 1978. Diel rhythm in activity of walleye, *Stizostedion vitreum*. Journal of Fish Biology 12: 593-599.
Kieffer, J.D., A.M. Wakefield and M.K. Litvak. 2001. Juvenile sturgeon exhibit reduced physiological responses to exercise. Journal of Experimental Biology. 204: 4281-4289.
Kiessling, A., S.S.O. Hung and T. Storebakken. 1993. Differences in protein mobilization between ventral and dorsal parts of white epaxial muscle from fed, fasted and refed white sturgeon (*Acipenser transmontanus*). Journal of Fish Biology 43: 401-408.
Kolok, A. 1991. Photoperiod alters the critical swimming speed of juvenile largemouth bass, *Micropterus salmoides*, acclimated to cold water. Copeia 4: 1085-1090.
Kryvi, H. 1977. Ultrastructure of the different fibre types in axial muscle of the sharks *Etmopterus spinax* and *Galeus melastomas*. Cell Tissue Research 184: 287-300.
Kryvi, H., P.R. Flood and D. Gulyaev. 1980. The ultrastructure and vascular supply of the different fibre types in the axial muscle of the sturgeon *Acipenser stellatus*, Pallas. Cell Tissue Res. 212: 117-126.
Kynard, B. 1998. Twenty-two years of passing shortnose sturgeon in fish lifts on the Connecticut River: What has been learned? *In* Fish Migration and Fish Bypasses. *Edited by* Kynard, B. Fishing News Books. pp. 255-266.
Lauder, G.V. and B.C. Jayne. 1996. Pectoral fin locomotion in fishes: testing drag-based models using three-dimensional kinematics. American Zoologist 36: 567-581.
Liu, H., Y. Sakurai, H. Munehara and K. Shimazaki. 1997. Diel rhythms of oxygen consumption and activity level of juvenile flounder *Paralichthys olivaceus*. Fisheries Science 63: 655-658.
Loew, E.R. and A.J. Sillman. 1998. An action spectrum for the light-dependent inhibition of swimming behaviour in newly hatched white sturgeon. Vision Research 38: 111-114.
Long, J.H. 1995. Morphology, mechanics and locomotion: the relation between the notochord and swimming motions in sturgeon. Environmental Biology of Fishes 44: 199-211.

Madison, D.M., R.M. Horrall, A.B. Stasko and A.D. Hasler. 1972. Migrating movements of adult sockeye salmon (*Oncorhynchus nerka*) in coastal British Columbia as revealed by ultrasonic tracking. Journal of the Fisheries Research Board of Canada 29: 1025-1033.
Maxime, V., G. Nonnotte, C. Peyraud, P. Williot and J.P. Truchot. 1995. Circulatory and respiratory effects of hypoxic stress in the Siberian sturgeon. Respiration Physiology 100: 203-212.
McKenzie, D.J., J.F. Piraccini, J.F. Steffensen, C.L. Bolis, P. Bronzi and E.W. Taylor. 1995. Effects of diet on spontaneous locomotor activity and oxygen consumption in Adriatic sturgeon (*Acipenser naccarii*). Fish Physiology and Biochemistry. 14: 341-355.
McKinley, R.S. and G. Power. 1992. Measurement of activity and oxygen consumption for adult lake sturgeon in the wild using radio-transmitted EMG signals. *In* Wildlife telemetry: remote monitoring and tracking of animals. Vol. *Edited by* I. G. Priede and S. M. Swift. Ellis Horwood., West Sussex, UK, pp. 307-318.
Mehner, T. and W. Wieser. 1994. Energetics and metabolic correlates of starvation in juvenile perch (*Perca fluviatilis*). Journal of Fish Biology 45: 325-333.
Nonnotte, G., V. Maxime, J.P. Truchot, P. Williot and C. Peyraud. 1993. Respiratory responses to progressive hypoxia in the sturgeon, *Acipenser baeri*. Respiration Physiology 91: 71-82.
Peake, S. 1999. Substrate preferences of juvenile hatchery-reared lake sturgeon, *Acipenser fulvescens*. Environmental Biology of Fishes 56: 367-374.
Peake, S., R.S. McKinley and D.A. Scruton. 2000. Swimming performance of walleye (*Stizostedion vitreum*). Canadian Journal of Zoology 78: 1686-1690.
Peake, S., F.W.H. Beamish, R.S. McKinley, D.A. Scruton and C. Katopodis. 1995. Swimming performance of lake sturgeon, *Acipenser fulvescens*. Canadian Technical Report of Fisheries and Aquatic Sciences No 2063. 26 pp.
Peake, S., R.S. McKinley and D.A. Scruton. 1997a. Swimming performance of various freshwater Newfoundland salmonids relative to habitat selection and fishway design. Journal of Fish Biology 51: 710-723.
Peake, S., F.W.H. Beamish, R.S. McKinley, D.A. Scruton and C. Katopodis. 1997b. Relating swimming performance of lake sturgeon, *Acipenser fulvescens*, to fishway design. Canadian Journal of Fisheries and Aquatic Sciences 54: 1361-1366.
Randall, D.J., D.J. McKenzie, G. Abrami, G.P. Bondiolotti, F. Natiello, P. Bronzi, L. Bolis and E. Agradi. 1992. Effects of diet on responses to hypoxia in sturgeon (*Acipenser naccarii*). Journal of Experimental Biology 170: 113-125.
Rome, L.C., I.H. Choi, G. Lutz and A. Sosnicki. 1992. The influence of temperature on muscle function in the fast swimming scup. I. Shortening velocity and muscle recruitment during swimming. Journal of Experimental Biology 163: 259-279.
Ruer, P.M., J.J. Cech and S.I. Doroshov. 1987. Routine metabolism of the white sturgeon, *Acipenser transmontanus*: effect of population density and hypoxia. Aquaculture 62: 45-52.
Sanderson, S.L., J.J. Cech and A.Y. Cheer. 1994. Paddlefish buccal flow velocity during ram suspension feeding and ram ventilation. Journal of Experimental Biology 186: 145-156.
Schwalme, K., W.C. Mackay and D. Lindner. 1985. Suitability of vertical slot and Denil fishways for passing north-temperate, non-salmonid fish. Canadian Journal of Fisheries and Aquatic Sciences 42: 1815-1822.
Secor, D.H. and T.E. Gunderson. 1998. Effects of hypoxia and temperature on survival, growth and respiration of juvenile Atlantic sturgeon, *Acipenser oxyrinchus*. Fishery Bulletin 96: 603-613.
Smiley, P.C. and G.R. Parsons. 1997. Effects of photoperiod and temperature on swimming performance of white crappie. Transactions of the American Fisheries Society 126: 495-499.
Thomas, S.L. and R.H. Piedrahita. 1997. Oxygen consumption rates of white sturgeon under commercial culture conditions. Aquaculture Engineering 16: 227-237.
Webb, P.W. 1971. The swimming energetics of trout. II. Oxygen consumption and swimming efficiency. Journal of Experimental Biology 55: 521-540.
Webb, P.W. 1986. Kinematics of lake sturgeon, *Acipenser fulvescens*, at cruising speeds. Canadian Journal of Zoology 64: 2137-2141.
Webb, P.W. 1998. Swimming. *In* The physiology of fishes. Vol. 2. *Edited by* D. H. Evans. CRC Press Inc., New York, pp. 3-24.
Webb, P.W. and R.S. Keyes. 1982. Swimming kinematics of sharks. Fishery Bulletin 80: 803-812.
Weihs, D. 1974. Energetic advantages of burst swimming of fish. Journal of Theoretical Biology 48: 215-229.
Wilga, C.D. and G.V. Lauder. 1999. Locomotion in sturgeon: Function of the pectoral fins. Journal of Experimental Biology 202: 2413-2432.

CHAPTER 8

STURGEON AND PADDLEFISH METABOLISM

THOMAS D. SINGER[1] AND JAMES S. BALLANTYNE[2]
[1]*University of Waterloo, Department of Biological Sciences, Waterloo, Ontario, Canada, N2L 3G1*
[2]*University of Guelph, Department of Zoology, Guelph, Ontario, Canada, N1G 2W1*

8.1 Introduction

The metabolism of Acipenseriformes is of interest to biologists in several disciplines. Metabolically, sturgeons and paddlefish bridge the evolutionary gap between the unusual metabolic adaptations of the osmoconforming marine chondrichthyans with their high levels of urea (Ballantyne 1997) and the osmoregulating teleosts. As representatives of an early stage in vertebrate evolution the Acipenseriformes provide insights into the acquisition of some metabolic (e.g. enzyme isoforms) and regulatory mechanisms (primarily hormonal) found in higher vertebrates. Thus this group can provide us with important clues to understand the evolution of the metabolic organization of the vertebrates. Within the group (Acipenseriformes) the distinctions are subtler. Thus, while the paddlefish and sturgeons diverged about 150 million years ago (Birstein and DeSalle 1998) they have many metabolic characteristics in common.

An understanding of the metabolism of sturgeon and paddlefish is particularly relevant now, due to the declining stocks of most species in this group. Virtually all species are endangered or threatened (Birstein et al. 1997), and efforts are in place to improve the husbandry for restocking or aquaculture. To this end, there is a need to improve diets and monitor the hormonal status to control reproduction, all of which require an understanding of the basic metabolic features of this group. Unfortunately, by comparison with other major fish groups (agnathans, elasmobranchs and teleosts), relatively little is known of this aspect of chondrostean biology. For this reason, although the focus of this book is the North American Acipenseriformes, we have taken a more global view where appropriate, since much of the metabolic information on Asian and European Acipenseriformes is also applicable to the North American species.

8.2 Digestion

Acipenseriformes are carnivores at all life stages. Adult paddlefish are filter feeders, and adult sturgeon may be piscivorous or feed on benthic invertebrates

G.T.O. LeBreton et al. (eds.), Sturgeons and Paddlefish of North America, 167–194.
© 2004 *Kluwer Academic Publishers. Printed in the Netherlands.*

(Buddington and Christofferson 1985). All larval forms are zooplanktivores. As diet changes through development, the relative importance of lipid, protein and amino acids also changes. In acipenserids, carbohydrates and lipids dominate in larval stages feeding on zooplankton, while protein becomes more important in the adult diet.

The digestive system is typical of a primitive carnivore. Overall the length of the digestive system is short (70-100% of body length) consisting of an esophagus, secretory forestomach, non-secretory stomach (pylorus), intestine, pyloric caecum, spiral valve (intestine), and rectum (Buddington and Christofferson 1985). The role of these segments in the digestion of lipid carbohydrate and protein are discussed below.

8.2.1 LIPID

Lipid digestion in sturgeon, as in other fish, depends on solubilization of lipid by bile salts, followed by enzymatic digestion by lipases. Bile enters the intestine via the bile duct, which connects the gall bladder to a point close to the pyloric sphincter (Buddington and Christofferson 1985). The main component of the bile (90%) is taurocholate with small amounts of tauroallocholate, the 5α-analog of taurocholate, and 5β-cyprinol sulfate (Haslewood 1967; Tammar 1974). Small amounts of bile alcohols of the 5α and 5β conformation, including 5β-cyprinol have been detected in Acipenseriformes (Haslewood 1967). In general the trend in the evolution of bile salts among the vertebrates has involved a reduction from 27 carbon 5α-alcohol sulfates to 24 carbon 5β-acids (Haslewood 1967). The sturgeons and paddlefish thus have a combination of more "primitive" bile constants (bile alcohols) as well as more "advanced" forms (bile acids such as taurocholate.)

Digestion of the emulsified lipids by the action of lipases occurs throughout the gut of adult white sturgeon *A.transmontanus* (Buddington and Christofferson 1985; Buddington and Doroshov 1986b). The capacity to digest lipids develops early with digestive lipase activity peaking at 20 days in lake sturgeon *A. fulvescens* (Buddington 1985).

Bacteria contribute to the lipid uptake by synthesizing volatile fatty acids from otherwise indigestible dietary components. Anaerobic bacterial fermentation takes place in the spiral valve where volatile fatty acids (acetate, propionate and butyrate) are produced (Callman and Macy 1984). The hydrogen gas produced as a by-product of this process in the spiral valve, is used in the swim bladder (Callman and Macy 1984).

8.2.2 PROTEIN

The essential amino acid requirements of sturgeon do not differ from those of teleosts (Kaushik et al. 1991). While the proportion of dietary protein required for optimal growth of sturgeon is similar to those of other fish (40% is optimal for growth in *A. transmontanus*) (Moore et al. 1988), the type of dietary protein does affect the growth of sturgeon (Stuart and Hung 1989). Protein digestion begins in the fore-stomach by the action of pepsin (25-35 units per mg protein) (Buddington and Christofferson 1985). This is aided by acid secretion in this region resulting in a pH of 3-4 (Buddington and Christofferson 1985). Trypsin, chymotrypsin and pepsin have been demonstrated in the alimentary canal of *A.transmontanus* (Buddington and Doroshov 1986b). In the more alkaline regions of the gut past the stomach, trypsin and carboxypeptidase B are the predominant proteases (Buddington and Christofferson 1985). Elastase and leucine aminopeptidase are also present but in low amounts

(Buddington and Christofferson 1985). Aminopeptidase is found in the brush border membranes of the small and large intestine (Hung et al. 1989).

Absorption of amino acids in the fore-stomach is low compared to that of the intestine and spiral valve (Buddington and Christofferson 1985). The spiral valve is the most important site for absorption of amino acids (Callman and Macy 1984). The postprandial changes in plasma amino acids have been examined and indicate that total plasma free amino acid levels peak about 8 hours after feeding (Ng et al. 1996), but ammonia excretion may begin as early as 6 hours after feeding depending on the diet (Dabrowski et al. 1987). The digestion induced elevation of metabolism (specific dynamic action) peaks about 12 hours after feeding (Dabrowski et al. 1987). Dietary free amino acids are not absorbed as efficiently as the same amount of dietary protein and it has been suggested that this is due to excessive losses of amino acids through the gills (Ng et al. 1996).

8.2.3 CARBOHYDRATE

Larval sturgeon may metabolize dietary carbohydrate better than juveniles or adults (Buddington and Doroshov 1986a). This is indicated by amylase activities that peak after 29 days post hatch in *A. fulvescens* (Buddington 1985). Amylase is found throughout the gut of *A. transmontanus* (Buddington and Doroshov 1986b). Although carbohydrate is likely not an important energy source except in larval sturgeon (Buddington and Chrisofferson 1985). Sturgeon may deal with dietary carbohydrate better than many teleost fish and this feature may be of an economic advantage for sturgeon aquaculture.

Although juvenile sturgeon have a poor ability to utilize dietary raw starch (Kaushik et al. 1989), certain carbohydrates (maltose and glucose) provide high rates of food conversion (Hung 1988). Dextrin, cornstarch and sucrose can also be utilized but lactose, fructose and cellulose result in poor energy retention (Hung 1988; Hung et al. 1990). A high carbohydrate diet results in elevated plasma glucose and liver glycogen (Fynn-Aikens et al. 1992). Maltose and glucose diets also result in high plasma glucose levels (Hung 1991).

8.3 Oxidative Metabolism

The physiological and metabolic responses to ambient oxygen levels have been examined in several species of sturgeons and paddlefish. Two types of response are possible. In one approach, oxygen regulators adjust their ventilation rates to compensate for changing oxygen levels, maintain a constant respiration rate, and thus their aerobic metabolism. Oxygen conformers on the other hand allow a decline in their respiration rates as environmental oxygen decreases, with concomitant reductions in metabolic rate and aerobic metabolism. Most species of sturgeon display oxygen conformation although the range of oxygen concentrations over which this occurs may vary with species and population (Burggren and Randall 1978; Ruer et al. 1987; Secor and Gunderson 1997). Reduced metabolism at low oxygen concentrations has been attributed to reduced activity in A. *transmontanus* (Crocker and Cech 1997). In sturgeon, which are oxygen conformers, aerobic metabolism may be replaced by anaerobic mechanisms. These will be dealt with in a subsequent section.

Studies of *Acipenser baeri* indicate this species is an oxygen regulator over a significant range of ambient oxygen levels (Nonotte et al. 1993). Similarly, paddlefish are oxygen regulators and maintain respiration rates constant over the range 150 to 90 mmHg (Burggren and Bemis 2001). Paddlefish are unusual in how close the lethal PO_2 is to the critical PO_2 at which respiration rates decline (Burggren and Bemis 2001). Thus paddlefish would appear to have little capacity for anaerobic metabolism.

8.3.1 OXYGEN DELIVERY

The delivery of oxygen to tissues depends on the characteristics of haemoglobin. Little is known of the properties of sturgeon haemoglobin although regulation of oxygen binding may be mediated by ATP and 2,3 DPG, as both of these have been reported in the erythrocytes of Acipenser (Crigg 1974). The hematocrit of North American species are given in Table 8-1, and indicate similar oxygen carrying capacities to those found in some teleost fish. There are no reports of the levels of myoglobin in tissues of sturgeon although this pigment must be present since sturgeon have discrete superficial regions of red muscle fibres (colour due to myoglobin) distinguishable from deeper white fibres as in teleost fishes (Radelli et al. 1999).

8.3.2 MITOCHONDRIAL METABOLISM

There have been no studies of the respiratory properties of isolated mitochondria from sturgeon or paddlefish, and there is only one study that has measured activities of enzymes of aerobic metabolism in sturgeon (Singer et al. 1990). Consistent with the lower metabolic rates of sturgeon, the levels of aerobic enzymes such as citrate synthase, in tissues of sturgeon are somewhat lower than those of more active teleost fish (Singer et al. 1990). Further investigations are required to investigate the metabolic organization of sturgeon and paddlefish mitochondria and these would certainly provide useful information about the substrate utilization of specific tissues.

Individual enzymes of glycolysis have been characterized and compared with other species. The properties of sturgeon muscle pyruvate kinase have been examined and indicate the enzyme is modulated by fructose 1,6 bisphosphate as in mammals (Randall and Anderson 1975). The sturgeon enzyme is more sensitive to temperature than the mammalian enzyme in that the Hill coefficient and the K_m decrease with decreasing temperature as is more typical of other poikilotherms (Randall and Anderson 1975). The molecular weight of the sturgeon PK is similar to that of the mammalian enzyme (56,000 Mw) (Anderson and Randall 1975).

The kinetic characteristics of another glycolytic enzyme, glyceraldehyde-3-phosphate dehydrogenase have been examined in European sturgeon and two binding sites for NAD and NADH have been identified as is typical of other organisms (Kelemen et al. 1975; Deparade et al. 1981). The cooperativity of the enzyme has been examined, and an induced fit model has been used to explain the negative cooperativity observed in this enzyme (Branlant et al. 1983). Pyruvate kinase has two electrophoretic forms in sturgeon, but the "muscle" form is not as highly specialized to function in muscle as in higher vertebrates (Guderley et al. 1983). Thus the tissue specific design of isoforms of glycolytic enzymes in Acipenseriformes is less complex than in higher vertebrates, and reflects their evolutionary position.

Table 8-1. Plasma parameters in North American Acipenseriformes. a = (Crocker and Cech 1998); b = (Jarvis 2001); c = (Barton et al 1998) ; d = (Grant et al. 1970); e = (Hung 1991) ; f = (Gillis and Ballantyne, 1996); g = (Fynn-Aikins et al. 1992); h = (Rasmussen 1980); i = (Singer et al. 1990); j = (McKinley et al. 1993)

Parameter	**A. transmontanus**	**A. fulvescens**	**A. brevirostrum**	**Polyodon spathula**
Lactate (mM)	0.31±0.31[a]		2.4±0.1[b]	2.89[c]
Glucose (mM)	4.24[g] 3.5±1.13[a]	3.5 - 4[f]	6.5±0.4[b]	1.10 female[d] 1.51 male[d] 3.23 - 4.80[c]
Free amino acids total (mM)		1.5 - 2.8[f]	3.30±0.17[b]	
Triacylglycerol (mg•dl^{-1})	596[g]			
Nonesterified Fatty acids (nmol•l^{-1})	34 (mg•dl^{-1})[g]	2490[i] 1719±261 spring[j] 915±57 summer[j] 1003±114 autumn[j]	1060.5±119.6[b]	
Protein (mg•ml^{-1})	1.69[g] 45[h]			162 female[d] 1.74 male[d]
Osmolarity (mOsm/•l^{-1})			273.2±1.4 [b]	258.1 female[d] 260.2 male[d]
Hematocrit (%)	23 a		23.5±0.9b	

8.4 Carbohydrate Metabolism

There is evidence that the carbohydrate metabolism of paddlefish and sturgeon differs somewhat from that of teleost fish in some aspects. Barton et al. (1998) found no significant changes in plasma glucose in response to a number of stressors, including handling, chasing and confinement unlike teleost fish where elevated plasma glucose is a relatively sensitive stress indicator (Iwama et al. 1995). Other studies report lower resting plasma glucose levels in paddlefish (Table 8-1) compared to most teleosts. This may reflect the continuous filter-feeding habits of paddlefish.

8.4.1 GLYCOLYSIS

Sturgeon have a low glycolytic capacity, based on a broad survey of key enzymes measured in five tissues of *A. fulvescens* (Singer et al. 1990). In particular, activities of both hexokinase (HK) and phosphofructokinase (PFK) levels were lower in lake sturgeon compared with most teleosts, suggesting that carbohydrates are not the preferred fuel in most tissues. The low activities of HK levels in the lake sturgeon were similar to those reported for the little skate *Raja erinacea*, a marine elasmobranch, while sturgeon PFK levels were even lower than in this elasmobranch (Moon and Mommsen 1987).

Individual enzymes of glycolysis have been characterized and compared with other species. The properties of sturgeon muscle pyruvate kinase have been examined and indicate the enzyme is modulated by fructose 1,6 bisphosphate as in mammals (Randall and Anderson 1975). The sturgeon enzyme is more sensitive to temperature than the mammalian enzyme in that the Hill coefficient and the K_m decrease with decreasing temperature as is more typical of other poikilotherms (Randall and Anderson 1975). The molecular weight of the sturgeon PK is similar to that of the mammalian enzyme (56,000 Mw) (Anderson and Randall 1975).

The kinetic characteristics of another glycolytic enzyme, glyceraldehyde-3-phosphate dehydrogenase have been examined in European sturgeon and two binding sites for NAD and NADH have been identified as is typical of other organisms (Kelemen et al. 1975; Deparade et al. 1981). The cooperativity of the enzyme has been examined, and an induced fit model has been used to explain the negative cooperativity observed in this enzyme (Branlant et al. 1983). Pyruvate kinase has two electrophoretic forms in sturgeon, but the "muscle" form is not as highly specialized to function in muscle as in higher vertebrates (Guderley et al. 1983). Thus the tissue specific design of isoforms of glycolytic enzymes in Acipenseriformes is less complex than in higher vertebrates, and reflects their evolutionary position.

8.4.2 GLUCONEOGENESIS

The capacity for synthesis of glucose and glycogen has not been extensively examined in Acipenseriformes. Based on measurements of enzyme activities, gluconeogenesis mainly occurs in the liver and kidney as in most other fish groups (Singer et al. 1990). The levels of enzymes of gluconeogenesis in the brain of sturgeon indicate significant capacity for gluconeogenesis in this tissue (Singer et al. 1990). The levels of gluconeogenic enzymes in sturgeon red muscle are somewhat lower than those reported for red muscle of other teleosts but higher than those reported for elasmobranchs (Singer et al. 1990; Moon and Foster 1995). Further research into the pathways for gluconeogenesis from lactate generated in muscle would provide valuable clues to the evolution of this process in the vertebrates.

8.4.3 PENTOSE PHOSPHATE PATHWAY

While the pentose phosphate pathway is important for a variety of metabolic functions such as production of NADPH for lipogenesis and provision of ribose sugars for nucleotide synthesis little work has been done to determine the tissue distribution of these functions. One study has demonstrated the enzymes of the pentose phosphate pathway in liver of *A. transmontanus* and indicates these are sensitive to the levels of D-glucose in the diet (Fynn-Aikens et al. 1992).

8.4.4 LACTATE METABOLISM

Based upon heart LDH:CS enzyme activity ratios, the lake sturgeon appears to have a lower anaerobic capacity compared with many other species of fish. In the red muscle LDH:CS was intermediate between that of a marine elasmobranch and several teleosts (Singer et al. 1990). This finding appears to contrast the known anoxia tolerance of sturgeon (Burggren et al. 1985). *A. transmontanus* is frequently found in deoxygenated benthic regions of lakes, in ice-covered waters and stagnant regions of estuaries, where an anaerobic metabolic strategy would be beneficial. Burggren and Randall (1978) have provided evidence that 1-kg *A. transmontanus* show evidence of metabolic depression under mild hypoxia and this would reduce the need for anaerobic production of lactate. The same study also revealed that under severe hypoxic exposure, there was no repayment of the oxygen debt. Likewise Cech et al. (1984) also found depressed metabolic rates reflected in lower growth rates of 1-6-g juvenile *A. transmontanus*. In contrast, a study conducted by Ruer et al. (1987) to determine the effect of population density and hypoxia on routine metabolism in *A. transmontanus*, showed no differences in metabolic rates for 1.7- 2.2-kg fish under normoxic and hypoxic conditions. Taken together this indicates sturgeon may rely on an inherently low metabolic rate and alternative fermentations of carbohydrates to survive anoxic conditions and avoid lactate acidosis. Further investigations of these pathways are needed.

Plasma lactate levels for several North American Acipenseriformes are given in Table 8-1. In juvenile paddlefish, plasma lactate levels were similar to those of many teleost fish (Table 8-1), and were shown to increase 2- 3-fold in response to handling stress (Barton et al. 1998). Lactate concentrations in the plasma returned to pre-stress resting levels after 24h, much like the response in teleosts (Barton et al. 1998). The mechanism for handling lactate loads (gluconeogenesis versus oxidation) has not been quantified.

The LDH isoform pattern has been established for the pallid sturgeon *Scaphirhynchus albus* and the paddlefish (Whitt et al. 1975). Three genes code for LDHs in the Acipenseriformes (Whitt et al. 1975) and the bichirs (Fisher et al. 1980) The LDH-C gene is thought to have arisen from the LDH-B gene, and in the Acipenseriformes its broad tissue expression is similar to that of the B gene (Whitt et al. 1975). In more advanced fishes, the tissue expression of the LDH-C gene is much more specific to a narrower range of tissues, such as the liver or the retina (Whitt et al. 1975). The Acipenseriformes thus display an early stage in the evolution of the LDH-C gene.

Analysis of the kinetics of LDHs from sturgeon indicate higher activation energy and a lower K_m for pyruvate than in other fish species (Klyachko and Ozernyuk 1998). The significance of this has not been established.

8.5 Lipid Metabolism

8.5.1 LIPID STORAGE

Lipids are stored in several tissues at high levels in sturgeon. Lovern (1932) suggested the peritoneal fat is the main fat store in *A. sturio*, although the lipid content of other tissues can be very high. Sturgeon may have muscle lipid content as high or higher than that of other fish that are considered fatty, such as salmon and mackerel (Krzynowek and Murphy 1987). Triacylglycerols (TAG) are the main storage lipid in the muscle of sturgeon and the level stored in muscle is related to the amount of lipid in the diet (Gershanovich and Kiselev 1993). Sturgeon fed high lipid diets preferentially deposit lipid in the liver and the digestive tract rather than in muscle (Medale et al. 1991). The red muscle of sturgeon contains much more lipid than white muscle (Decker et al. 1991). This may indicate that the red muscle of sturgeon has a high capacity for lipid oxidation similar to the situation in teleosts but unlike the elasmobranch condition where extrahepatic lipid oxidation is minimal (Ballantyne 1997). The lipid content of the white muscle of North American species of

Acipenseriformes is given in Table 8-2. The lipid content of paddlefish white muscle is lower than that of sturgeon (Table 8-2). The white muscle lipid level in cultured paddlefish is 3 $gm·kg^{-1}$, while that of red muscle is much higher (Decker et al. 1991). In the European species, *A. baeri*, wild caught fish have about 20 % of dry muscle mass as lipid (Medale et al. 1991). In the same species the lipid content of the liver of cultured sturgeon (*A. baeri*) is 60 – 74 % of the dry mass (Medale et al. 1991) rivalling that of elasmobranchs and gadids (Krzynowek and Murphy 1987). When expressed in terms of wet weight, a comparison of three European species reported muscle lipid ranged from 45- 106 $g·kg^{-1}$ wet wt (Badiani et al. 1997). The lipid content of *A. naccarii* liver varies from 20 to 32 % of the wet weight and muscle from 3 to 7 % depending on diet (McKenzie et al. 1997).

The requirement for specific essential fatty acids indicates the growth of cultured *A. transmontanus* is not affected by isonitrogenous, isocaloric diets differing widely in the type of fatty acids in the lipid component (Xu et al. 1993). In spite of this, dietary fatty acid type (n3 or n6) does reflect the n3/n6 ratio in muscle, liver and brain (Xu et al. 1993). Indeed, cultured and wild *A. oxyrinchus* can be distinguished based on fatty acid composition due to their dietary differences (Chen et al. 1995). Similar effects were noted in European sturgeon (McKenzie et al. 1997). The n3/n6 ratios of sturgeon are high in seawater and low in freshwater (Table 8-2), similar to those of teleosts.

8.5.2 LIPID TRANSPORT

Lipids of Acipenseriformes are transported in the plasma by a variety of lipoproteins as in other fish species. The levels of very low density lipoproteins (VLDL), low density lipoproteins (LDL) and high density lipoproteins (HDL) have been examined in three species (Babin and Vernier 1989). VLDL levels range from 348-726 $mg·dl^{-1}$, LDL levels from 23 – 510 and HDL from 204-649 $mg·dl^{-1}$ (Babin and Vernier 1989). The VLDL and LDL levels are in the range reported for elasmobranch and teleost fish (Babin and Vernier 1989). The cholesterol contents of the LDL and HDL of sturgeon are similar (17-20% of total lipid) although the total HDL content is 5 times that of the LDL (Lizenko et al. 1995). The HDL levels are thus lower than those reported for most teleosts but considerably higher than those reported for elasmobranchs (Babin and Vernier 1989). Total triacylglycerol levels in the plasma of North American species are given in Table 8-1.

Plasma nonesterified fatty acids (NEFAs) are the most metabolically active lipid fraction in the blood of vertebrates and have been measured in several species of North American Acipenseriformes (Table 8-1). The levels of plasma NEFAs in lake sturgeon have been shown to vary seasonally, with levels measured in the spring significantly higher than those found in summer (McKinley et al. 1993). These studies indicate levels of plasma NEFAs similar to those of some teleost fish, and much higher than those reported for elasmobranchs (Ballantyne 1997). This suggested that unlike the elasmobranchs, which lack a plasma NEFA carrier (albumin), the Acipenseriformes can transport NEFAs via an albumin-like protein. Indeed, the plasma of a sturgeon has been shown to display a band with similar mobility to that of albumin (Rasmussen 1980). Serum protein levels of 45$mg·ml^{-1}$ of sturgeon (Table 8-1) are higher than those recorded in dogfish shark (30$mg·ml^{-1}$), lower than those from teleosts and about the same as Australian lungfish (Rasmussen 1980). This may be due to the presence of albumin, the most abundant plasma protein. Further studies into the nature of the plasma nonesterified fatty acid binding protein would provide interesting insights into the evolution of this lipid transport mechanism in the vertebrates.

Table 8-2. Proximate composition and lipid characteristics of white muscle of North American Acipenseriformes. a = (Badiani et al. 1997); b = (Hung et al. 1993); c = (Jarvis 2001); d = (Krzynowek and Murphy 1987); e = (Chan et al. 1997); f = (Decker et al. 1991); g = (Beamish et al. 1996).

Trait	*A. trans-montanus*	*Brevi-rostrum*	*Oxyrinchus*	*Acipenser fulvescens*	*Polyodon spathula*
Moisture	755.5±5.4 [a] 756-784 [b]	782±4.0[c]		777.5±16.8[g]	838.0±4.0
Protein	195.7±2.8 [a] 127-137 [b]	182±1.0 [c]		182.9±9.8 [g]	159.0±1.4
Lipid	44.9±4.1 [a] 52-76 [b]	21±5.0 [c]		33.1±11.3 [g]	2.7±2.1
Ash	11.4±0.3 [a] 25-31 [b]			11.0±0.8 [g]	10.0±0
Cholesterol	0.614±0.016[a]				
Energy value MJ kg^{-1}	4.99±0.17 [a]				
% Saturated fatty acids	26.9±0.29 [a]			24.56±3.81[e]	28.5
% Mono-unsaturated fatty acids	45.4±0.58 [a]			23.73±4.05[e]	56.1
% Poly-unsaturated fatty acids	21.8±0.60 [a]		29.2-42.8 [d]	14.44±2.14[e]	15.3
n3/n6	4.21 ±0.08 [a]			0.66 [e]	

8.5.3 LIPID CATABOLISM

Based on measurements of the enzymes of lipid catabolism fatty acids can be oxidized in several tissues of sturgeon, including the liver, kidney, heart, red muscle and brain (Singer et al. 1990). This distinguishes the sturgeon from the more primitive elasmobranchs that lack the capacity for significant lipid catabolism outside of the liver (Ballantyne 1997). Relatively high levels of plasma NEFAs occur in sturgeon (Table 8-1) and these would provide the metabolic substrates for lipid catabolism in these tissues. Little is known of the regulation of lipid catabolism in Acipenseriformes, although plasma NEFAs have been shown to be responsive to season and nutritional status in wild lake sturgeon (McKinley et al. 1993).

8.5.4 LIPOGENESIS

The synthesis of lipid de novo has not been directly investigated in Acipenseriformes. It is likely that most de novo lipogenesis occurs in the liver since that tissue has the highest activities of the enzymes involved in supporting lipid synthesis (Fynn-Aikins et al. 1992). The levels of these enzymes are responsive to dietary conditions. In particular, levels of enzymes in pathways supplying the NADPH needed for lipogenesis in liver (glucose-6-phosphate dehydrogenase, 6-phosphogluconate dehydrogenase, isocitrate dehydrogenase and malic enzyme) are elevated in sturgeon fed high carbohydrate diets (Fynn-Aikins et al. 1992). Correspondingly, elevated plasma TAG levels are observed indicating increased lipogenesis under these conditions although plasma NEFAs were not affected (Fynn-Aikins et al. 1992). In addition to de novo synthesis, sturgeon can elongate and desaturate dietary linoleic and linolenic acids (Xu et al. 1993; Agradi et al. 1993; Chen et al. 1995).

8.5.5 CHOLESTEROL METABOLISM

Little is known of the cholesterol metabolism of Acipenseriformes. The serum cholesterol levels of sturgeon are high ($400 mg \cdot 100 ml^{-1}$) but similar to those of some fish species (Rasmussen 1980). Plasma cholesterol levels in paddlefish are in the same range as those reported for teleost fish with females having markedly lower levels than males (Grant et al. 1970). Muscle cholesterol content in three species ranged from 0.6-0.7 $g \cdot kg^{-1}$ wet weight (Badiani et al. 1997). The brain of the *Acipenser ruthenus* contains typical proportions of cholesterol (Tamai et al. 1994) and the cholesterol content of sturgeon roe (caviar) is similar to that of other fish roe (Iwasaki and Harada 1964).

8.5.6 PHOSPHOLIPID METABOLISM

There have been few studies of the metabolism of phospholipids of Acipenseriformes. Some inferences may be made from the studies of the phospholipids found in various tissues. The lipids in the brain of *Acipenser ruthenus* contain typical proportions of phosphatidylcholine (PC), phosphatidylethanolamine (PE), phosphatidylinositol (PI), phosphatidylserine (PS) and sphingomyelin compared with that of most vertebrates (Tamai et al. 1994). The brain gangliosides contain more than 67% 18:0 with 16:0 and 18:1 making up another 28 % (Tamai et al. 1994). The predominant fatty acids in brain glycosphingolipids are 22:1 and 24:1 accounting for about 75% of the total (Tamai et al. 1994). The galactosyl glycerides are predominantly of the diacyl form with alkylacyl components making up 16% of the glyceride (Tamai et al. 1994).

Plasmalogens were estimated in the gills of *A. transmontanus* at 85 $\mu mol \cdot g^{-1}$ or 6.5% by weight (Nevenzel et al. 1985). PE from brains of Eurasian sturgeon is about 50% in the plasmalogen form while PC is mostly in the diacyl form (Kruglova 1984). The classes (saturated, monoenes and polyenes) of fatty acids of brain plasmalogen and diacyl PE are similar to those of teleost fishes (Kruglova 1984). The fatty aldehydes of the plasmalogens of PE in the brains of sturgeon are about 35% saturated and 65% unsaturated similar to that of teleost fish (Kruglova 1984).

The phospholipid composition of the eggs of *A. stellatus* have been examined and it has been suggested that phospholipids provide some of the energy for development of the eggs since the proportion of phospholipids declined post fertilization (Abrosimova et al. 1999). The limited studies of the metabolism of phospholipids indicate sturgeon can synthesize PC from dietary choline but choline must be supplied in the diet (Hung and Lutes 1988). This may be an important role for dietary choline in sturgeon (Hung 1989).

8.6 Ketone Body Metabolism

In higher vertebrates, ketone bodies are important energy sources during starvation especially in tissues such as the brain that normally rely on carbohydrate for energy. In teleost fish, ketone bodies are of limited importance in starved or fed states but may be important in more primitive fishes such as elasmobranchs even in the fed state (Ballantyne 1997). The only investigation of ketone body metabolism in sturgeon reported enzyme activities of β-hydroxybutyrate dehydrogenase (βHBDH) in all tissues examined (Singer et al. 1990). Other ketone body enzymes including succinyl CoA, ketotransferase and acetoacetyl CoA thiolase were also detected in tissues such as the liver, kidney, heart, red muscle and brain indicating these substrates may be used as energy sources in sturgeon tissues, unlike in most teleost fish (Singer et al. 1990). Sturgeon thus more closely resemble elasmobranch fishes, which use ketone bodies extensively in extrahepatic tissues (Driedzic and Hart 1984; Ballantyne 1997).

8.7 Nitrogen Metabolism

Fish utilize a wide range of mechanisms to deal with the end products of nitrogen metabolism. Most teleosts are ammonotelic and excrete ammonia across the gills, while most elasmobranchs are ureotelic. The North American sturgeon *A. transmontanus* has been shown to be non-urea-retaining (as presumably are all Acipenseriformes), in contrast to elasmobranchs (Rasmussen 1980).

The nitrogen metabolism of acipenserids appear to be more similar to that of teleosts in that the main form of nitrogen excretion is ammonia via the gill (urea represents only about 17% of the nitrogen excretion) (Medale et al. 1991). Most nitrogen is excreted at the gills of sturgeon as is the case in teleosts and urinary nitrogen excretion is correspondingly low at about 2% of the total (Medale et al. 1991). The proportion of ammonia to urea excreted by the sterlet (*Acipenser ruthenus*) ranges from 1.4 to 2.6-fold in fed versus starved fish respectively (Gershanovich and Pototskij 1992). Ammonia toxicity in *A. brevirostrum* has been determined for fingerlings in low Cl^- water with the 96h LC50 at 11.3±8.17 mg/L nitrite nitrogen which is low compared with levels of 84.0-812.0 mg/L from eels measured in 100% seawater (Fontenot et al. 1998).

8.7.1 AMINO ACID METABOLISM

Based upon a survey of key enzymes, lake sturgeon have the ability to utilize amino acids as oxidative substrates (Singer et al. 1990). Moderate levels of glutamate dehydrogenase, alanine aminotransferase, and aspartate aminotransferase were detected in the brain, red muscle, heart, kidney and liver, with the potential to utilize glutamate and alanine highest in the kidney compared with other tissues. This ability is clearly reflected in the digestive enzyme complement in *A. transmontanus* with high trypsin and pepsin secretions in adults (Buddington 1985, Buddington and Doroshov 1986b).

Sturgeon, like many fish, have a high protein requirement for growth compared with other domestic animals. Juvenile *A. transmontanus* require dietary protein to be approximately 40% of the diet for proper growth (Moore et al. 1988). Interestingly, diets containing high amounts of free amino acids resulted in poor growth of juvenile *A. transmontanus* (Ng et al. 1996).

Plasma glutamine levels in sturgeon are low (Kaushik et al. 1994; Gillis and Ballantyne 1996; Jarvis and Ballantyne, 2003) but higher than those reported for most elasmobranch fishes. The higher levels of circulating glutamine found in sturgeons may relate to the lack of a need for glutamine for urea synthesis as occurs in elasmobranchs. Similarly, the ratio of liver to brain glutamine synthetase activity of sturgeon is typical of non-urea-retaining fish (Webb and Brown 1980).

Total plasma free amino acid levels for North American species are given in Table 8-1. The circulating levels of some essential amino acids are sensitive to dietary levels of those amino acids (Kaushik et al. 1994). In this context, the effects of starvation on plasma amino acids of *A. fluvescens* have been examined (Gillis and Ballantyne 1996). After 60 days of food deprivation, elevations of plasma amino acids (both essential and non-essential) occur (Gillis and Ballantyne, 1996). Glutamine and lysine by contrast rose in the food-deprived group after 10 days but the significance of this has not been established (Gillis and Ballantyne 1996). In food-deprived *A. transmontanus*, the levels of plasma branched-chain amino acids (leucine, isoleucine and valine) all fell significantly compared with fed fish (Swallow 1985). During short-term starvation arginine oxidation increased in sturgeon (Fauconneau et al. 1986). There is some evidence that during short-term fasting, protein synthesis decreases (Fauconneau et al. 1986).

8.8 Hormonal Control of Metabolism

The hormonal regulation of metabolism of sturgeon and paddlefish has not been extensively investigated, but there is sufficient information available to comment on some of the main endocrine pathways.

8.8.1 CORTISOL

The role of cortisol in fishes is as a mediator of stress relations (Mommsen et al. 1999) as is the case in higher vertebrates. Elasmobranchs however, use a different steroid 1-α-hydroxycorticosterone (Idler and Truscott 1967). Since sturgeon and paddlefish are the extant fish group most closely allied to the elasmobranchs, the nature of the stress hormone in this group was a fascinating evolutionary question. Early studies (Idler and O'Halloran 1970; Idler and Sangalang 1970; Sangalang et al. 1971) established the nature and site of synthesis (interrenal tissue) of cortisol. Subsequently, the measurement of cortisol along with cortisone and corticosterone in the plasma of the paddlefish (Idler and Truscott 1980) and other studies (Barton et al. 1998) established cortisol as the main stress steroid in Acipenseriformes.

Plasma levels of cortisol in paddlefish are similar to those of freshwater teleosts (Grant et al. 1970). As in teleost fish, paddlefish display elevated plasma cortisol in response to stress but the highest levels of cortisol that could be induced are lower than those of teleost fish (Barton et al. 1998). Some stressors, such as hypercapnia in *A. transmontanus,* cause cortisol levels to rise (Crocker and Cech 1998). Based on their own work Barton et al. (1998) concluded chondrostean fish are less responsive to physical stress than teleosts.

The consequences of elevated cortisol levels are important in an aquaculture context. Measurements of plasma cortisol levels have been used to establish the health of cultured sturgeon and low plasma cortisol levels have been correlated with poor gamete quality in Russian sturgeon (Semenkova et al. 1999). The establishment of the normal ranges of plasma cortisol levels for individual species should be established to aid in management and aquaculture.

8.8.2 SEX HORMONES

The role of sex hormones in modulating metabolism of Acipenseriformes has not been examined, although considering the extensive metabolic changes known to occur in teleost fish during reproduction the effects must also be substantial in sturgeons and paddlefish. The endocrine control mechanisms of reproduction are similar in teleosts and sturgeon (Doroshov et al. 1997). All of the relevant hormones have been measured in the plasma. Follicle stimulating hormone (FSH), luteinizing hormone (LH), estradiol, progesterone and testosterone have been determined in *A. persicus* (Safi et al. 1999). In addition, testosterone has been measured in the plasma of Atlantic sturgeon, *A. oxyrinchus* (Sangalang et al. 1971).

The hormones mediating sex hormone release have been examined. Physiological evidence for the role of 2 forms of gonadotropin have been reported in *A. transmontanus* (Moberg et al. 1995; Doroshov et al. 1997), although 4 forms have been reported in other sturgeon species (Goncharov et al. 1983; Kuznetzov et al. 1983). Circulating levels of gonadotropin are <1-2 ng/ml in immature *A. transmontanus* (Doroshov et al. 1997). The gonadotropin releasing hormone peptide sequence of sturgeon is identical to that of the mammalian hormone as is the case in other primitive fish (Lescheid et al. 1995; Sherwood et al. 1991). It has been suggested that this hormone arose in an organism ancestral to the bony fish, but only after the divergence of the chondrichthyans (Sherwood et al. 1991). Thus the development of hormonal control of reproduction appears to have advanced early in vertebrate evolution.

8.8.3 MELATONIN

As an important mediator of circadian rhythms, melatonin has an ancient history in vertebrates and invertebrates. There are no studies on the effects of photoperiod on plasma melatonin in Acipenseriformes, although melatonin has been detected in the gastrointestinal tissues of *A. fulvescens* (Bubenik and Pang 1997).

8.8.4 GROWTH HORMONE

Growth hormone is an important mediator of growth rate and the efficiency of metabolism. The amino acid sequences of sturgeon growth hormone have been determined and conform to the expected relationships between this group, the elasmobranchs and the teleosts (Rubin and Dores 1994).Two forms of growth hormone have been determined in *A. guldenstadti* and these differ in three amino acids (Yasuda et al. 1992). The binding of sturgeon growth hormone to hepatic receptors has been determined and indicates that there are high and low affinity binding sites similar to the situation in higher vertebrates (Nicholl et al. 1986). Future studies should focus on establishing if the metabolic effects of growth hormone in Acipenseriformes are similar to those observed in teleosts.

8.8.5 GLUCAGON

Glucagon is an important hormone in mediating carbohydrate and lipid metabolism in fish. It is not known if the glucagon in Acipenseriformes has a similar metabolic effect to that observed in teleosts. As is the case for most teleost fish, paddlefish have two forms of glucagons- (G1 and G2), and these differ by a single amino acid (Nguyen et al. 1994). The glucagons of paddlefish most closely resemble those of the more advanced gar rather than those of other fish (Nguyen et al. 1994). The potency of paddlefish G-1 in eliciting glycogenolysis in teleost hepatocytes is greater than that of G-2 (Nguyen et al. 1994). Future investigations should be focused on establishing the metabolic actions of glucagon in Acipenseriformes.

8.8.6 INSULIN

Insulin is involved in regulating energy metabolism in vertebrates. The pancreas is a diffuse organ found in various tissues and throughout the body cavity (Buddington and Christofferson 1985). Paddlefish display two forms of insulin that resemble the insulin of the gar more closely than that of other fish (Nguyen et al. 1994). The two forms of insulin in paddlefish differ by a single amino acid (Nguyen et al. 1994). Nothing is known of the metabolic effects of insulin in Acipenseriformes.

8.8.7 CATECHOLAMINES

Plasma catecholamines have not been reported for North American species of sturgeon, but studies of the European species *A. baeri* indicate an increase in catecholamines with hypoxia (Maxime et al. 1995) and hypercapnia (Crocker and Cech 1998).

8.9 Toxicant Metabolism

Considering the endangered status of sturgeon and paddlefish, it is somewhat surprising that an understanding of their ability to deal with environmental challenges from man-made toxicants has not been extensively examined. Many of the enzymes involved in detoxification have been detected in sturgeon. Cytochrome P-450, NADPH-cytochrome c reductase, benzo(a)-pyrene hydrolase, 7-ethoxycoumarin-O-deethylase, epoxide hydrolase, UDP-glucuronyltransferase (UDPGT) and glutathione S-transferase have been reported in the liver of a European sturgeon *A. baeri* at levels similar to those of rainbow trout, except for UDPGT which was ten-fold lower (Perdu-Durand and Cravedi 1989). The highest levels of NADPH-cytochrome c reductase were found in the kidney of *A. baeri* and significant activity was also detected in the gill and liver (Perdu-Durand and Cravedi 1989). The pH optima of all of the above enzymes were similar to those reported for other fish and mammals (Perdu-Durand and Cravedi 1989).

It has been suggested that the low levels of UDPGT in liver correlate with the low capacity for detoxification by glucuronidation in sturgeon, and this may make them more susceptible to the toxic effects of phenols (Perdu-Durand and Cravedi 1989). The liver of white sturgeon can hydroxylate aniline but at rates lower than that of numerous teleost species (Buhler and Rasmussen 1968). Similarly, the ability of sturgeon liver to dealkylate aminopyrine is low but in the range of some teleost species (Buhler and Rasmussen 1968). Thus it seems that if anything, the capacity for detoxification of various compounds is lower in Acipenseriformes than in teleosts. Certainly, further investigation into the ability of sturgeon and paddlefish to deal with toxicants is required.

8.10 Environmental Effects on Metabolism

A variety of environmental factors such as temperature, oxygen availability and salinity can have substantial effects on the metabolism of fishes. All of these parameters are relevant to the metabolism of Acipenseriformes especially in an aquaculture context where optimal growth conditions of temperature, salinity and oxygen are desirable.

8.10.1 TEMPERATURE

The optimal temperatures for growth and reproduction for several North American Acipenseriformes are given in Table 8-3. In general, the species from warmer waters have higher temperature preferences for growth. The ability of sturgeon to adapt to temperature change has only been examined in a few studies. As is the case in other fish species, the capacity for a homeoviscous response of membrane phospholipids only develops after hatching in white sturgeon *A. transmontanus* (Buddington et al. 1993). By contrast there are no studies to indicate if adaptive enzymatic responses occur in sturgeon when challenged by changing temperature.

The effects of temperature on endocrine function has also not been extensively examined. One study indicates that a lower temperature (17°C) reduced plasma cortisol and chloride levels in freshwater *A. naccarii* compared to that of controls (25°C) (Cataldi et al. 1998). This hormonal response suggests hormonal mediation of enzyme systems in response to changing temperature.

Temperature does not seem to affect the nature of the nitrogen compounds excreted, perhaps indicating that there is little effect of temperature on amino acid metabolism (Gershanovich and Pototskij 1995). Both ammonia and urea excretion rate increase with temperature from 12 to 28°C and the ratio of ammonia to urea remains constant over that range (Gershanovich and Pototskij 1995). The interactive effect of temperature and hypoxia indicate poorer survival and growth at higher temperatures and low oxygen levels than at low temperature and hypoxia (Secor and Gunderson 1997).

Sturgeon are frequently cultured at higher temperatures to enhance growth. The impact of these conditions on maturation and egg quality (especially lipid components) have not been examined.

Table 8-3. Optimal temperatures for growth and reproduction in North American Acipenseriformes.

Species	Optimal Growth (°C)	Upper lethal	Spawning	Reference
Polyodon spathula	16-18	ND	13-16	Barton et al. 1998
Acipenser transmontanus	23-26	ND	ND	Hung et al. 1993
A. fulvescens	20	23 (some mortalities)	ND	Fajfer et al. 1999
A. oxyrinchus (small, 0.3 gm)	19	ND	ND	Kelly and Arnold 1999
A. oxyrinchus (large, 60 gm)	15	ND	ND	Kelly and Arnold 1999

8.10.2 SALINITY

North American sturgeon and paddlefish demonstrate a wide range of salinity tolerances. The salinity tolerance associated with each species corresponds to one of three spawning and feeding migration patterns: anadromous, freshwater amphidromous or potamodromous (as described by Bemis and Kynard 1997). The life history patterns and corresponding salinity preferences for each life cycle stage for North American sturgeon and paddlefish are summarized in Table 8-4.

Fully anadromous sturgeon (e.g. *Acipenser transmontanus*, *A. medirostrum*, and *A. oxyrhinchus*) spawn in freshwater rivers, while both juveniles and adults forage in ocean waters, with some juveniles returning to estuarine areas to feed in summer. There is some evidence that both *Acipenser transmontanus* and *A. oxyrhinchus* spawn in brackish water (McEnroe and Cech 1985; Smith 1985). Interestingly, *A. medirostrum* is rarely found in freshwater and spends most of its time in the brackish water estuaries of larger rivers (Houston 1988). The sole freshwater amphidromous species, *Acipenser brevirostrum*, spawns in freshwater rivers while most juveniles and adults forage in the brackish water of estuaries, and only rarely do adults enter full strength seawater (Kynard 1997). It appears that latitudinal differences in adult *Acipenser brevirostrum* seawater residency may be related to opportunities for foraging, with increased seawater use in northern stocks directly linked to enhanced foraging abundances in northern versus southern estuaries (Kynard 1997).

The remaining North American Acipenseriformes are potamodromous (e.g. *Acipenser fulvescens, Scaphirhynchus albus, S. platorynchus, S. suttkusi* and *Polyodon spathula*) with spawning occurring in freshwater rivers while juvenile and adults forage within the same freshwater river or within adjacent freshwater lakes. However, there are several observations of *A. fulvescens,* in regions of elevated salinities in the St. Lawrence River and Hudson Bay (Harkness and Dymond 1961; Vladykov and Greeley 1963). Likewise evidence of salinity tolerance in paddlefish comes from incidental reports of fish captured in brackish water by shrimp trawlers, and the movement of tagged paddlefish into brackish water (Graham 1997). In contrast there are no reports of the shovelnose sturgeons (*Scaphirhynchus albus, S. platorynchus, or S. suttkus*) outside of freshwater (Keenlyne 1997).

Table 8-4. Anadromy and corresponding salinity tolerance of North American Acipenseriformes.

Species	Common name	Anadromy	Salinity Tolerance		
			Juvenile	Adult	Spawning
Acipenser oxyrinchus	Atlantic Sturgeon	Anadromous (Smith and Dingley 1984)	Freshwater	Full seawater 34 ppt	Possibly spawn in brackish water
Acipenser brevirostrum	Shortnose Sturgeon	Freshwater Amphidromous (Kynard 1997)	Limited to estuaries	Adult use of seawater varies with latitude	Freshwater
Acipenser fulvescens	Lake sturgeon	Potamodromous (LeBreton and Beamish 1998)	Freshwater (seawater tolerance = 15ppt)	Freshwater	Freshwater
Scaphirhynchus albus	Pallid Sturgeon	Potamodromous (Simons et al. 2001)	Freshwater	Freshwater	Freshwater
Scaphirhynchus platorynchus	Shovel-nose sturgeon	Potamodromous (Keenlyne 1997)	Freshwater	Freshwater	Freshwater
Scaphirhynchus suttkusi	Alabama Sturgeon	Potamodromous (Simons et al. 2001)	Freshwater	Freshwater	Freshwater
Acipenser transmontanus	White sturgeon	Anadromous (Semakula and Larkin 1968)	Freshwater	Full seawater 34 ppt	Possibly spawn in brackish water
Acipenser medirostrum	Green Sturgeon	Anadromous (Houston 1988)	Freshwater	Rarely found in freshwater	Freshwater
Polyodon spathula	Paddlefish	Potamodromous (Graham 1997)	unknown	Incidental evidence to survive in brackish water	Freshwater

While the wide range of life history patterns and seawater tolerances of these ancient fish make them ideal to study the adaptive significance of salinity tolerance as well as the evolution of osmoregulatory metabolism, the ability to tolerate seawater has been tested in only a small number of sturgeon species. Juvenile *A. fulvescens*, exposed to a range of salinities could not maintain homeostasis above 15-ppt. seawater, and could not tolerate salinities exceeding 25 ppt (LeBreton and Beamish 1998). The ability of sturgeon to tolerate elevated salinities appears to be size dependent. However, the underlying mechanisms have yet to be understood (Gershanovich et al. 1991); LeBreton and Beamish 1998). Comparisons of juvenile and adult *A. transmontanus* showed a gradual increase in salt tolerance, with increasing body weight in adults. The largest of the North American species, they are able to tolerate salinities of 34 ppt (McEnroe and Cech 1985). Likewise, adult *A. oxyrinchus,* which grow to a similar size as *A. transmontanus,* can tolerate salinities of 34 ppt (Altinok et al. 1998). Salinity tolerance has also been examined in the Adriatic sturgeon, *A. naccarii*, an anadromous species that spends most of its time in brackish water. The highest salinity tolerated by *A. naccarii* adults, which grow to only a maximum of 25kg, is 20ppt, while the larvae can only tolerate salinity isoosmotic to plasma (Cataldi et al. 1999).

The physiological and biochemical mechanism behind seawater tolerance and acclimation of Acipenseriformes remains poorly understood. Preliminary studies have demonstrated that sturgeon share some of the main features of osmotic regulation with teleosts. Sturgeon, paddlefish and teleosts all utilize the same fundamental organ systems to maintain near constant plasma osmolarity hyperosmotic to freshwater and hypoosmotic to seawater. As in teleosts the gills of sturgeon contain large numbers of chloride cells, which are the primary site of salt secretion (Altinok et al. 1998). Following seawater exposure there are changes in both the number and size of the gill lamellae chloride cells in teleosts (see Jurss and Bastrop 1995 for review). Similarly, seawater and brackish water adapted *A. oxyrinchus* have significantly larger and a higher number of gill lamellar chloride cells compared with freshwater fish (Altinok et al. 1998). By contrast in the Adriatic sturgeon, *A. naccarii,* the opposite effect is observed with brackish water sturgeon having fewer chloride cells compared with freshwater fish (Cataldi et al. 1995). The distribution of chloride cells to non-gill osmoregulatory organs in sturgeon differs from that of teleosts. While the opercular epithelium of certain teleosts (e.g. tilapia, *Oreochromis mosambicus*, killifish *Fundulus heteroclitus*) contain numerous chloride cells (Fosket et al. 1983; Karnaky 1986) the opercular membrane, pseudobranch and spiracle from *A. oxyrinchus* contain few or no chloride cells (Altinok et al. 1998).

Like teleosts, all sturgeon maintain plasma ion levels hypo-osmotic to seawater and hyperosmotic to freshwater. Preliminary studies suggest that euryhaline sturgeon such as *A. transmontanus* maintain plasma ion levels lower than teleosts in both fresh and seawater. Urist and van de Putte (1967) reported sodium levels in freshwater adapted *A. transmontanus* of approximately 123 mM Na^+ compared with levels of 140-160 mM Na^+ in most freshwater teleosts, while seawater adapted sturgeon maintained 130 mM Na^+ versus 160-180 mM Na^+ for most marine teleosts. Separate studies indicate sodium fluxes of sturgeon in seawater are similar to those of marine teleosts (Potts and Rudy 1972). However, sodium fluxes are higher in freshwater than those of freshwater teleosts and sturgeon are more permeable to sodium than freshwater teleosts (Potts and Rudy 1972). It has been suggested that regulation of plasma ions is slower than that of euryhaline teleosts (McEnroe and Cech 1987). Overall it appears that sturgeons are less capable of tolerating rapid seawater exposure, and require extended periods of time to adjust to changes in salinity compared with teleosts. These important differences may explain why sturgeon are confined to major river systems with large estuaries and not associated with smaller rivers, which are inhabited by certain teleosts that can tolerate abrupt salinity changes.

Energetically it is clear that osmoregulation is metabolically costly, especially for juvenile sturgeon. The growth of shortnose sturgeon *A. brevirostrum*, reared for 10 weeks at four salinities, was impeded under high salinity (20ppt) with highest growth rates in freshwater (Jarvis et al. 2001). Likewise, young-of-the-year Adriatic sturgeon reared in 11 ppt brackish water had 30% higher standard metabolic rate and 17% lower growth rate compared with sturgeon reared in freshwater (McKenzie et al. 2001). Together these studies suggest that osmoregulatory costs are lowest in freshwater during juvenile stages of development in sturgeon. This contradicts the idea that osmoregulatory costs would theoretically be minimized in brackish water when ion concentrations are isoosmotic to plasma levels, as suggested by earlier studies of teleosts (Canagataram 1959; Rao 1968).

Adaptation to seawater in all sturgeon likely involves the use of amino acids as intracellular solutes. Muscle total free amino acid levels rise with abrupt transfer to higher salinity in a sturgeon hybrid but remain constant if the salinity change is gradual (Gershanovich et al. 1991). Certain amino acids such as β-alanine and glycine increase with increasing salinity regardless of the time course of acclimation (Gershanovich et al. 1991), and are important as osmolytes in many species of fish and other organisms (Ballantyne and Chamberlin 1994). Although taurine is typically an important amino acid in osmotic relations, it was found to decline in muscle with increased salinity, regardless of the time-course of acclimation (Gershanovich et al. 1991).

Other amino acids such as alanine and glutamate increase in a transitory fashion then decline, but only if introduction to higher salinity is abrupt (Gershanovich et al. 1991). Part of this response may be the use of these amino acids as energy sources for adjustments that are made in ion levels at the new salinity. Salinity acclimation in *A brevirostrum* resulted in no change in the activities of glutamate dehydrogenase in any tissue (Jarvis and Ballantyne 2003). There is some evidence that salinity acclimation has other effects on nitrogen metabolism, since increased salinity results in elevated urea excretion rates and reduced ammonia excretion (Gershanovich and Pototskij 1995). It is not known if this implies elevated urea cycle activity at high salinities, or is simply due to arginase activity.

The hormonal responses to salinity change likely involve cortisol, although the few studies available did not examine levels in the critical early stages (first 24 hours) and thus may have missed the "cortisol spike" observed in teleost fish in response to high salinity. So even though plasma cortisol levels are the same in sturgeon adapted to salinities from 5 to 640 mOsm and respiration rates are similar (McKenzie et al. 1999), it is likely cortisol played a role in the early stages of this adaptation. Evidence for this comes from measurements of gill Na^+,K^+-ATPase that increased more than 3 fold over that range (McKenzie et al. 1999). Gill and intestine Na^+,K^+-ATPase were unaffected by salinity acclimation to 20 ppt in *A. brevirostrum* (Jarvis 2001). Gill Na^+,K^+-ATPase activity is typically increased by cortisol in fish (McCormick 1996). Gill H^+-ATPase activity was similar in sturgeon *A. naccarrii* acclimated to salinities ranging from 5 to 640 mOsm (McKenzie et al. 1999).

Adjustments in lipid metabolism with salinity change have not been extensively examined in sturgeon. The activities of 3-hydroxyacyl CoA dehydrogenase, an enzyme of fatty acid catabolism, are unaffected by salinity acclimation in *A. brevirostrum* (Jarvis and Ballantyne, 2003). In teleost fish there is usually a shift towards higher levels of n3 series fatty acids at higher salinities (Henderson and Tocher 1987), although the n3/n6 ratio of freshwater sturgeon is similar to that of freshwater teleosts in liver, gonad, muscle and carcass total lipid (Garcia-Gallego et al. 1999). The n3/n6 ratio of plasma non-esterified fatty acids (NEFAs) was unaffected by salinity acclimation in *A. brevirostrum* (Jarvis and Ballantyne 2003).

It is clear much more work needs to be done, especially with North American species to establish the biochemical basis for salinity acclimation. A better understanding of salinity regulation in acipenserids would also be of great benefit in establishing better diets and protocols for the culture and introduction of sturgeon into seawater. As a group the Acipenseriformes lend themselves perfectly to elucidating the adaptive significance of salinity tolerance. With recent advancements in our understanding of the molecular basis of salt transport in teleosts (see Marshall and Singer 2002) studies of sturgeon species may provide insight into the question of whether early Acipenseriformes were potamodromous or anadromous.

8.10.3 HYPOXIA

Sturgeon have evolved mechanisms for maintaining oxygen uptake in situations such as bottom feeding where oxygen availability is compromised. They can alternate between buccal and opercular water intake while feeding, to maintain oxygen uptake (Burggren and Randall 1978). The growth of *A. transmontanus* is reduced at environmental O_2 levels of 90 torr (Cech et al. 1984). Although some sturgeon are oxygen conformers and thus reduce their energy needs as oxygen levels decline, some of the responses to low oxygen, typical of teleost fish, are observed. Plasma glucose (Maxime et al. 1995) and lactate levels rise (Randall et al. 1992) as in other fish species. Plasma catecholamines and cortisol are also elevated during hypoxia, (Maxime et al. 1995). During recovery from periods of hypoxia, or during hypoxia below 40 mmHg, lactate is released into the circulation (Nonnotte et al. 1993).

Dietary lipid and vitamin E influence regulation of metabolism in hypoxia (Randall et al. 1992). It was suggested that this is due to an altered prostaglandin complement mediated by differences in dietary n3 fatty acids. Enrichment of the diet with n3 fatty acids alters the response to hypoxia but does not improve hypoxia tolerance (Randall et al. 1992). Elevated dietary n3 improves the resistance of the sturgeon heart to hypoxia (Agnisola et al. 1996).

Interestingly, some of the regulatory elements in adaptation to hypoxia have been investigated in sturgeon. The role of prostaglandins in mediating the hypoxic response has been implied. The particular type of fatty acid (n3 or n6) and the amount of vitamin E (a vitamin protecting polyunsaturated fatty acids from oxidative damage) found in the diet have been found to affect the response to hypoxia in sturgeon (Randall et al. 1992). Further investigations of this type are needed to establish the particular eicosanoid pathways involved in the response to hypoxia.

8.11 Evolutionary Considerations

The evolutionary position of sturgeons and paddlefish between the elasmobranchs and the teleost fishes provides a potentially useful "missing link" in the evolution of the aquatic vertebrates. Anatomically Acipenseriformes retain many "primitive" features including a heterocercal tail and a spiral valve in the intestine. Some aspects of the anatomy of neurosecretory systems such as the caudal system also indicate an intermediate position for the Acipenseriformes between the elasmobranchs and the teleosts (Waugh et al. 1995). Similarly there is growing evidence to suggest an intermediate metabolic position.

The number of isoforms of various proteins found in Acipenseriformes is also indicative of their evolutionary position. As indicated above, the number and type of isoforms of LDH indicate an intermediate position between the chondrichthyans and the more advanced bony fishes (Fisher et al. 1980). Similar trends are observed for the isoforms of glyceraldehyde-3-phosphate dehydrogenase and creatine kinase (Fisher et al. 1980).

Some aspects of the endocrinology of the Acipenseriformes more closely resemble those of the elasmobranchs than those of the teleosts. This applies not only to the sequence similarities but also to the forms of hormones found. For example, the hormone scyliorhinin I, originally identified in elasmobranchs has been identified in sturgeon (Wang et al. 1999a) but not in teleosts (Wang et al. 1999b). Scyliorhinin II has been found in elasmobranchs but not in Acipenseriformes (Wang et al. 1999b). The sturgeons have growth hormones with sequences that more closely resemble elasmobranchs than those of teleosts (Yasuda et al. 1992).

Molecular analysis of the primary structure of the peptide hormone (urotensin II) produced in caudal neurosecretory system, supports this view (Waugh et al. 1995). There is only one form of urotensin II in *A. ruthenus* (McMaster et al. 1992), and if this is typical of all Acipenseriformes, it represents an intermediate position between the elasmobranchs and the teleosts.

The evolution of the postranslational processing of some hormones, such as α-MSH, indicates that Acipenseriformes have a more "primitive" process involving fewer acetylations if compared to that of higher vertebrates (Keller et al. 1994). There is a lower level of diacetylated compared to monoacetylated forms in both sturgeon and paddlefish, and this is lower than reported for the other fishes such as the bowfin (*Amia)*, the gar (*Lepisosteus)*, the lungfish (*Neoceratodus)* and mammals (Keller et al. 1994).

The gene coding for MSH-related hormones and β-endorphins (proopiomelanocortin; POMC gene) in Acipenseriformes exists in two forms indicating a duplication similar to that .found independently in the tetraploid salmonids (Danielson et al. 1999). The rates of evolution of these genes differ between the Acipenseriformes and the teleosts, with the teleost rate being very rapid by comparison (Danielson et al. 1999).

Some aspects of the metabolism of Acipenseriformes indicate the loss of certain pathways in the evolution of the teleost fishes. Sturgeon have the ability to synthesize ascorbic acid in the kidney (Dabrowski 1994; Moreau and Dabrowski 1996; Maeland and Waagbo 1998) similar to that of elasmobranchs. In contrast, teleost fish lack this ability and must rely on dietary ascorbate..

Although limited, studies of the metabolism of Acipenseriformes provide valuable insights into the evolution of the vertebrates. Much more needs to be done to establish further details of the differences and similarities to the elasmobranchs and teleosts with respect to regulatory mechanisms and modes of gene expression.

Since Acipenseriformes may exist in 4n, 8n or 16n ploidy levels (Birstein and DeSalle 1998), future studies of isoform complexity should take this into account in interpreting the metabolic complexity of the group. Assuming the ancestral ploidy level is 4n, species such as *A. oxyrhinchus* and *A. sturio* may better reflect the ancestral metabolic condition. The multiple ploidy levels present in the Acipenseriformes provide a unique opportunity to examine the evolution of metabolic complexity as they relate to genome duplication.

8.12 References

Abrosimova, N.A., S.S. Abrosimov, and A.A. Biryukova. 1999. Lipid composition characteristics of stellate sturgeon, *Acipenser stellatus*, eggs and incubation success under conventional hatchery conditions. J. Appl. Ichthyol. 15: 220-223.

Agnisola, C., D.J. McKenzie, E.W. Taylor, C.L. Bolis and B. Tota. 1996. Cardiac performance in relation to oxygen supply varies with dietary lipid composition in sturgeon. Am. J. Physiol. 271: R417-R425.

Agradi, E., G. Abrami, G. Serrini, D. McKenzie, C. Bolis, and P. Bronzi. 1993. The role of dietary n-3 fatty acid and vitamin E supplements in growth of sturgeon (*Acipenser naccarii*). Comp. Biochem. Physiol. 103A: 187-195.

Altinok, I., S.M. Galli, and F.A. Chapman. 1998. Ionic and osmotic regulation capabilities of juvenile Gulf of Mexico sturgeon, *Acipenser oxyrinchus desotoi.* Comp. Biochem. Physiol. 120A: 609-616.

Anderson, P.J. and R.F. Randall. 1975. Comparison of the subunit and primary structure of the pyruvate kinases from rabbit and sturgeon muscles. Biochem. J. 145: 575-579.

Babin, P.J. and J. Vernier. 1989. Plasma lipoproteins in fish. J.Lipid Res. 30: 467-489.

Badiani, A., S. Stipa, N. Nanni, P.P. Gatta, and M. Manfredini. 1997. Physical indices, processing yields, compositional parameters and fatty acid profile of three species of cultured sturgeon (Genus *Acipenser*). J. Sci. Food Agric. 74: 257-264.

Ballantyne, J.S.1997. Jaws, the inside story. The metabolism of elasmobranch fishes. Comp. Biochem. Physiol. 118B: 703-742.

Ballantyne, J.S. and M.E. Chamberlin.1994. Regulation of cellular amino acid levels. In: *Cellular and molecular physiology of cell volume regulation* (Strange, K., Ed.) CRC Press Inc., Boca Raton. pp. 111-122.

Barton, B.A., A.B. Rahn, G. Feist, H. Bollig, and C.B. Schreck.1998. Physiological stress responses of the freshwater chondrostean paddlefish (*Polyodon spathula*) to acute physical disturbances. Comp. Biochem. Physiol. 120A: 355-363.

Beamish, F.W.H., J.A. Jebbink, A. Rossiter and D.L.G. Noakes. 1996. Growth strategy of juvenile lake sturgeon (*Acipenser fulvescens*) in a northern river. Can. J. Fish. Aquat. Sci. 53:481-489.

Bemis,W.E. and B. Kynard. 1997. Sturgeon rivers: an introduction to acipenseriform biogeography and life history. Env. Biol. Fishes 48:167-183.

Birstein, V.J., W.E. Bemis and J.R.Waldman. 1997. The threatened status of acipenseriform species: a summary. Env. Biol. Fishes 48: 427-435.

Birstein,V.J. and R. Desalle. 1998. Molecular phylogeny of Acipenseridae. Mol. Phylogen. Evol. 9:141-155.

Branlant, G., B. Eiler, J.F. Biellmann, H.P.Luzt and P.L. Luisi. 1983. Applicability of the induced-fit model to glyceraldehyde-3-phosphate dehydrogenase from sturgeon muscle. Study of the binding of oxidised nicotinamide adenine dinucleotide and nicotinamide 8-bromoadenine dinucleotide. Biochemistry 22: 4437-4443.

Bubenik, G.A. and S.F. Pang. 1997. Melatonin levels in the gastrointestinal tissues of fish, amphibians and a reptile. Gen. Comp. Endocrinol. 106: 415-419.

Buddington, R.K. 1985. Digestive secretions of lake sturgeon, *Acipenser fulvescens* during early development. J. Fish. Biol. 26: 715-723.

Buddington,R.K. and J.P. Christofferson. 1985. Digestive and feeding characteristics of the chondrosteans. In. North American sturgeons: biology and aquaculture potential. F.P. Binkowski and S.I. Doroshov (Eds). Dr. W. Junk Publishers, Dordrecht, Netherlands.

Buddington, R.K. and S.I. Doroshov. 1986a. Development of digestive secretions in white sturgeon juvenile (*Acipenser transmontanus)*. Comp. Biochem. Physiol. 83A: 233-238.

Buddington, R.K. and S.I. Doroshov. 1986b. Digestive enzyme complement of white sturgeon (*Acipenser transmontanus)*. Comp. Biochem. Physiol. 83A: 561-567.
Buddington, R.K., J.R. Hazel, S.I. Doroshov and J. Van Eenennaam. 1993. Ontogeny of the capacity for homeoviscous adaptation in white sturgeon (*Acipenser transmontanus)*. J. Exp. Zool. 265: 18-28.
Buhler, D.R. and M.E. Rasmusson. 1968. The oxidation of drugs by fishes. Comp. Biochem. Physiol. 25: 223-239.
Burggren, W.W. and D.J. Randall. 1978. Oxygen uptake and transport during hypoxic exposure in the sturgeon *Acipenser transmontanus*. Resp. Physiol. 34:171-183.
Burggren,W., K. Johansen and B. McMahon. 1985. Respiration in phyletically ancient fishes. In: Evolutionary Biology of Primitive Fishes. R.A. Foreman, A. Gorbman, J.M. Dodd and R. Olsson (Eds). Plenum Press, New York, pp. 217-252.
Burggren, W.W. and W.E. Bemis. 2001. Metabolism and ram gill ventilation in juvenile paddlefish, *Polyodon spathula* (*Chondrostei: Polyodontidae*). Physiol. Zool. 65: 515-539.
Callman, J.L. and J.M. Macy. 1984. The predominant anaerobe from the spiral intestine of hatchery –raised sturgeon (*Acipenser transmontanus*) a new *Bacteroides* species. Arch. Microbiol. 140: 57-65.
Canagataram, P. 1959. Growth rates of fishes in different salinities. J. Fish. Res. Board Can. 16: 121-130.
Cataldi,E., E. Ciccotti, P. Di Marco, O. Di Santo, P. Bronzi,P. and S. Cataudella. 1995. Acclimation trials of juvenile Italian sturgeon to different salinities: morphophysiological descriptors. J. Fish Biol. 47: 609-618.
Cataldi, E., C. Barzaghi, P. Di Marco, C. Boglione, L. Dini, D.J. McKenzie and S. Cataudella. 1999. Some aspects of osmotic and ionic regulation in Adriatic sturgeon *Acipenser naccarii.* I. Ontogenesis of salinity tolerance. J. Appl. Ichthyol. 15: 57-60.
Cataldi, E., P. DiMarco, A. Mandich and S. Cataudella. 1998. Serum parameters of Adriatic sturgeon *Acipenser naccarii* (Pisces: Acipenseriformes): effects of temperature and stress. Comp. Biochem. Physiol. 121A: 351-354.
Cech, J.J., S.J. Mitchell and T.E. Wragg. 1984. Comparative growth of juvenile white sturgeon and striped bass: effects of temperature and hypoxia. Estuaries 7: 12-18.
Chan, H.M., M. Trifonopoulos, A. Ing, O. Recveur, and E. Johnson.. 1997. Consumption of freshwater fish in Kahnawake: risks and benefits. Environmental Research 80A: 5213-5222.
Chen, I.C., F.A. Chapman, C.I. Wei, K.M. Portier and S.F. O'Keefe. 1995. Differentiation of cultured and wild sturgeon (*Acipenser oxyrinchus desotoi*) based on fatty acid composition. J. Food Sci. 60: 631-635.
Crigg, G. C. 1974. *Chemical Zoology Volume VIII Deuterostomians, cyclostomes and fishes* (Florkin, M. and Scheer, B.T., Eds.) Academic Press Inc., New York. pp. 331-368.
Crocker,C.E. and J.J. Cech. 1997. Effects of environmental hypoxia on oxygen consumption rate and swimming activity in juvenile white sturgeon, *Acipenser transmontanus*, in relation to temperature and life intervals. Env. Biol. Fishes 50: 383-389.
Crocker,C.E. and J.J. Cech 1998. Effects of hypercapnia on blood-gas and acid-base status in the white sturgeon, *Acipenser transmontanu*s. J. Comp. Physiol. 168B: 5-60.
Dabrowski, K. 1994. Primitive Actinopterygian fishes can synthesize ascorbic acid. Experientia 50: 745-748.
Dabrowski, K., S.J. Kaushik and B. Fauconneau. 1987. Rearing of sturgeon (*Acipenser baeri* Brandt) larvae III. Nitrogen and energy metabolism and amino acid absorption. Aquaculture 65: 31-41.
Danielson, P.B., J. Alrubaian, M. Miller, J.M. Redding and R.M. Dores. 1999. Duplication of the POMC gene in the paddlefish (*Polyodon spathula*): analysis of the γ-MSH, ACTH and, and β-endorphin regions of ray-finned fish POMC. Gen. Comp. Endo. 116: 164-177.
Decker, E.A., A.D. Crum, S.D. Mims and J.H. Tidwell. 1991. Processing yields and composition of paddlefish (*Polyodon spathula*), a potential aquaculture species. J. Agr. Food Chem. 39: 686-688.
Deparade, M.P., K. Gloggler and W.E. Trommer. 1981. Isolation and properties of glyceraldehyde-3-phosphate dehydrogenase from a sturgeon from Caspian sea and its interaction with spin-labelled NAD^+ derivatives. Biochim. Biophys. Acta 659: 422-433.
Doroshov, S.I., G.P.Moberg, J.P. Van Eenennaam. 1997. Observations on the reproductive cycle of cultured white sturgeon *Acipenser transmontanus*. Env. Biol. Fishes 48: 265-278.
Driedzic, W.R. and T. Hart. 1984. Relationship between exogenous fuel availability and performance by teleost and elasmobranch hearts. J. Comp. Physiol. 154B: 593-599.
Fajfer, S., L. Meyers, G. Willman, T. Carpenter and M.J. Hansen. 1999. Growth of juvenile lake sturgeon reared in tanks at three densities. North American Journal of Aquaculture 61: 331-335.
Fauconneau, B., P. Aguirre, K. Dabrowski and S.J. Kaushik. 1986. Rearing of sturgeon (*Acipenser baeri*) larvae. 2. Protein metabolism influence of fasting and diet quality. Aquaculture 51: 117-131.
Fisher, S., J.B. Shaklee, S.D. Ferris and G.S. Whitt. 1980. Evolution of five multilocus isozyme systems in the chordates. Genetica 52/53: 73-85.
Fontenot, Q.C., J.J. Isely and J.R. Tomasso. 1998. Acute toxicity of ammonia and nitrite to shortnose sturgeon fingerlings. Prog. Fish-Culturist 60: 315-318.

Foskett,J.K., H.A. Bern, T.E. Machen and M. Conner. 1983. Chloride cells and the hormonal control of teleost fish osmoregulation. J. Exp. Biol. 106:255-281
Fynn-Aikins, K., S.S.O. Hung, W. Liu and H. Li. 1992. Growth, lipogenesis and liver composition of juvenile white sturgeon fed different levels of D-glucose. Aquaculture 105: 61-72.
Garcia-Gallego, M., A. Sanz, A. Domezain and M. De La Higuera. 1999. Age-size influences on tissue-lipid quality of the sturgeon *Acipenser naccarii* from intensive culture. J. Appl. Ichthyol. 15: 261-264.
Gershanovich, A.D. and G.A. Kiselev. 1993. Growth and haematological response of sturgeon hybrids Russian sturgeon (*Acipenser guldenstadt* Brandt) x beluga (*Huso huso* L.) to protein and lipid contents in diet. Comp. Biochem. Physiol. 106A: 581-586.
Gershanovich, A.D. and I.V. Pototskij. 1992. The peculiarities of nitrogen excretion in sturgeons (*Acipenser ruthenus*) (Pisces, Acipenseridae)-I. The influence of ration size. Comp. Biochem. Physiol. 103A: 609-612.
Gershanovich, A.D. and I.V. Pototskij. 1995. The peculiarities of non-faecal nitrogen excretion in sturgeons (Pisces; Acipenseridae) - 2. Effects of water temperature, salinity and pH. Comp. Biochem. Physiol. 111A: 313-317.
Gershanovich, A.D., G.A.Vaitman, S.S. Vladimirsky and T.E. Rubtsova. 1991. Changes in chemical composition of muscle in young hybrids between Russian sturgeon *Acipenser guldenstadti brandt* x beluga *Huso huso* L. (Pisces; Acipenseriformes) under different levels of salinity. Comp. Biochem. Physiol. 100A: 667-673.
Gillis, T.E. and J.S. Ballantyne. 1996. The effects of starvation on plasma free amino acid and glucose concentrations in lake sturgeon, *Acipenser fulvescens*. J.Fish Biology 49: 1306-1316.
Goncharov, B.F., A.A. Kuznetzov and E. Burzawa-Gerard. 1983. Pituitary gonadotropic hormone from a chondrostean fish, starred sturgeon (*Acipenser stellatus* pall.) IV. Differences in biological action of isoforms and absence of sex-specific forms. Gen. Comp. Endocrinol. 49: 375-382.
Graham,K. 1997. Contemporary status of the North American paddlefish, *Polyodon spathula*. Env. Biol. Fishes 48: 279-289.
Grant, B.F., P.M. Mehrle and T.R. Russell. 1970. Serum characteristics of spawning paddlefish (*Polyodon spathula)*. Comp. Biochem. Physiol. 37: 321-330.
Guderley, H., L. Hamel and J. Lafond, J. 1983. Close resemblance between muscle pyruvate kinase from a primitive vertebrate, the river sturgeon *Acipenser fulvescens*, and the ancestral type K isozyme. J. Comp. Physiol. 153B: 247-256.
Harkness, W.J.K. and J.R. Dymond. 1961. The lake sturgeon. The history of its fishery and problems of conservation. Ont. Dep. Lands Forests, Fish Wildlife Br.121p.
Haslewood, G.A.D. 1967. Bile salt evolution. J. Lipid Res. 8: 535-550.
Henderson,R.J. and D.R. Tocher. 1987. The lipid composition and biochemistry of freshwater fish. Prog. Lipid Res. 26: 281-347.
Houston, J.J. 1988. Status of the green sturgeon, *Acipenser medirostris*, in Canada. Can. Field-Nat. 102:286-290.
Hung, S.S.O. 1988. Present knowledge of white sturgeon nutrient requirements. Aquaculture Magazine (November/December): 50-52.
Hung, S.S.O. 1989. Choline requirement of hatchery-produced juvenile white sturgeon (*Acipenser transmontanus)*. Aquaculture 78: 183-194.
Hung, S.S.O. 1991. Carbohydrate utilization by white sturgeon as assessed by oral administration tests. J. Nutr. 121: 1600-1605.
Hung, S.S.O., F.K. Fynn-Aikins, P.B. Lutes and R. Xu. 1989. Ability of juvenile white sturgeon (*Acipenser transmontanus*) to utilize different carbohydrate sources. J. Nutr. 119: 727-733.
Hung, S.S.O., J.M. Groff, P.B. Lutes and F.K. Fynn-Aikins. 1990. Hepatic and intestinal histology of juvenile white sturgeon fed different carbohydrates. Aquaculture 87: 349-360.
Hung, S.S.O. and P.B.Lutes. 1988. A preliminary study on the non-essentiality of lecithin for hatchery-produced juvenile white sturgeon (*Acipenser transmontanus*. Aquaculture 68: 353-360.
Hung, S.S.O., P.B. Lutes, A.A. Shqueir and F.S. Conte. 1993. Effect of feeding rate and water temperature on growth of juvenile white sturgeon (*Acipenser transmontanus)*. Aquaculture 115: 297-303.
Idler, D.R. and M.J. O'Halloran. 1970. Steroids of a chondrostean: identification of interrenal tissue in the American Atlantic sturgeon, *Acipenser oxyrinchus* Mitchill, by histological and histochemical methods. J. Endocrinol. 48: 621-626.
Idler, D.R. and G.B. Sangalang. 1970. Steroids of a chondrostean: in-vitro steroidogenesis in yellow bodies isolated from kidneys and along the posterior cardinal veins of the American Atlantic sturgeon, *Acipenser oxyrinchus* Mitchill. J. Endocrinol. 48: 627-637.
Idler, D.R. and B. Truscott. 1967. 1alpha-hydroxycorticosterone: synthesis in vitro and properties of an interrenal steroid in the blood of cartilaginous fish (Genus *Raja*). Steroids 9: 457-477.

Idler, D.R. and B. Truscott. 1980. *Evolution of vertebrate endocrine systems* (Pang, P.K.T. and Epple, A., Eds.) Graduate studies Texas Tech University, Texas. pp. 357-373.

Iwama, G.K., J.D. Morgan and B.A. Barton. 1995. Simple field methods for monitoring stress and general condition of fish. Aquacult. Res. 26: 273-282.

Iwasaki, M. and R. Harada. 1984. Cholesterol content of fish gonads and livers. Bull. Jap. Soc. Sci. Fish. 50: 1623

Jarvis, P.L. 2001. Effects of salinity on the growth and metabolism of shortnose sturgeon, *Acipenser brevirostrum*. M.Sc. Thesis. Department of Zoology, University of Guelph, Guelph, Ontario Canada.

Jarvis, P.L., J.S. Ballantyne and W.E. Hogans. 2001. The influence of salinity on the growth of juvenile shortnose sturgeon. North American Journal of Aquaculture 63: 272-276.

Jarvis, P.L. and J.S. Ballantyne. 2003. Metabolic responses to salinity acclimation in juvenile shortnose sturgeon *Acipenser brevirostrum*. Aquaculture 219: 891-909.

Jurss, K. and R. Bastrop. 1995. The function of mitochondria-rich cells (chloride cells) in teleost gills. Rev. Fish Biol. Fisheries 5:235-255.

Karnaky, K.J. 1986. Structure and function of the chloride cells of *Fundulus heteroclitus* and other teleosts. Am. Zool. 26: 209-224.

Kaushik, S.J., J. Breque and D. Blanc. 1991. Requirements for protein and essential amino acids and their utilization by Siberian sturgeon (*Acipenser baeri*). In: Acipenser (Williot, P., Ed.) Cemagref Publ., 25-39.

Kaushik, S.J., J. Breque and D. Blanc. 1994. Apparent amino acid availability and plasma free amino acid levels in Siberian sturgeon (*Acipenser baeri)*. Comp. Biochem. Physiol. 107A: 433-438.

Kaushik, S.J., P. Luquet, D. Blanc and A. Paba. 1989. Studies on the nutrition of Siberian sturgeon, *Acipenser baeri*. I. Utilization of digestible carbohydrates by sturgeon. Aquaculture 76: 97-107.

Keenlyne, K.D. 1997. Life history, latitudinal patterns, and status of the shortnose sturgeon. Env. Biol. Fishes 48: 291-298.

Kelemen, N., N. Kellershohn and F. Seydoux. 1975. Sturgeon glyceraldehyde-3-phosphate dehydrogenase. Formation of binary complexes with coenzymes and substrates. Eur. J. Biochem. 57: 69-78.

Keller, H., M. Redding, G. Moberg and R.M. Dores. 1994. Analysis of the post-translational processing of the a-MSH in the pituitaries of the chondrostean fishes, *Acipenser transmontanus* and *Polyodon spathula.* Gen. Comp. Endocrinol. 94: 159-165.

Kelly, J.L. and D.E. Arnold. 1999. Effects of ration and temperature on growth of age-0 Atlantic sturgeon. North Am. J. Aqua. 61: 51-57.

Klyachko, O.S. and N.D. Ozernyuk. 1998. Functional and structural properties of lactate dehydrogenase from embryos of different fishes. Comp. Biochem. Physiol. 119B: 77-80.

Kruglova, E.E. 1984. Plasmalogen forms of phospholipids in the sturgeon brain. J. Evol. Biochem. Physiol. 20:97-102.

Krzynowek, J. and J. Murphy. 1987. Proximate composition, energy, fatty acid, sodium and cholesterol content of finfish, shellfish, and their products. *National Oceanic Atmosphere Administration Technical Report* 55: 1-48.

Kuznetzov, A.A., B.F. Goncharov and E. Burzawa-Gerard. 1983. Pituitary gonadotropic hormone from a chondrostean fish, starred sturgeon (*Acipenser stellatus* pall.). Gen. Comp. Endocrinol. 49: 364-374.

Kynard,B. 1997. Life history, latitudinal patterns, and status of the shortnose sturgeon, *Acipenser brevirostrum*. Env. Biol. Fishes 48: 319-334.

LeBreton, G.T.O. and F.W.H. Beamish. 1998. The influence of salinity on ionic concentrations and osmolarity of blood serum in lake sturgeon, *Acipsenser fulvescens*. Env. Biol. Fishes 52: 477-482.

Lescheid, D.W., J.F.F. Powell, W.H. Fischer, M. Park, A. Craig, O. Bukovskaya, I.A. Barannikova and N.M. Sherwood. 1995. Mammalian gonadotropin-releasing hormone (GnRH) identified by primary structure in Russian sturgeon, *Acipenser gueldenstaedti*. Reg. Peptides 55: 299-309.

Lizenko,Y.I., V.S. Sidorov, V.I. Luk'yanenko, T.I. Regerand, S.D. Gut'yanova, T.S. Vasil'yeva and S.A. Taksheyev. 1995. General characterization of lipid composition of the lipoproteins of the blood serum of sturgeons (Acipenseridae). J. Ichthyol. 35: 263-270.

Lovern, J.A. 1932. CCXXXV. Fat metabolism in fishes. II. The peritoneal, pancreatic and liver fats of the sturgeon (*Acipenser sturio)*. Biochem. J. 26: 1985-1988.

Maeland, A. and R. Waagbo. 1998. Examination of the qualitative ability of some cold water marine teleosts to synthesize ascorbic acid. Comp. Biochem. Physiol. 121A: 249-255.

Marshall,W.S. and T.D. Singer. 2002. Cystic fibrosis transmembrane conductance regulator in teleost fish. Biochem. Biophys. Acta 1556: 16-27.

Maxime, V., G. Nonnotte, C. Peyraud, P. Williot and J.P. Truchot. 1995. Circulatory and respiratory effects of an hypoxic stress in the Siberian sturgeon. Resp. Physiol. 100: 203-212.

McCormick, S.D. 1996. Effects of growth hormone and insulin-like growth-factor I on salinity tolerance and gill Na^+,K^+-ATPase in Atlantic salmon (*Salmo salar*): interactions with cortisol. Gen. Comp. Endocrinol. 101: 3-11.
McEnroe, M. and J.J. Cech. 1985. Osmoregulation in juvenile and adult white sturgeon, *Acipenser transmontanus*. Env. Biol. Fishes 14: 23-30.
McEnroe, M. and J.J. Cech. 1987. Osmoregulation in white sturgeon: life history aspects. Am. Fish. Soc. Symp. 1: 191-196..
McKenzie,D.J., G. Piraccini, N. Papini, C. Galli, P. Bronzi, C.G. Bolis and E.W. Taylor. 1997. Oxygen consumption and ventilatory reflex responses are influenced by dietary lipids in sturgeon. Fish Biochem. Physiol. 16:365-379.
McKenzie, D.J., E. Cataldi, P. DiMarco, A. Mandich, P. Romano, S. Ansferri, P. Bronzi and S. Cataudella, S. 1999. Some aspects of osmotic and ionic regulation in Adriatic sturgeon *Acipenser naccarii*. II Morpho-physiological adjustments to hyperosmotic environments. J. Appl. Ichthyol. 15: 61-66.
McKenzie, D.J., E. Cataldi, P. Romano, S.F. Owen, E.W. Taylor and P. Bronzi. 2001. Effects of acclimation to brackish water on the growth, Respiratory metabolism and swimming performance of young-of-the-year Adriatic sturgeon *Acipenser nacarii*. Can. J. Fish. Aquat. Sci. 58: 1104-1112.
McKinley, R.S., T.D. Singer, J.S. Ballantyne and G. Power. 1993. Seasonal variation in plasma non-esterified fatty acids of lake sturgeon (*Acipenser fulvescens*) in the vicinity of hydro-electric facilities. Can. J. Fish. Aquat. Sci. 50: 2440-2447.
McMaster, D., M.A. Belenky, A.L. Polenov and K. Lederis. 1992. Isolation and amino acid sequence of urotensin II from the sturgeon *Acipenser ruthenus*. Gen. Comp. Endocrinol. 87: 275-285.
Medale, F., D. Blanc and S.J. Kaushik. 1991. Studies on the nutrition of Siberian sturgeon *Acipenser baeri*. II. utilization of dietary non-protein energy by sturgeon. Aquaculture 93: 143-154.
Moberg,G.P., J.G. Watson, S. Doroshov, H. Papkoff and R.J. Pavlick. 1995. Physiological evidence for two sturgeon gonadotrophins in *Acipenser transmontanus*. Aquaculture 135:27-39.
Mommsen,T.P., M.M. Vijayan and T.W. Moon. 1999. Cortisol in teleosts: dynamics, mechanisms of action, and metabolic regulation. Rev. Fish Biol. Fisheries 9: 211-268.
Moon, T.W. and G.D. Foster. 1995. Tissue carbohydrate metabolism, gluconeogenesis and hormonal and environmental influences. In: *Biochemistry and molecular biology of fishes, Volume 4. Metabolic biochemistry* (Hochachka, P.W. and Mommsen, T.P., Eds.) Elsevier Science, Amsterdam. pp. 65-100.
Moon, T.W. and T.P. Mommsen. 1987. Enzymes of intermediary metabolism in tissues of the little skate, *Raja erinacea*. J.Exp. Zool. 244: 9-15.
Moore, B.J., S.S.O. Hung and J.F. Medrano. 1988. Protein requirement of hatchery-produced juvenile white sturgeon (*Acipenser transmontanus*. Aquaculture 71: 235-245.
Moreau, R. and K. Dabrowski. 1996. The primary localization of ascorbate and its synthesis in the kidneys of acipenserid (Chondrostei) and teleost (Teleostei) fishes. J. Comp. Physiol.B. 166: 178-183.
Nevenzel, J.C., A. Gibbs and A.A. Benson. 1985. Plasmalogens in the gill lipids of aquatic animals. Comp. Biochem. Physiol. 82B: 293-297.
Ng, W.K., S.S.O. Hung and M.A. Herold. 1996. Poor utilization of dietary free amino acids by white sturgeon. Fish Physiol. Biochem. 15: 131-142.
Nguyen, T.M., T.P. Mommsen, S.M. Mims and J.M. Conlon. 1994. Characterization of insulins and proglucagon-derived peptides from a phylogenetically ancient fish, the paddlefish (*Polyodon spathula)*. Biochem. J. 300: 339-345.
Nicoll, C.S., J.F.Tarpey, G.L. Mayer, and S.M. Russell. 1986. Similarities and differences among prolactins and growth hormones and their receptors. Am. Zool. 26: 965-983.
Nonnotte, G., V. Maxime, J.P. Truchot, P. Williot and C. Peyraud. 1993. Respiratory responses to progressive ambient hypoxia in the sturgeon, *Acipenser baeri*. Resp. Physiol. 91: 71-82.
Perdu-Durand, E.F. and J.P. Cravedi. 1989. Characterization of xenobiotic metabolizing enzymes in sturgeon (*Acipenser baeri*). Comp. Biochem. Physiol. 93B: 921-928.
Potts, W.T.W. and P.P. Rudy. 1972. Aspects of osmotic and ionic regulation in the sturgeon. J. Exp. Biol. 56: 703-715.
Radelli, G., C. Domeneghini, S. Arrigi, F. Vaini and F. Mascarello. 1999. Histochemical and immunohistochemical investigation of muscle fibres in the sturgeon (Chondrostei; Acipenser*)*. J. Appl. Ichthyol. 15: 87-91.
Randall, D.J., D.J. McKenzie, G. Abrami, G.P. Bondiolotti, F. Natiello, P. Bronzi, L. Bolis and E. Agradi. 1992. Effects of diet on responses to hypoxia in sturgeon (*Acipenser naccarii*). J. Exp. Biol. 170: 113-125.
Randall, R.F. and P.J. Anderson. 1975. Purification and properties of the pyruvate kinase of sturgeon muscle. Biochem.J. 145: 569-573.
Rao, G.M.M. 1968. Oxygen consumption of rainbow trout (*Salmo gairdneri*) in relation to activity and salinity. Can. J. Zool. 46: 781-786.

Rasmussen, L.E. 1980. Comparative distribution of nitrogenous compounds in *Acipenser transmontanus* and *Hydrolagus colliei*: enzyme kinetics and isozyme ratios alterations. Comp. Biochem. Physiol. 67B: 665-677.
Rubin, D.A. and R.M. Dores. 1994. Cloning of a growth hormone from a primitive bonyfish and its phylogenetic relationships. Gen. Comp. Endocrinol. 95: 71-83.
Ruer, P.M., J.J. Cech and S.I. Doroshov. 1987. Routine metabolism of the white sturgeon, *Acipenser transmontanus* :effect of population density and hypoxia. Aquaculture 62: 45-52.
Safi, S.H., G.A. Takami, I. Nowrouzian, M. Mahmoodi and S. Bokaei. 1999. Evaluation of sturgeon follicle-stimulating hormone, luteinizing hormone, estradiol, progesterone and testosterone in *Acipenser persicus* serum to identify fertile broodstock. J. Appl. Ichthyol. 15: 196-198.
Sangalang, G.B., M. Weisbart and D.R. Idler. 1971. Steroids of a chondrostean: corticosteroids and testosterone in the plasma of the American Atlantic sturgeon, *Acipenser oxyrinchus* Mitchill. J. Endocrinol. 50: 413-421.
Secor, D.H. and T.E. Gunderson. 1997. Effects of hypoxia and temperature on survival, growth and respiration of juvenile Atlantic sturgeon, *Acipenser oxyrinchus*. Fishery Bulletin 96: 603-613.
Semakula, S.N. and P.A. Larkin. 1968. Age, growth, food, and yield of the white sturgeon (*Acipenser transmontanus*) of the Fraser River, British Columbia. J. Fish. Res. Bd. Canada 25: 2589-2602.
Semenkova, T.B., L.V. Bayunova, A.A. Boev and V.P. Dyubin. 1999. Effects of stress on serum cortisol levels of sturgeon in aquaculture. J. Appl. Ichthyol. 15: 270-272.
Sherwood, N.M., S. Doroshov and V. Lance. 1991. Gonadotropin-releasing hormone (GnRH) in bony fish that are phylogenetically ancient: reedfish (*Calamoichthys calabaricus*), sturgeon (*Acipenser transmontanus*), and alligator gar (*Lepisosteus spatula*. Gen. Comp. Endocrinol. 84: 44-57.
Simons,A.M., R.M. Wood, L.S. Heath, B.R. Kuhajda and R.L. Mayden. 2001. Phylogenetics of Scaphirhynchus based on mitochondrial DNA sequences. Trans. Am. Fish. Soc. 130: 359-366.
Singer, T.D., V.G. Mahadevappa and J.S. Ballantyne. 1990. Aspects of the energy metabolism in the lake sturgeon, *Acipenser fulvescens*: with special emphasis on lipid and ketone body metabolism. Can. J. Fish. Aquat. Sci. 47: 873-881.
Smith, T.I.J. 1985. The fishery, biology and management of Atlantic sturgeon, *Acipenser oxyrinchus*, in North America. Env. Biol. Fishes 14: 61-72.
Smith, T.I.J. and E.K. Dingley. 1984 Review of biology and culture of Atlantic (*Acipenser oxyrinchus*) and shortnose sturgeon (*A. brevirostrum*). J. World Maricul. Soc. 15: 210-218.
Stuart, J.S. and S.S.O. Hung. 1989. Growth of juvenile white sturgeon (*Acipenser transmontanus*) fed different proteins. Aquaculture 76: 303-316.
Swallow, R.L. 1985. The effect of food deprivation on the plasma free amino acid levels of sturgeon *Acipenser transmontanus*. Env. Biol. Fishes 14: 93-95.
Tamai,Y., H. Kojima, K. Takayama-Abe. 1994. Lipids and myelin proteins in the brains of coelacanth (*Latimeria chalumnae*), lungfish (*Lepidosiren paradoxa* and *Protopterus aethiopicus*), bichir (*Polypterus senegalus*), and sturgeon (*Acipenser ruthenus*) (Osteichthyes): phylogenetic implications. Can. J. Fish. Aquat. Sci. 51:1265-1272.
Tammar, A.R. 1974. *Chemical Zoology. Volume VIII. Deuterostomes, cyclostomes and fishes* (Florkin, M. and Scheer, B.T., Eds.) Academic Press Inc., New York. 595-612.
Urist, M.R. and K.A. Van de Putte. 1967. Comparative biochemistry of the blood of fishes: identification of fishes by the chemical composition of serum. In: Sharks skates and rays. Gilbert,P.W. R.F. Mathewson and D.P. Rall (Eds). The Johns Hopkins Press. Baltimore pp. 271-285.
Vladykov, V.D. and J.R. Greeley. 1963. Order Acipenseroidei. In: Fishes of the western North Atlantic. Mem. Sears Found. Mar. Res. 1: 24-60.
Wang,Y., B.A.Barton, L. Thim, P.F. Nilsen and J.M. Conlon. 1999a. Purification and characterization of galinin and scylrhinin I from the hybrid sturgeon, *Scaphirhynchus platorhynchus* x *Scaphirhynchus albus* (Acipenseriformes.) Gen. Comp. Endocrinol. 113: 38-45.
Wang, Y., B.A. Barton, P.F. Nilsen and J.M. Conlon. 1999b. Tachykinins (Substance P and neuropeptide γ) from the brains of the pallid sturgeon, *Scaphirhynchus albus* and the paddlefish, *Polyodon spathula* (Acipenseriformes). Gen. Comp. Endocrinol. 116:21-30.
Waugh,D., J. Youson, S.D. Mims, S. Sower and J.M. Conlon. 1995. Urotensin II from the river lamprey (*Lampetra fluviatilis*), the sea lamprey (*Petromyzon marinus*) and the paddlefish (*Polyodon spathula*. Gen. Comp. Endocrinol. 99:323-332.
Webb, J.T. and G.W. Brown. 1980. Glutamine synthetase: assimilatory role in liver as related to urea retention in marine chondrichthyes. Science 208: 293-295.
Whitt, G.S., J.B. Shaklee and C.L. Markert. 1975. Evolution of the lactate dehydrogenase isozymes of fishes. In: Isozymes IV. Genetics and evolution. Markert, C.L. (Ed.) Academic Press, New York pp. 381-400.
Xu, R., S.S.O. Hung and J.B. German. 1993. White sturgeon tissue fatty acid compositions are affected by dietary lipids. J. Nutr. 123: 1685-1692.

Yasuda, A., K. Yamaguchi, T. Noso, H. Papkoff, A.L. Polenov, C.S. Nicoll and H. Kawauchi. 1992. The complete amino acid sequence of growth hormone from sturgeon (*Acipenser guldenstadti)*. Biochim. Biophys. Acta 1120: 297-304.

CHAPTER 9

GROWTH, BIOENERGETICS AND AGE

G.T.O. LEBRETON[1] AND F.W. (BILL) H. BEAMISH[1,2]
[1]*University of Guelph, Guelph, Ontario, N1G 2W1, Canada*
[2]*Department of Biology, Faculty of Science, Burapha University, Bangsaen, Chonburi, 20131, Thailand*

9.1 Introduction

Growth, bioenergetics, and age are notable characteristics in North American species of sturgeon and paddlefish. Sturgeon are well known for their large body size and longevity (Figures 8-1 & 2, Table 8-1). While their longevity and size have contributed to their fame, their slow rates of growth and maturation in concert with a general loss of habitat (Kempinger 1996), have also made them susceptible to even moderate levels of exploitation (Fortin et al. 1996, Beamesderfer et al. 1995, Hoffnagle & Timmons 1989). As a result, all species have suffered dramatic reductions in abundance throughout North America.

Growth is a fundamental characteristic of all living things. Defined as a change in magnitude, mass, energy or proximate components (Weatherley & Gill 1987), growth offers an easily observable and quantifiable metric that may be compared among populations and species. Before a full discussion about growth in sturgeon can begin, an understanding of basic bioenergetic principles relating to growth is useful. To grow, an organism must acquire energies from its environment through ingestion of food. Only a portion of the ingested food is available for activities such as growth, maintenance of bodily functions and other activities such as swimming, breeding and foraging. The indigestible portion of ingested food is lost as feces with other losses resulting from metabolic excretions and waste heat. Growth, as the somatic addition of proteins, fats and carbohydrates, is therefore inextricably linked to a number of processes and occurs only when ingested energies exceed those lost to excretion and expended for metabolism. (Brett & Groves 1979).

Growth in fishes differs from growth in birds and mammals as it is highly variable. This characteristic, known as plasticity, (Weatherly 1990) occurs in response to both intrinsic and extrinsic factors and their operation on thermodynamic processes in fish bodies (Weisberg 1993). Intrinsic factors incorporate variables such as age, sex, maturity and pathological condition (Brett 1979). For example, absolute growth of large fishes fed well is much greater than that for small fishes that are also fed well; however, the relative growth is greater for small fish. It should be clear that intrinsic factors also influence other activities such as metabolism. For example standard or

G.T.O. LeBreton et al. (eds.), Sturgeons and Paddlefish of North America, 195–216.
© 2004 *Kluwer Academic Publishers. Printed in the Netherlands.*

basal metabolic rate, the measured metabolic rate of a fasting and resting individual, is much lower in larger individuals than small after adjustment to a common weight.

Extrinsic factors may be divided into abiotic and biotic components. Abiotic components include temperature (Lutes et al. 1990, Hung et al. 1993), latitude (Power & McKinley 1997), oxygen, carbon dioxide (Burggren & Randall 1978, Crocker & Cech 1996), pH, salinity, conductivity (Fortin et al. 1996), and light. Biotic components include food availability and competition (Fajfer et al. 1999, Jodun et al. 2002). Of the abiotic factors, temperature is extremely important as it operates to pace or regulate the thermodynamics involved in food processing and metabolism (Weatherly 1990). When sturgeon are fed to satiety, growth increases with temperature to a maximum or optimal measure (Fry 1947). If temperatures are further increased, fish growth declines as respiration and maintenance are elevated beyond the limits that nutrient assimilation pathways can support. Slight further increases may result in thermal death, the result of protein coagulation in cells and a breakdown of metabolic function (Brett 1979).

Environmental factors can be divided into four categories depending upon how they influence growth (Brett 1979). Controlling factors, such as temperature, influence rates of thermodynamic reactions. Other environmental factors such as pH, also a controlling factor, may reduce growth by requiring the individual to redirect energies towards homeostasis and away from somatic tissue expansion (Brett 1979). Limiting factors influence chemical reactions required for growth at specific concentrations. Low oxygen, for example, imposes limits on metabolic pathways curtailing growth. Masking factors override or modify the influence of other factors. Finally, directive factors guide or cue activities to particular changes in the environment. Light, and related photoperiod, generally acts as a directive factor, cueing the internal endocrine rhythms of fishes in response to alterations in season.

Though these abiotic, extrinsic factors influence growth, it should be noted that growth is primarily dependent on the intake and assimilation of food substances that fuel growth processes (Brett 1979). Therefore, the amount that an organism can grow is directly related, not only to the quantity and quality, but also availability of food resources. For this reason biotic factors such as intra- and interspecific competition can influence growth by altering food availability.

The species of sturgeon in North America offer an excellent opportunity to study the influence of larger scale environmental factors such as latitude, longitude and salinity on growth, however, effects are often complicated by the interactions of many factors acting concurrently as well as the stage of ontogenetic development. While sturgeon are exciting and economically (ecologically) important, their large size and scarcity can make them difficult subjects of research, especially in the laboratory. In field studies the assignment of age is complicated by difficulties encountered during annuli interpretation, a lack of validation studies, and a poor understanding of how seasonal temperature change influences annuli development in the hard tissues of these species (Morrow et al. 1998).

Even though growth research is hindered by numerous factors, science has made progress. Much data has been collected and published, over extensive periods of time, on the growth of North American sturgeon. Recently, new techniques used to construct growth chronologies, and to elicit other ecological information locked within the hard tissues of these species, promise to provide extensive amounts of growth data in the future (Veinott et al. 1999, LeBreton & Beamish 2000B). Aquaculture and the markets for sturgeon meat and caviar have greatly fuelled efforts to understand sturgeon growth. In these culture settings environmental variables, feeding rates, and levels of competition have been carefully manipulated and studied to enhance the growth of these species. The objective of this chapter is to summarize current knowledge of growth, bioenergetics, and age of North American sturgeon and paddlefish. Improved understanding of the parameters will contribute to the management and conservation of wild populations of these threatened and endangered species.

Fig. 9-1. Several examples of large sturgeon and paddlefish specimens. (A) Dr. Douglas Peterson (Right) with lake sturgeon (Acipenser fulvescens) from the mouth of the Manistee River, Michigan. Photo credit: Paul Vecsei. (B) Paddlefish (Polyodon spathula) from Wisconsin River. Photo credit: U.S. Fish and Wildlife Service. (C) Mike Ruggles from Montana Fish, Wildlife and Parks holding a large pallid sturgeon (Scaphirhynchus albus). Photo credit: Gerald Mestl. (D) Shortnose sturgeon (Acipenser brevirostrum) from the Kennebacasis River, Saint John, New Brunswick Photo credit: Michael Browne and Alisha Giberson. (E) White sturgeon caught by Oregon Fish & Wildlife Service working in conjunction with the Washington Fish & Wildlife Service on the Columbia River below the Dalles Dam. Photo credit: Paul Vecsei. (F) Fisheries biologist John North with a large green sturgeon (Acipenser medirostris) taken from the Columbia River estuary. Photo credit: Paul Vecsei.

9.2 Growth and Environmental Factors

9.2.1 WATER TEMPERATURE

Temperature is the main controlling factor influencing growth in fishes, and for this reason, it is especially of interest to those culturing sturgeon in North America. Numerous hatcheries have been established throughout the continent to facilitate restoration of sturgeon populations and supply aquaculture operations (Mason et al. 1992). From these facilities, information has been gathered relating growth and temperature for a variety of species. These studies are most often conducted on larval or juvenile individuals, as these are easier to accommodate and rear than adults and demonstrate greater relative growth (Cui & Hung 1995).

Fish growth increases with temperature to a maximum following which growth rates are precipitously curtailed until a lethal temperature is reached. When investigating the impact of temperature on growth it is important to note that interactions between abiotic and biotic factors serve to complicate the relation between growth and temperature. For example, variations in ambient temperature influences, not only growth-related thermodynamic processes, but also metabolic rates. For example, as temperature rises, so does the rate of digestion and therefore the amount of food that can be consumed (Brett 1979).

In general, optimum temperatures for growth in sturgeon appear to lie between approximately 20 and 24°C with smaller fish preferring the higher temperatures (Figure 9-2). Paddlefish appear to respond in a similar fashion to temperature but data are scarcer. In both sturgeon and paddlefish, studies of temperature effects on growth are complicated by differences in diet, meal size and feeding frequency (Hung et al. 1993, Kelly & Arnold 1999, Kroll et al. 1994, Lin et al. 1997).

Temperature studies have favoured white sturgeon because this species is very important to North American aquaculture. White sturgeon larvae were found to grow more rapidly at 18.4 than 14.7°C in a study by Lutes et al. (1990). In another study, growth of small white sturgeon (0.6g) was similar at 20 and 25°C and higher than that at 15°C (Cech et al 1984). In larger, 30g fish, growth and feeding efficiency were higher at 23°C than at either 20 or 26°C (Hung et al. 1993). In this study, oxygen concentration at 26°C was lower than at 23°C which may have resulted in reduced body weight index and feeding efficiency as sturgeon are known oxygen conformers and will reduce their metabolism during low oxygen regimes (Burggren & Randall 1978). In another study, growth rates of sturgeon of about 30g were highest at 20°C when maintained on a fixed meal size of 1.5-2.0% of their live body weight per day (Hung & Lutes 1987). A growth model developed by Cui & Hung in 1995 indicated that white sturgeon under 1.2kg decrease in weight below approximately 10°C. This investigation also noted that the optimum temperature for growth decreased with increasing body size. For example, the model suggests that 1000g fish would demonstrate a maximum expected growth rate of 0.9% body weight per day at 18°C while 50g fish demonstrated a maximum expected growth rate of 3.4% body weight per day at 26°C. Interestingly, the authors noted that there are several limitations to this model, as more information is needed on how temperature, size, and diet impact the feeding-growth relationship in sturgeon, particularly for large individuals.

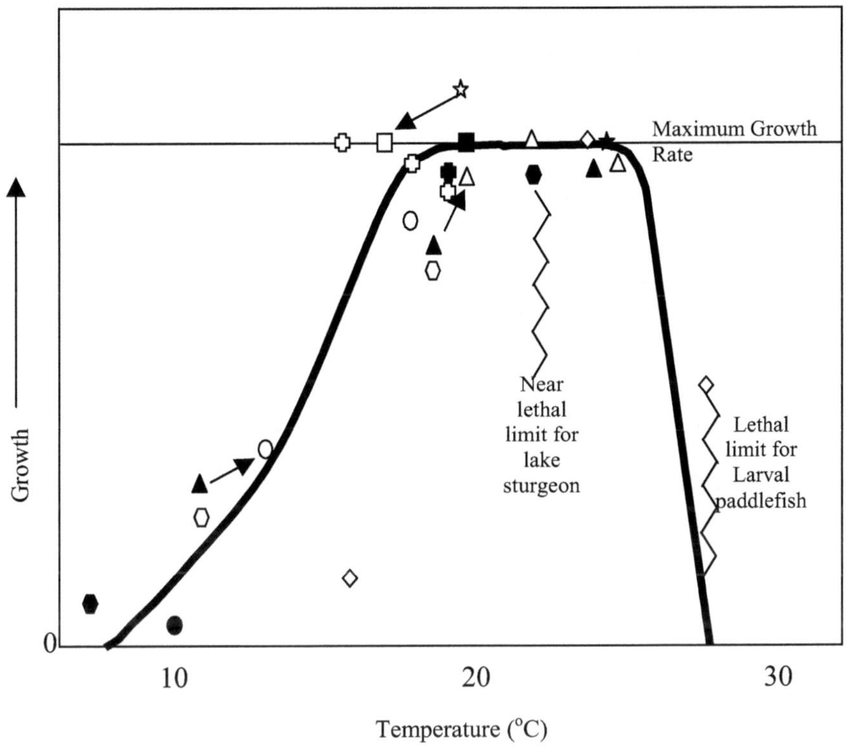

Fig. 9-2. Growth as a function of temperature for North American species of sturgeon and paddlefish. Maximum growth rate as largest change in body size over determined period of time. This figure assumes that the reported maximum growth rates reported all occurred at optimum rations.

Symbol legend

○ *Larval white sturgeon (Lutes et al. 1990)*
△ *30g white sturgeon (Hung et al. 1993)* ■ *30g white sturgeon (Hung and Lutes 1987)*
□ *250 g white sturgeon (Hung et al. 1989)* ▲ *0.5-0.6g white sturgeon (Cech et al. 1984)*
● *1200g white sturgeon (Cui and Hung 1995)* ☆ *1000g modeled white sturgeon (Cui and Hung 1995)*
★ *50g modeled white sturgeon (Cui and Hung 1995)* ⬡ *Juvenile lake sturgeon (Fajfer et al. 1999)*
⬢ *Lake sturgeon (Wehrly et al. 1995)* ✚ *60 g Atlantic sturgeon (Kelly and Arnold 1999)*
✚ *0.3g Atlantic sturgeon (Kelly and Arnold 1999)* ◇ *Larval paddlefish (Kroll et al. in 1992)*

The change in relative body weight per day in juvenile lake sturgeon (106-day old) was found to increase with increasing water temperatures between 14 and 20°C (Fajfer et al. 1999). Wehrly reported similar results in 1995 in the same species reared at temperatures between 7 and 23°C. Fajfer et al. 1999 reported that growth more than doubled from 1.9 to 4.9% body weight per day between 15 and 20°C when fish were fed to satiation twice daily. Interestingly, Wehrly (1995) noted 45% mortality in lake sturgeon reared at 23°C suggesting that this temperature is near the upper tolerance limit for this species.

Growth in Atlantic sturgeon follows a similar pattern to that for white sturgeon in relation to temperature and body size. Based on lab trials at 15, 17 and 19°C, Kelly & Arnold in 1999 found that a six-month-old, 60g Atlantic sturgeon, demonstrated most rapid instantaneous growth at 15°C and 1.5% body weight/day ration. One-month old, 0.3g fish grew most rapidly at 19°C and 7% ration. Greater growth rates in fish at lower temperatures were suggested to be attributed to a reduction in the energy required for maintenance allowing for a greater percentage of available energy being directed to somatic components (Brett 1979).

Little information is present in the literature regarding the influences of temperature on the North American paddlefish, *Polyodon spathula*. However, in a paper published by Kroll et al. in 1992 it was reported that larvae reared on formulated diets, fed to ad libitum every hour, and kept at temperatures of 16, 20, 24, and 28°C, demonstrated the highest and lowest growth rates at 24 and 16°C, respectively. It was suggested in this paper that if a constant temperature was to be used for rearing of larval paddlefish a range of 18 to 20°C is suitable for the first thirty days of life.

9.2.2 LATITUDE

Lake sturgeon occupy a large portion of central North American; across 24° of latitude, from northern Manitoba south to Arkansas, and 41° of longitude, from the St. Lawrence River west to Alberta (Power & McKinley 1997). Numerous populations occur throughout this range living under a myriad of environmental regimes. In North America some more northerly regions are located on the granitic Canadian Shield, therefore water conductivity, alkalinity, and pH are lower than in other areas to the south which may result in lowered fish growth due to reduced system productivity (Fortin et al. 1996). Power & McKinley (1997) related differences in length, mass, and condition factor for 16 populations of lake sturgeon from throughout their ecological range to temperature. They expressed temperature as degree days greater than 5°C that they called thermal opportunity for growth. This investigation found that in northern regions where thermal opportunities for growth were low, individuals were shorter, weighed less and had a lower condition factor than elsewhere. Similar results were reported by Fortin et al. in 1996, from an investigation that related growth and condition of 32 different lake sturgeon populations with latitude and air temperature. Results indicated a negative relationship between growth rates and latitude, while mean annual air temperature provided the best predictor for growth as measured by mean total length between the ages of 23-27 years of age. This relationship is evident when growth in a population from the La Grande Rivière at 54° latitude is compared with that in Lake Wisconsin at 43° latitude. These populations exhibited mean total lengths between the ages of 23 and 27 years of 769mm and 1491mm, respectively.

Interestingly, the study by Power & McKinley (1997) noted that lake sturgeon growth rates, when adjusted for the population's thermal opportunity for growth, demonstrated a significant inverse relationship with latitude. For example, fish located at approximately 54° north demonstrated growth rates of approximately 0.055mm/degree-day while those located at approximately 43° grew 0.035mm/degree-day. It was proposed that this remarkable relationship, termed a "latitudinal countergradient", may be explained by improved food sources in northern populations, genetic variations, or differing thermal regimes. For example, it was postulated that temperatures for growth in southern regions may become too high during the summer resulting in reduced growth rates (Power & McKinley 1997). However, the latitudinal countergradient may also be the result of other factors indirectly operating on growth. For example, photoperiod or length of daylight in more northerly regions is greater than that experienced in the south during the summer. This would allow for a longer foraging period per day and therefore greater food intake per day by individuals from these northern populations.

Latitudinal variation is also known to influence the growth of shortnose sturgeon (Dadswell et al. 1984 in Kynard 1997). Southern individuals grow more rapidly and mature at younger ages than fish in the north but fail to achieve the larger sizes of their more northerly counterparts. Similarly, shovelnose sturgeon from the northern limits of their distribution in Montana reach sizes up to 7kg while more southerly specimens are expected to achieve only 4.5kg. (Carlander 1969 in Keenlyne 1997).

Latitude has also been found to influence growth of paddlefish in a similar manner as it does sturgeon. Hoffnagle & Timmons (1989) compared growth variations among six paddlefish populations. The same inverse relationship between growth rate and latitude that is seen in sturgeon seems to apply to paddlefish. First year growth rates in paddlefish in Old Hickory Reservoir, Tennessee, at approximate latitude of 36°, were reported to be 305 mm/year while in Yellowstone River in Montana at approximately 47° latitude they were only 137 mm/year.

9.2.3 HABITAT

Growth rates in all fishes vary with habitat. Habitat encompasses a myriad of environmental parameters including water temperature, food sources (foraging habitat), stream flow, oxygen and carbon dioxide levels, pH, salinity, and access to other habitats required during different seasons or life stages. These factors interact to produce optimal habitats for various species of fish and therefore impact fish growth. For example, alterations of water flow within river systems may change the area of foraging habitat available, reduce food intake and curtail growth (McKinley et al. 1993). Sturgeon in North American have suffered extensive loss of habitat during the last century from the impoundment of river systems. These changes have impacted sturgeon several ways, closing their migratory routes, altering seasonally available prey and excluding sturgeon from estuarial systems.

Beamesderfer et al. (1995) suggested that impoundment of habitat, and the division of one original population into several smaller populations, can cause different growth rates within the smaller groups. This investigation compared biomass, length, growth rate, and condition factor in white sturgeon from impounded and unimpounded sections of the Columbia River. The unimpounded sections were suggested to contain the better habitats for all life stages of white sturgeon. Correspondingly, populations in these sections demonstrated growth parameters that were equal to or greater than those in impounded regions. Similar results were reported by Devore et al. (1995) who found that white sturgeon from the lower Columbia River grew better than impounded populations from the Snake River and Fraser River. It was suggested that growth rates in the unimpounded sections of river systems were greater because these sturgeon have access to both marine and freshwater based prey systems. Access to these systems is believed to provide an alternate food source during periods of food scarcity. Also, lower growth rates, poor condition factors, and smaller size of females at maturity in impounded sections of the Columbia River was suggested to be due to elevated rates of intraspecific competition (Beamesderfer et al. in 1995). These factors combine to lower the potential production of reservoir impounded populations relative to unimpounded populations.

A similar pattern was reported for an impounded population of shortnose sturgeon (Taubert 1980a in Kynard 1997) in the Connecticut River. Growth rates for this population were the lowest for all populations surveyed, and may have been associated with their denied access to foraging opportunities in the estuary. In contrast, growth rates of lake sturgeon in impounded sections of the St. Lawrence River did not differ from those for free ranging populations in the system (Johnson et al. 1998). This investigation suffered to some extent from limited information but concluded that habitat changes associated with the construction of the St. Lawrence had no significant impact on sturgeon growth. It is possible that lake sturgeon, being a species found almost completely in freshwater, does not require access to foraging sites in estuaries and therefore is not impacted by the St. Lawrence dams.

However, hydroelectric dams and impoundments have been found to impact growth of lake sturgeon in Northern Ontario through their deleterious effect on prey. Water levels below hydroelectric dams are extremely variable in response to the demand for electricity. During periods of low demand, water levels downstream from hydroelectric dams may be low enough to expose large areas of substrate and destroy resident benthic invertebrate populations reducing foraging opportunities for sturgeon. McKinley et al. (1993) compared the nutritional condition of lake sturgeon above and below hydroelectric dams and found that those upstream had a more elevated nutritional status than those below the hydroelectric facilities, based on their plasma nonesterified fatty acid concentrations. It was postulated that the higher nutritional status of the upstream fish was due to the more abundant and diverse prey found upstream. Lower nutritional status of the sturgeon below the impoundments was thought to be due to reduced foraging opportunities as a consequence of daily fluctuations in water flow downstream from the facilities.

While dams and impoundments forcibly alter sturgeon habitat, sturgeon sometimes select habitats that impair their optimum growth potential. For example, Gulf of Mexico sturgeon (*Acipenser oxyrinchus desotoi*) a subspecies of the Atlantic sturgeon, have been known to lose weight during their annual stay in freshwater (Gu et al. 2001). This species migrates from the Gulf of Mexico into the Suwannee River each summer where it does not feed, based on carbon isotope analysis, but which it uses as a thermal refuge. Ground water springs in these regions provide water temperatures that are cooler, and thought to be more favourable, than found in other sections of the river and estuarial areas.

9.2.4 CHEMICAL FACTORS

9.2.4.1 Carbon Dioxide

In many aquaculture settings elevated levels of carbon dioxide (CO_2) may become a concern. High blood concentrations of CO_2 reduce the affinity between red blood cells and oxygen, thereby reducing a fish's capacity to perform its essential activities such as swimming and feeding. High densities of fish, the injection of oxygen, and water recirculation can result in elevated levels of carbon dioxide or hypercapnia. Elevated levels of CO_2, a byproduct of aerobic respiration can result in reduced arterial pH in the blood of fish.

A study by Crocker & Cech (1996) examined the influence of elevated levels of CO_2 on the growth of white sturgeon. They found that white sturgeon growth was significantly decreased by CO_2 concentrations between 45-75mgl^{-1}. Such levels are similar to those reported in tanks where sturgeon are raised in high densities. White sturgeon which were grown in water with elevated levels of CO_2 (pH=7.0) changed in average mass from 4.0g to 5.6g during the same period of time. Interestingly, when these fish were cultured in water with a low pH (7.1) and normal concentrations of carbon dioxide, growth was not significantly depressed relative to the control population. These results indicate that elevated levels of carbon dioxide, not reduced blood pH, are responsible for lower growth rates. Reduced growth in fish cultured in waters containing high concentrations of carbon dioxide appears to be the result of reduced foraging time due to more in erratic swimming behaviour, which reduced food intake and lead to poorer growth rates.

9.2.4.2 Salinity

In freshwater, sturgeon regulate body water concentration by excretion of excess water across the kidneys. In salt water, the process of osmosis drives water out of the fish's body, which sturgeon and other fishes replace by swallowing seawater and excreting chloride ions across their gills. Maintenance of homeostasis in fresh and salt water requires energy, the amount varying with the physiological distance from the isosmotic salinity. There should be no energy required for osmotic regulation at the isosmotic salinity allowing for enhanced growth performance. This, in combination with the demand for improved aquaculture production in sturgeon, has stimulated interest in the effects of salinity on growth (LeBreton & Beamish 1999). While there is a substantial amount of data available on the salinity tolerances of various species of sturgeon there has been little work done on the impact that salinity has on growth in North American species. The single investigation by Jarvis et al. (2001), surprisingly, found growth performance of juvenile shortnose sturgeon, when subjected to different salinities, was highest in freshwater (0%) and not at 10%, their isosmotic salinity, as was expected.

9.3 Growth and Biological Factors

9.3.1 SOMATIC GROWTH CHARACTERISTICS

There are several body characteristics common to North American sturgeon distinguishing them from other fishes. Their relatively large, cylindrical, torpedo-shaped bodies are easily recognized and have remained relatively unchanged for hundreds of millions of years. This body form, and how it changes with age, can provide some insight into the growth of these fish. For example, consistent relationships between total length and weight have been demonstrated among several lake sturgeon populations (Beamish et al. 1996). It was found that the slope of this logarithmic relationship was close to 3.3 and indicates that lake sturgeon demonstrates a disproportionate increase in mass with length. This relationship also results in an increase in condition factor with total length. A similar relationship was found for shovelnose sturgeon and paddlefish by Morrow et al. (1998) and Reed et al. (1992), respectively. Slopes of this logarithmic relation were 2.80 and between 2.5-3.5 for shovelnose sturgeon and paddlefish, respectively. Pallid sturgeon increase slowly in mass to a fork length of 600mm and then increase more rapidly to 1200mm (Kallemeyn 1983). Harkness & Dymond (1961) noted that lake sturgeon increase rapidly in length yet little in weight during their first 5 years. Following this period, length increments decrease while mass gain increases. In general, as both sturgeon and paddlefish age, relative change in body length decreases while relative change in mass continues. It is probable that this continued increase in body mass late in life, while length remains unchanged, is the result of late gonad development in these species.

It should be noted however that condition factor in sturgeon is highly variable among populations. Condition factor for lake sturgeon of 50cm in length varied from 0.28 to 0.54 (Beamish et al. 1996) and that for small white sturgeon reared in laboratory conditions from 0.30-0.44 (Hung & Lutes 1987).

Body composition of lake sturgeon also fluctuates with growth. As body length and mass increase, somatic water content decreases. This is inversely matched by an increase in lipid concentration (Beamish et al. 1996). For example, the lipid content of a 100g sturgeon is approximately 2.8% of body mass. With an increase in mass to 500g, lipids comprise approximately 5.4% of the body weight. This gradual and disproportionate accumulation of lipid with age and growth ultimately provides the material source for gonadogenesis.

Sturgeon and paddlefish in North America display a wide variation in body mass at age the comparison of which is further complicated by differences in environmental quality and exploitation (Power & McKinley 1997). Morrow et al. in 1998 compared von Bertalanffy growth equations among shovelnose, white and shortnose sturgeon. The growth coefficient or slope for shovelnose sturgeon was 0.213, while for white and shortnose sturgeon they were 0.040 and 0.047-0.063, respectively. Shovelnose sturgeon were found to attain a length of approximately 68cm at age 10 years. At this same age shortnose sturgeon attain fork lengths of approximately 450mm. The growth coefficient for shovelnose sturgeon was notably higher than that for either white or shortnose sturgeon, as expected, because this faster growing species reaches maturity between ages 5 and 7 years while the white and shortnose sturgeon mature only after 10 and 25 years of age.

Not surprisingly, species that achieve comparatively small body size but have rapid growth rates are better able to withstand exploitation. For example, shovelnose sturgeon, which reach maturity between the ages of 5 and 7 years of age, and a maximum size of 1m, only become vulnerable to commercial fishing gear after they exceed these ages. For this reason the shovelnose sturgeon, which is currently not listed as a US federal endangered species, can withstand fishing pressures that other species cannot (Morrow et al. 1998).

9.3.2 SEX

Sex can govern growth in sturgeon and paddlefishes. Mature male lake sturgeon are typically smaller than females of the same age (Fortin et al. 1992, Johnson et al. 1998). Sexual dimorphism in size is reported also in mature Atlantic sturgeon, again with females being the larger at comparable age (Bain 1997). Interestingly, this paper states that female Atlantic sturgeons are believed to continue to grow once they have matured and moved into marine waters, whereas males do not. Alexander et al. (1987) noted sexual dimorphism in paddlefish growth only following maturity that occurs at age 8 when total length approaches one meter. After this age females were significantly heavier than males. Within the Alabama River drainage basin, female paddlefish are heavier, longer, and have a larger girth and longer snout lengths than males of the same age (Lein & DeVries, 1998). Data collected from paddlefish sampled in the Lake Kentucky region of Kentucky and Tennessee show there was no significant difference between sexes in the length-age relationship for age classes 2-12 (Hoffnagle & Timmons 1989). Similar results were found by Reed et al. 1992 on paddlefish, aged 1-14, sampled in Louisiana.

9.3.3 RATION

While temperature is considered the main rate controlling factor influencing growth, ration is considered to be the driving force (Stauffer 1973). As ration is increased from zero, growth increases from a negative value, signifying that the amount of food ingested is insufficient to meet the animal's maintenance requirements to zero at which point maintenance has been achieved. At this ration size energy taken into the body is sufficient to maintain the fish's metabolic demands but not to fuel somatic additions. As ration is further increased, a point is reached at which the ratio of growth to ration is maximized, known as optimum ration. With further additions to ration, the ratio of growth to ration begins to decrease to a point of maximum ration where further increases in ration do not result in increased growth.

Ration is dependent on the available food source quantity and quality. Sturgeon are generally characterized as indiscriminate, bottom-feeding carnivores that consume small prey that are available in their habitat. For example, a major component of lake sturgeon diet is insect larvae; mainly chironomids, Diptera, Ephemeroptera, Trichoptera, and Neuroptera and molluscs (Scott & Crossman 1973a, Kempinger 1996). However, occasionally small fish including sculpin and sticklebacks are consumed by lake sturgeon (Houston 1987). The diet of shortnose sturgeon in the Hudson River includes a similar list or organisms (Bain 1997). Atlantic sturgeon in saltwater consume amphipods, isopods, shrimps, molluscs and fish (Scott & Crossman 1973b).

White sturgeon from the lower Columbia River are known to move from the river system to the marine environment based on prey availability. In marine environments they feed on eulachon *Thaleichthys pacificus*, northern anchovy *Engraulis mordax*, American Shad *Alosa sapidissima*, moribund salmonids *Oncorhynchus spp*., and invertebrates (DeVore et al. 1995). Young white sturgeon, when reared in captivity, take artificial foods when they begin exogenous feeding. It is however important to note that these fish do not grow as well as similar individuals fed natural diets (Lutes et al. 1990). Interestingly, the slower growth in these fish is not due to a lower consumption of food as, on inspection, gastrointestinal tracts are generally full. It is possible that these results reflect insufficient knowledge about the dietary requirements of juvenile sturgeon.

Paddlefish are primarily zooplanktivorous (Michaletz et al. 1982). The feeding behaviour of these fish is known to differ with age. Young fish tend to select and capture individual prey organisms while adults, greater than 120mm in size, are filter feeders. However, paddlefish will accept formulated diets (see chapter on aquaculture). This differs from white sturgeon that, if allowed to imprint on natural foods, will not accept formulated diets afterwards (Kroll et al. 1994).

The amount of food ingested obviously has a major influence on growth. Hung et al. (1993) demonstrated that an increase in feeding rate from 2.0 to 2.5% body weight/day resulted in significant mass increases in 30g white sturgeon at both 23 and 26°C over an 8 week period. Temperature and ration can also combine to influence growth in sturgeon. For example, white sturgeon at 23°C demonstrated feed efficiency that linearly decreased as feeding rate was increased. However, at 26°C feeding efficiency decrease when feeding rate was elevated from 2.5 to 3.0% body weight per day (Hung et al. 1993). In general this study indicated that the feed, protein, and energy required for maintenance at either 23 or 26°C was below 2% body weight per day if the diet provided contained 41% protein and 3.2kcal metabolizable energy per gram. Moore et al. (1988) demonstrated that the dietary protein requirements for 154-300g white sturgeon were approximately 40% crude protein in feed. These requirements are similar to that of many other fishes, for example rainbow trout, which require approximately 36% protein, but lower than Japanese eel, small mouth bass, and plaice which require 50% protein. Interestingly, as with most fish, white sturgeon have very high protein requirements relative to other cultured mammals and birds (12-27%) (Stuart & Hung 1989). A more detailed discussion about the influence of ration on growth is outlined in the chapter dealing with aquaculture in this volume.

In white sturgeon deprived of food, weight loss was most pronounced during the first two weeks. Growth rates during these periods have been reported at -0.52% body weight/day. Following these first two weeks, negative growth rates dropped to -0.37 and -0.29% body weight/day. (Hung et al. 1997). In this study, weight loss of the viscera of juvenile white sturgeon was greater than that of the carcass. Under conditions of food deprivation, lipids are more readily metabolized and replaced by water (Beamish et al. 1996). In general, proximate composition of young lake sturgeon (65g) from Northern Ontario was reported to be approximately 82% water, 12% protein, 3% lipid, and 3% inorganic content (Beamish et al. 1996). Over a 10-week period of food deprivation, protein content was also reduced but not nearly so much as lipid. Thus, lipid content was reduced by 84% whereas protein was only reduced by 9%. Similar results were found in plasma constituents with lipids and glucose being significantly reduced while protein concentrations remained unchanged. This study also suggested that sturgeon may have some gluconeogenic capability to convert muscle protein to glucose. Interestingly, whole body protein concentrations for lake sturgeon, 12-15%, is an intermediate value falling between more primitive fishes, such as lamprey, 10-13%, and teleosts, at about 16%. Similarly, the inorganic composition of the lake sturgeon body is also intermediate to lamprey, <1%, which are entirely cartilaginous, and bony fish, 5%, which possess relatively higher concentrations of calcified tissues.

Temporal changes in proximate body composition of lake sturgeon collected from northern Ontario indicate lipid depletion over winter and accumulation during summer (Beamish et al. 1996). Energy concentrations within both muscle and the whole body of lake sturgeon collected in March, near the end of winter were lower than those measured during the summer or fall. These changes reflect the impact of winter on the energy resources of lake sturgeon living at the northern extremes of their distribution and the obvious importance of accumulating energy reserves prior to entering this long, presumably, largely non trophic period. These results concur with the findings of several other investigations that indicate rapidly growing white sturgeon have a higher lipid content and a lower water content than more slowly growing fish (Stuart & Hung 1989, Hung et al. 1993). Lipid concentrations in white sturgeon were found to increase when feeding rates were elevated from 2.5 to 3.0% body weight per day when fish were held at 23°C. Similar increases in lipid content were found when the same species was held at 26°C and feeding rate was increased from 3.0 to 3.5% body weight per day. Body inorganic contents of white sturgeon were also found to be significantly influenced by ration and temperature treatments however there was a discernable trend noted within the data. Protein and water content did not discernibly differ when ration was changed from 2.0 to 3.5% body weight per day at either 23°C or 26°C.

Ration and growth are also related to the size of the organism. As in other fish species, relative growth rates in small sturgeon are greater than in large individuals. Small fish are able to process relatively more food due to elevated metabolic rates. Further, the capacity of their stomach relative to body mass is greater than that in large fish (Brett 1979). Cui & Hung in 1995 found relative food consumption by small white sturgeon (50g) was double that by large fish (1kg) at 3.69 and 1.74% body weight/ day, respectively. This same relationship can also be seen in Atlantic sturgeon (60g) which grew most rapidly when fed a ration of 1.5% body weight per day (Kelly & Arnold 1999). Smaller fish (0.3g) were observed to grow most rapidly at a 7% body weight per day ration. Kelly & Arnold (1999) noted that the interaction of temperature and ration were significant in smaller fish (0.3g) but this was not found to be significant in experiments conducted with larger individuals (60g). Interpretation of these results suggests that biotic (ration) and abiotic (temperature) factors are inextricably coupled such that ration is the force driving growth, temperature is the rate controlling force, and size of the individual is the scaling factor which adjusts rates for the individual. Therefore, a 4°C difference in temperature is not as significant a change in the thermal environment for a 60g individual as is that same thermal alteration for a 0.3g fish. Also, as an individual ages growth slows and optimum temperatures become less apparent. As sturgeon are known to achieve very large body sizes the interaction of temperature and ration on adults may be expected to be very much reduced.

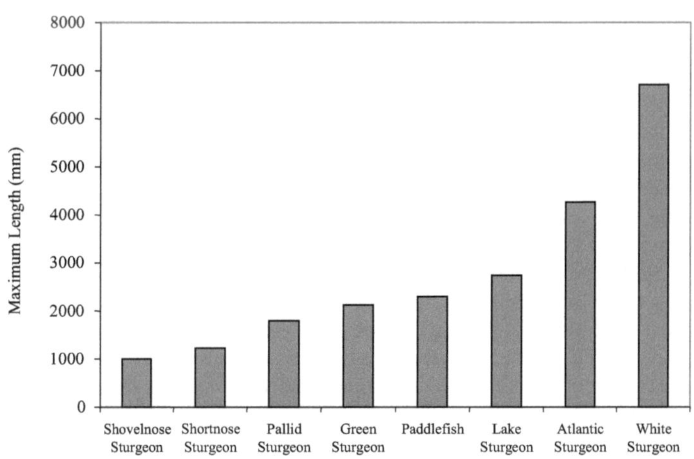

Fig. 9-3. Maximum reported total lengths for North American sturgeon and paddlefish. Sources: Shovelnose: (Page and Burr 1991), Shortnose (Birstein 1993), Pallid (Page and Burr 1991), Green (Eschmeyeret al. 1983), Paddlefish (Hochleithner and Gessner 1996), Lake (Scott and Crossman 1973), Atlantic (Scott and Crossman 1973), White (Scott and Crossman 1973).

9.4 Age, Aging, and the Use of Hard Tissues in Ecological Investigations

Of great importance to the study of growth in wild sturgeon populations is the accurate and precise determination of an individual's age. Many calcified structures including clavicles, cleithra, opercules, medial nuchals, dorsal scutes and pectoral fin rays have been investigated for their use in aging of individual sturgeon (Brennan & Cailliet 1989). While saccular otoliths (sagittae) are commonly used in the age interpretation of other species of fish, they have been difficult to interpret in sturgeon and are generally not used in age and growth investigations.

Fin rays have been most widely used to measure age and growth in many species of acipenserids because they are relatively easy to interpret, can be collected without killing the individual, have been confirmed to document the true annual age through capture-recapture studies, and can be correctly interpreted by novice readers (Rossiter et al. 1995, Brennan & Cailliet, 1989, 1991). Cross sections of these bones (Figure 9-4) demonstrate one pair of opaque and translucent bands or rings associated with winter and summer growth, respectively, in temperate populations. Growth rings are generally wide during early years of growth and become more narrowly spaced towards the outer edge of the fin ray during the later years. Young fish are generally not difficult to age from the cross section of the pectoral fin ray, however, older fish do present numerous problems due to the narrow spacing of the rings, the presence of double annuli and false annuli, difficulty in determining the location of the first annuli, and an inability to distinguish between annuli (Keenlyne & Jenkins 1993). Young lake sturgeon, especially those under the age of 15, were found to demonstrate the most consistent interpretation among different interpreters (Rossiter et al. 1995). It has been reported that considerable disagreement over the age of an individual sample can occur in white sturgeon (Brennan & Cailliet 1989). In lake sturgeon, agreement among independent age interpretations was never below 80% (Rossiter et al. 1995). In 200 shovelnose sturgeon collected in Mississippi, overall aging agreement was found to be just over 30% (Morrow et al. 1998), with agreement being inversely correlated with age of sturgeon. Difficulties in the age determination of sturgeon require that all age estimates be verified, that is, the repeatability of a numerical interpretation, independent of age, must be ensured (Brennan & Cailliet 1989). This is conducted by repeated agings of the same calcified structure by multiple researchers.

Validating the assumption that each pair of opaque and translucent growth annuli correspond with one year's age in each species can be achieved by capture and recapture experiments. Validation is defined in Brennan & Cailliet (1989) as the determination of the true temporal meaning of a growth increment and is considered to be analogous to accuracy. Aging can be validated by intramuscular injections of oxytetracycline (OTC), which is deposited in hard tissues, and serves as a marker. Typically, OTC-injected fish are also tagged, and the first ray from one pectoral fin removed to provide an estimate of age at the time of capture. When fish are recaptured, the first fin ray from the opposite pectoral fin is removed and viewed under ultraviolet light. Marked fish deposit a band of OTC on the fin ray at the time of capture that fluoresces under ultraviolet light (Rossiter et al. 1995). The time span between captures can then be determined from the number of growth rings developed and compared to researchers' records. These investigations have demonstrated that one growth ring is deposited throughout the course of one calendar year in both white sturgeon (Brennan & Cailliet 1991) and lake sturgeon (Rossiter et al 1995). It should be noted that recapture experiments can be hampered by the shedding of tags in sturgeon. Devore et al. (1993) noted that 48 months after tagging, white sturgeon in the Columbia River shed 71% of implanted tags.

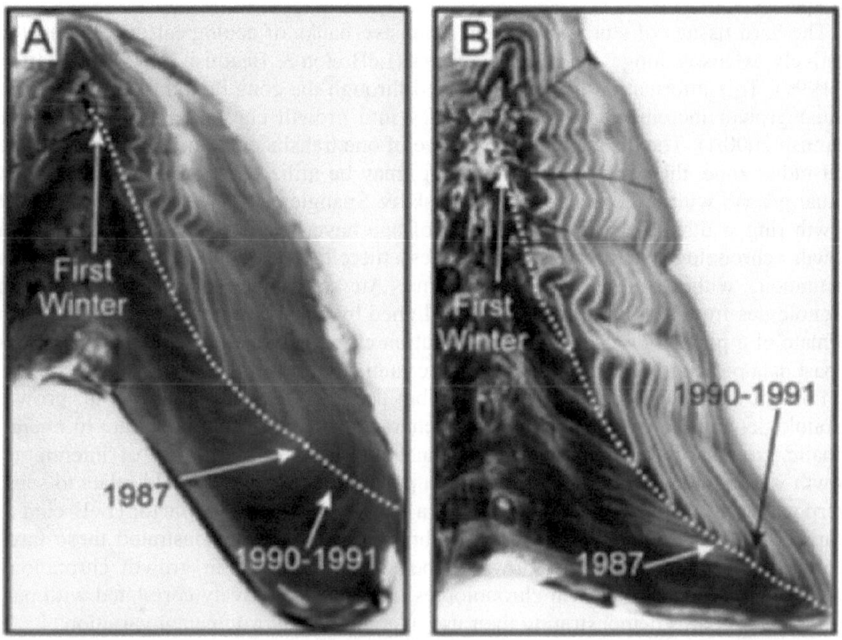

Fig. 9-4. Cross sections of lake sturgeon pectoral fin rays. (A) 1973 year-class, age = 23; (B) 1977 year-class, age = 19. Years indicated by arrows are periods of synchronous growth variation. Dotted lines indicate the radius over which increment widths were measured.

Growth annuli, as mentioned, are composed of one narrow translucent band and a wider opaque portion (Brennan & Cailliet 1989). The translucent band is formed during winter growth and contains less protein and is more mineralized than the opaque, wider, summer growth section (Chilton & Beamish 1982). Interestingly, while the annual nature of the growth rings in sturgeon pectoral fin rays has been determined for several species, this assumption cannot be applied to all species or even populations within a species. For example, Morrow et al (1998) found two pairs of opaque and translucent bands per year in the fin rays of shovelnose sturgeon which they related to high summertime temperatures.

In paddlefish, dental bones have been used to estimate ages from individuals. Within this species, halo bands, areas of blurred demarcation, around annuli can make aging difficult. Halo bands are most commonly found in individuals sampled from unstable environments (Lein & DeVries 1998). As some population of paddlefish generally feed in the spring and fall, halo bands may be representative of periods of slow growth or "checks" that occur during summer when the individual is not feeding.

Furthermore, paddlefish in more southern latitudes are known to feed throughout the winter and demonstrate elevated growth during these periods. Therefore annuli present in the dental bones of these individuals may be representative of slowed growth in the summer due to depressed dissolved oxygen and elevated temperatures (Lein & DeVries 1998). For these reasons age validation studies on paddlefish populations need to be conducted to more fully enhance our ability to interpret age from the bone tissues of this species.

The hard tissues of sturgeon contain extensive banks of ecological data due to the relatively extensive longevity of these species (LeBreton & Beamish 2000A, Veinott et al. 1999). This information may be extracted through the compilation of the widths of annual growth increments from hard tissues into growth chronologies (LeBreton & Beamish 2000A). The distance from the edge of one translucent zone to the edge of the next radial zone, the width of a growth ring, may be utilized as a relative measure of annual growth within the individual (Cyterski & Spangler 1996). A temporal series of growth ring widths, for which the effects of age have been removed, is defined as a growth chronology. Growth chronologies therefore document relative growth fluctuations within an individual over time. An assemblage of individual growth chronologies from a group of individuals, aligned by calendar year and averaged, is an estimate of a population chronology. Population chronologies offer ecologists archives of past data pertaining to population dynamics and environmental variation.

For growth rings from sturgeon to be used in the development of growth chronologies they must demonstrate three characteristics. These must relate to overall somatic growth of the individual, they must demonstrate synchrony of interannual growth variation among individuals within a population, and they should relate to some interpopulation and interannual variation of a known environmental factor (LeBreton & Beamish 2000A). Lake sturgeon growth chronologies have demonstrated these three characteristics and have been shown to be suitable for use in growth chronology development. Sturgeon growth chronologies have been positively correlated with past temperature records demonstrating their use in archiving environmental variation.

When developing sturgeon growth chronologies it is most important to address aging errors. Errors due to misinterpretation of growth rings are difficult to detect due to the highly subjective nature of age interpretation. Aging errors will shift temporal data incorrectly in time thereby disrupting any correlation in growth variation that may be detectable among individuals within a population. To minimize errors of this nature it is recommended that each individual tissue sample be independently assessed for age a minimum of three times. This allows the aging error of the sample to be calculated (Beamish & Fournier 1981). By minimizing the aging error of samples used in the development of population growth chronologies it may be possible to minimize noise among growth chronologies and maximize the common signal (LeBreton et al. 1999).

Sturgeon growth chronologies have been shown to exhibit negative correlation with tree ring growth chronologies from neighbouring populations of trees (LeBreton & Beamish 2000B). These correlations are likely due to the influence of temperature on growth in both organisms. Within temperate ecosystems, relatively warm summers may reduce growth in trees by causing excessive transpiration. Conversely, lake sturgeon in these areas may experience elevated growth as water temperatures remain in an optimum thermal range for long periods of time during warm summers (LeBreton & Beamish 2000B).

Sturgeon growth chronology investigations are not the only means by which ecological data is being extracted from sturgeon bone tissue. As bone tissue grows it will include in it a variety of rare elements and isotopes present in the individual's environment at the time of tissue development. For example, strontium (Sr) levels are much higher in salt water than fresh water. Strontium can be easily substituted for calcium in the bones of sturgeon during periods of marine residence. For this reason strontium concentrations, when sampled from the cross section of the fin ray of white sturgeon, can be used to track the movements of sturgeon between fresh and salt water (Veinott et al. 1999). Sampling of these delicate structures can be conduced using laser ablation to obtain tiny sample volumes required to compare concentrations of elements with annuli position. Loss of elements due to activities such as spawning and migration tend to complicate research of this nature.

Table 9-1. Maximum age for North American species of sturgeon and paddlefish

Species	Maximum Age	Source
Atlantic Sturgeon	60	Scott and Crossman 1973b
Lake sturgeon	154	Scott and Crossman 1973a
Paddlefish	30	Purkett 1963
Pallid Sturgeon	40	Hochleithner and Gessner 1996
Shortnose Sturgeon	67	Hochleithner and Gessner 1996
Shovelnose Sturgeon	27	Christenson and Hatzenbeler 1996
White Sturgeon	80+	Scott and Crossman 1973

9.5 Conclusion

While science has made extensive progress to enhance our understanding of the growth of sturgeon and paddlefish in North America there is still a great deal that remains unknown. This lack of knowledge, especially that pertaining to wild organisms, is likely to continue as numbers dwindle from human insults on populations and habitat.

The value of sturgeon growth research is being acknowledged, primarily for its economic value in caviar and meat production. In addition, their large sizes, late maturation, and the relatively extensive longevity of these species make them unique organisms for use in growth studies. It has been shown that long lived individuals contain extensive records of environmental change within their calcified tissues and hence serve as monitors of environmental quality. While research into the extraction of this data has only begun, it does present some promising findings.

Finally, the importance of sturgeon growth research may be, as yet, unrealized. Humans share the characteristics of elevated positioning within the food chain, extensive longevity, and late maturation with sturgeon and paddlefish. In addition we also live in a highly variable environment, constantly subjected to insults and injuries from human society. It is possible that the populations of sturgeon and paddlefish in North America may serve as test cases, or ecological mine canaries, for human populations of the future.

9.6 References

Alexander, C.M., A.I. Myhr III, & L.J. Wilson. 1987. An assessment of the harvest potential of paddlefish stocks in Watts Bar Reservoir, Tennessee. Proceedings of the Annual Conference Southeastern Association of Fish and Wildlife Agencies 39:45-55.

Bain, M. 1997. Atlantic and shortnose sturgeons of the Hudson River: common and divergent life history attributes. Environmental Biology of Fishes. 48: 347-358.

Beamesderfer, R.C., T.A. Rien, & A.A. Nigro. 1995. Differences in the Dynamics and Potential Production of Impounded and Unimpounded White Sturgeon Populations in the Lower Columbia River. Transactions of the American Fisheries Society. 124:857-872.

Beamish, R.J., & Fournier, D.A. 1981. A method for the comparing the precision of a set of age determinations. Can. J. Fish. Aquat. Sci. 38: 982–983.

Beamish, F.W.H., J. Jebbink, A. Rossiter, & D.G.L. Noakes. 1996. Growth strategy of juvenile lake sturgeon in a northern river. Can. J. Fish. Aquat. Sci. 53: 481-489.

Birstein, V.J. 1993. Sturgeon and Paddlefishes: Threatened fishes in need of conservation. Conservation Biology. 7:773-787.

Brennan, J.S. & G.M. Cailliet. 1989. Comparative age-determination techniques for white sturgeon in California. Transactions of the American Fisheries Society. 118:

Brennan, J.S. & G.M. Cailliet. 1991. Age determination and validation studies of white sturgeon, *Acipenser transmontanus,* in California. Pages 209-234 *in* P. Williot, editor. Acipenser. Centre National du Machisme Agricole du Génie Rural des Eaux et des Forêts, Bordeaux, France.

Brett, J.R. 1979. Environmental factors and growth. Pages 599-675 *in* W.S. Hoar, D.R. Randall, and J. R. Brett, editors. Fish Physiology, Volume 8. Academic Press, New York.

Brett, J. R. & Groves, T. D. D. 1979. Physiological energetics. Pages 279-344 *in* Hoar, W.S.; Randall D.J.; Brett, J.R. eds. Fish Physiology, VIII; Bioenergetics and Growth. Academic Press, New York, 1979.

Burggren, W.W. & D.J. Randall. 1978. Oxygen uptake during hypoxic exposure in the sturgeon, *Acipenser transmontanus.* Respiratory Physiology 34:171-183.

Cech, J.J. Jr., Mitchell, S.J. & Wragg, T.E., 1984. Comparative growth of juvenile white sturgeon and striped bass: effects of temperature and hypoxia. Estuaries, 7:12: 12-18.

Chilton, D.E., & R.J. Beamish. 1982. Age determination methods for fishes studied by the ground-fish program at the Pacific Biological Station. Canadian Special Publication of Fisheries and Aquatic Sciences 60.

Crocker, C.E. & J.J. Cech. 1996 The effects of hypercapnia on the growth of juvenile white sturgeon, *Acipenser transmontanus.* Aquaculture. 147, 293-299.

Cui, Y. & S.S.O. Hung. 1995. A prototype feeding-growth table for white sturgeon. Journal of Applied Aquaculture. 5(4) 25-34.

Cyterski, M.J. & G.R. Spangler. 1996. Development and utilization of a population growth history of Red Lake walleye, *Stizostedion vitreum*. Environmental Biology of Fishes.

DeVore, J.D., B.W. James, C.A. Tracy, and D.A. Hale. 1993. Dynamics and Potential Production of White Sturgeon in the Columbia River Downstream from Bonneville Dam. Bonneville Power Administration, Project Number 86-50, Contract Number DE-AI79-86BP63584.

DeVore, J.D., B.W. James, C.A. Tracey, & D.A. Hales. 1995. Dynamics and Potential Production of White Sturgeon in the Unimpounded Lower Columbia River. Transactions of the American Fisheries Society. 124:845-856.

Fajfer, S., L. Myers, G. Willman, T. Carpenter, & M. J. Hansen. 1999. Growth of juvenile lake sturgeon reared in tanks at three densities. North American Journal of Aquaculture. 61:331-335.

Fortin, R., P. Dumont, & S. Guénette. 1992. Biologie, exploitation, modélisation et gestion des populations d'esturgeon jaune (*Acipenser fulvescens*) dans 14 réseaux de lacs et de rivieres du Québec. Québec, Ministere du Loisir, de la Chasse et de le Pêche, Service de l'amenagement et de l'exploitation de la faune et Service de la faune aquatique, Montreal et Quebec, Que.

Fortin, R., P. Dumont, & S. Guénette. 1996. Determinants of growth and body condition of lake sturgeon (*Acipenser fulvescens*). Canadian Journal of Fisheries and Aquatic Sciences. 53:1150-1156.

Fry, F.E.J. 1947. Effects of the environment on animal activity. University of Toronto Biological Sciences Series 55, Ontario Fisheries Research Lab. 68:1-62.

Gu, B., D.M. Schell, T. Frazer, M. Hoyer, & F.A. Chapman. 2001. Stable carbon isotope evidence for the reduced feeding of Gulf of Mexico sturgeon during their prolonged river residence period. Estuarine, Coastal, and Shelf Science. 53: 275-280.

Harkness, W.J.K. & J.R. Dymond. 1961. The lake sturgeon: the history of its fishery and problems of conservation. Ontario Department of Lands and Forests, Fish and Wildlife Branch, Toronto.

Hochleithner, M. and J. Gessner. 1996. The sturgeon and Paddlefishes of the World. Biology and Aquaculture. AquaTech Publications. Kitzbuehel. 207pp.
Hoffnagle, T.L. & T.J. Timmons. 1989. Age, growth, and catch analysis of the commercially exploited paddlefish population in Kentucky Lake, Kentucky-Tennessee. North American Journal of Fisheries Management 9:316-326.
Houston, J.J. 1987. Status of the lake sturgeon, *Acipenser fulvescens*, in Canada. Canadian Field Naturalist101:171-185.
Hung, S.S.O. & P.B. Lutes, 1987. Optimum feeding rate of hatchery-produced juvenile white sturgeon (*Acipenser transmontanus*): at 20°C. Aquaculture, 65: 307-317.
Hung, S.S.O., P.B. Lutes, F.S. Conte, & T. Storebakken. 1989. Growth and feed efficiency of white sturgeon (*Acipenser transmontanus*) sub-yearlings at different feeding rates. Aquaculture 80:147-153.
Hung, S.S.O, P.B Lutes, A.A Shiqueir, & F.S. Conte, 1993. Effects of feeding rate and water temperature on growth of juvenile white sturgeon (*Acipenser transmontanus*). Aquaculture, 115: 297-303.
Hung, S.S.O., W. Liu, H. Li, T. Storebakken, & Y. Cui. 1997. Effects of starvation on some morphological and biochemical parameters in white sturgeon, *Acipenser transmontanus*. Aquaculture 151: 357-366.
Jarvis, P.L., J.S. Ballantyne, & W.E. Hogans. 2001. The Influence of Salinity on the Growth of Juvenile Shortnose Sturgeon. North American Journal of Aquaculture 63:272-276.
Jodun, W.A., M.J. Millard, & J. Mohler. 2002. The effect of rearing density on growth, survival, and feed conversion of juvenile Atlantic sturgeon. North American Journal of Aquaculture 24: 10-15.
Johnson, J.H., D.S. Dropkin, S.R. LaPan, J.E. McKenna, & R.M. Klindt. 1998. Age and growth of lake sturgeon in the upper St. Lawrence River. Journal of Great Lakes Research. 24(1): 474-478.
Kallemeyn, L.W. 1983. Status of the pallid sturgeon *Scaphirhynchus albus*. Fisheries (Bethesda) 8(1):3-9.
Keenlyne, K.D. & L.G. Jenkins. 1993. Age at sexual maturity of the pallid sturgeon. Transactions of the American Fisheries Society. 122:393-396.
Keenlyne, K.D. 1997. Life history and status of the shovelnose sturgeon, *Scaphirhynchus platorynchus*. Environmental Biology of Fishes 48: 291-298.
Kelly, J. & D.E. Arnold. 1999. Effects of ration and temperature on growth of Age-0 Atlantic sturgeon. North American Journal of Aquaculture. 61: 51-57.
Kempinger, J.J. 1996. Habitat, Growth, and Food of Young Lake Sturgeons in the Lake Winnebago System, Wisconsin. North American Journal of Fisheries Management 16: 102-114.
Kroll, K.J., J.P. Van Eenennaam, & S.I. Doroshov. 1992. Effects of water temperature and formulated diets on growth and survival of larval paddlefish. Transactions of the American Fisheries Society. 121:538-543.
Kroll, K.J., J.P. Van Eenennaam, S.I. Doroshov, J. Linares, E.J. Hamilton, & T. R. Russell. 1994. Growth and survival of Paddlefish Fry Raised in the Laboratory on Natural and Artificial Diets. The Pregressive Fish-Culturist 56-169-174.
Kynard, B. 1997. Life history, latitudinal patterns, and status of shortnose sturgeon, *Acipenser brevirostrum*. Environmental Biology of Fishes 48: 319-334.
LeBreton, G. T. O. and Beamish, F. W. H. (1998). The influence of salinity on ionic concentrations and osmolarity of blood serum in lake sturgeon, *Acipenser fulvescens*. Env. Biol. Fish. 52, 477–482.
LeBreton, G.T.O., F.W.H. Beamish & R.G. Wallace. 1999. Lake sturgeon (*Acipenser fulvescens*) growth chronologies. Can. J. Fish. Aquat. Sci. 56:1752-1756.
LeBreton, G.T.O. & F.W.H. Beamish. 2000A. Suitability of lake sturgeon growth rings in chronology development. Transactions of the American Fisheries Society 129(4): 1018-1030.
LeBreton, G.T.O. & F.W.H. Beamish. 2000B. Interannual growth variation in fish and tree rings. Can. J. Fish. Aquat. Sci. 57:2345-2356.
Lein, G.M. & D.R. DeVries, 1998. Paddlefish in the Alabama River drainage: Population characteristics and the adult spawning migration. Transactions of the American Fisheries Society 127:441-454.
Lin, J., Y. Cui, S.S.O. Hung, & S. Shiau. 1997. Effect of feeding strategy and carbohydrate source on carbohydrate utilization by white sturgeon (*Acipenser transmontanus*) and hybrid tilapia (*Oreochromis niloticus* X *O. aureus*). Aquaculture 148: 201-211.
Lutes, P.B., S.S.O. Hung, & F.S. Conte. 1990. Survival, growth, and body composition of white sturgeon larvae fed purified and commercial diets at 14.7°C and 18.4°C. The Progressive Fish-Culturist. 52: 192-196.
Mason, T.W., J.P. Clugston, & A.M. Foster. 1992. Growth of Laboratory-Held Gulf of Mexico Sturgeon (*Acipenser oxyrinchus desotoi*) The Progressive Fish Culturust 54:59-61.
McKinley, R.S., T.D. Singer, J.S. Ballantyne, & G. Power. 1993. Canadian Journal of Fisheries and Aquatic Sciences. 50:2440-2447.
Michaletz, P.H., C.F. Rabeni, W.W. Taylor, & T.R. Russell. 1982. Feeding ecology and growth of young-of-the-year paddlefish in hatchery ponds. Transactions of the American Fisheries Society.
Moore, B.J., S.S.O. Hung, & J.F. Medrano. 1988. Protein requirements of hatchery-produced juvenile white sturgeon (*Acipenser transmontanus*). Aquaculture, 71: 235-245.

Morrow, J.V., J.P. Kirk, K.J. Killgore, & S. G. George. 1998. Age, growth, and mortality of shovelnose sturgeon in the Lower Mississippi River. North American Journal of Fisheries Management. 18: 725-730.
Page, L.M., & B.M. Burr. 1991. A Field Guide to Freshwater Fishes North America North of Mexico. Peterson Field Guide Series. Houghton Mifflin and Company. Boston. 432 pp.
Power, M. & R.S. McKinley. 1997 Latitudinal variation in lake sturgeon size as related to the thermal opportunity for growth. Transactions of the American Fisheries Society. 126:549-558.
Purkett, C.A. 1963. The paddlefish fishery of the Oage River and Lake of the Ozarks, Missouri. Trans. Am. Fish. Soc. 92:125-129.
Reed, B.C., W.E. Kelso, & A, Rutherford. 1992. Growth, Fecundity, and Mortality of Paddlefish in Louisiana. Transactions of the American Fisheries Society. 121:378-384.
Rossiter, A., D.L.G. Noakes, & F.W.H. Beamish. 1995. Validation of Age Estimation for the Lake Sturgeon. Transactions of the American Fisheries Society. 124:777-781.
Scott, W.B., & Crossman, E.J. 1973a. Freshwater Fishes of Canada. Fish. Res. Board Can. Bulletin Number 184:82-89.
Scott, W.B., & Crossman, E.J. 1973b. Freshwater Fishes of Canada. Fish. Res. Board Can. Bulletin Number 184:92-96.
Stauffer, G.D. 1973. A growth model for salmonids reared in hatchery environments. Doctoral dissertation. University of Washington, Seattle.
Stuart , J.S. & S.S.O. Hung. 1989. Growth of juvenile white sturgeon (*Acipenser transmontanus*) fed different proteins. Aquaculture, 76:303-316.
Veinott, G. T. Northcote, M. Rosenau, & R.D. Evans. 1999. Concentrations of strontium in the pectoral fin rays of the white sturgeon (*Acipenser transmontanus*) by laser ablation sampling - inductively coupled plasma - mass spectrometry as an indicator of marine migrations. Can. J. Fish. Aquatic. Sci. 56:1981-1990.
Weatherly, A.H. 1990. Approaches to understanding fish growth. Transactions of the American Fisheries Society. 119:662-672
Weatherley, A.H. & H.S. Gill. 1987. The Biology of fish growth. Academic Press. London and New York.
Wehrly, K.E. 1995. The effect of temperature on the growth of juvenile lake sturgeon, *Acipenser fulvescens*. Michigan Department of Natural Resources. Fisheries Division, Research Report 2004, Lansing.
Weisberg, S. 1993. Using hard-part increment data to estimate age and environmental effects. Canadian Journal of Fisheries and Aquatic Sciences. 50: 1229-1237.

CHAPTER 10

GENETICS OF NORTH AMERICAN ACIPENSERIFORMES

MICHAEL R. ROBINSON AND MOIRA M. FERGUSON
Department of Zoology, University of Guelph, Guelph Ontario, N1G 2W1 Canada

10.1 Introduction

The Acipenseriformes are an ancient order of fishes that contain 25 to 27 extant species worldwide, all of which are native to the Northern Hemisphere (Bemis et al. 1997). There are many conflicting ideas and hypotheses surrounding the genetics and evolutionary history of Acipenseriformes, especially in the areas of cytology and phylogenetics. In spite of the many controversies apparent in the literature, researchers generally agree on a few salient points concerning the genetics of these fishes. First, the genomic organization of Acipenseriformes is highly complex (Wirgin et al. 1997). Their karyotypes consist of an extremely large number of chromosomes, which for some sturgeons is estimated to be as high as 500 (Blacklidge and Bidwell 1993). Second, all Acipenseriformes are polyploid, and as a group Acipenseriformes exhibit the highest levels of polyploidy known in fishes (Vasiliev 1999). Polyploidy events are believed to have been a major driving force in the evolution and subsequent radiation of taxa within this group (Birstein and DeSalle 1998; Vasiliev 1999). Third, as a group Acipenseriformes are monophyletic and evolutionarily ancient, and form the sister group of all extant Neopterygii (Bemis et al. 1997).

This chapter is an introduction to and a review of the concepts and controversies surrounding the genetics and evolution of North American sturgeon and paddlefish. This review will be structured into sections pertaining to specific aspects of acipenseriform genetics. A brief conclusion and synthesis with speculations on the future direction of research in sturgeon genetics concludes the chapter.

10.2 Genomic Organization

Early studies of the Acipenseriformes focused mainly on the karotypes of the European sturgeons (see reviews in Birstein and Vasiliev 1987; Fontana et al. 2001). The karyotypes of these fishes are characterized by large numbers of chromosomes, about half of which are very small microsomes approximately 1µm long. The microchromosomes are functional and have a highly defined secondary structure (Fontana et al. 1999). The remaining chromosomal complement is typically made up of

G.T.O. LeBreton et al. (eds.), Sturgeons and Paddlefish of North America, 217–230.
© 2004 *Kluwer Academic Publishers. Printed in the Netherlands.*

large metacentric and submetacentric macrochromosomes, usually ranging from 2 to 5μm in length (Birstein et al. 1997). The large numbers of microchromosomes present were often left uncounted by early cytologists, or were confused with other cellular artifacts resulting in a high degree of variability in the early estimates of sturgeon chromosome numbers (Van Eenennaam et al. 1998). Aneuploidy (chromosome numbers different than exact multiples of the haploid complement) in cell cultures, superimposed chromosomes, and chromosome fragmentation from the hypotonic shock needed to separate them in a preparation are all cited as possible sources of variability in the reported karyotypes for various species of Acipenseriformes (Fontana et al. 1997; Van Eenennaam et al. 1998).

Although modern karyological methods have resulted in more accurate chromosomal counts for many sturgeon species, there remains a great deal of controversy surrounding the ploidy-level interpretation of these data (Blacklidge and Bidwell 1993; Birstein et al. 1997, but see Fontana et al. 1997; Fontana et al. 2001). Ploidy levels in Acipenseriformes are normally inferred either through karyotyping (Ohno et al. 1969; Dingerkus and Howell 1976; Fontana 1994; Fontana et al. 1997; Van Eenennaam et al. 1998) or through estimates of cellular DNA content and genome size using flow cytometry (Birstein et al. 1993; Blacklidge and Bidwell 1993). Although recent studies of sturgeon transmission genetics (Ludwig et al. 2001; Pyatskowit et al. 2001; McQuown et al. 2002; Rodzen and May 2002) have provided an additional source of evidence for inference of sturgeon ploidy levels, ploidy evolution in sturgeons will likely continue to be a contentious issue in the sturgeon genetics community.

Geneticists divide extant sturgeons into three generalized groups based on their actual or hypothesized karyotypes; 1), species with approximately 120 chromosomes , 2) species with approximately 240 chromosomes, and 3) species with around 500 chromosomes (Table 10-1, see Birstein et al. 1997; Fontana 2002 for reviews). It should be noted that the 500 chromosome group has been putatively defined based on DNA content and not on actual karyotypes (Blacklidge and Bidwell 1993).

Basic assumptions on the ploidy levels of these groups vary among authors. Some researchers propose a tetraploid-octoploid relationship between the 120 and 240 chromosome groups (Birstein and Vasiliev 1987; Blacklidge and Bidwell 1993; Birstein et al. 1997), while others hypothesize that they actually reflect a diploid-tetraploid relationship (Fontana 1994; Fontana et al. 1997; Vasiliev 1999). Since karyotypic evidence for the 500 chromosome group is lacking, they are typically excluded from such comparisons. Early karyotyping efforts with European sturgeons (see Birstein and Vasiliev 1987 for review) lacked the fluorescence-based methods needed to visualize secondary structures in resolved chromosomes, such as banding patterns and nucleolar organizing regions (NORs). These kinds of structures are important for ploidy-level inferences because they can help distinguish between homologous chromosomes and chromosomes that are only superficially similar based on overall morphology (Fontana 1994; Fontana et al. 1997). In lieu of being able to recognize NOR regions, researchers based their estimates of ploidy on the number of active nucleoli in metaphase nuclei, which they assumed would be directly related to the overall ploidy level of the organism (Birstein and Vasiliev 1987; Birstein et al. 1997). The numbers of active nucleoli ranged from 2 to 3 in the 120 chromosome group, and from 6 to 8 in the 240 chromosome group (Birstein et al. 1997), thus superficially supporting a tetraploid-octoploid relationship between these two groups.

Table 10-1. Known genome size (pg DNA / cell), approximate chromosome numbers (2n), and inferred or hypothesized ploidy levels of North American sturgeon and paddlefish species.

Group	Taxa	Common Name	Genome size	Chromosome no.	Ploidy[a,g]
I	*P. spathula*	American paddlefish	4.9[a]	120[f]	2n / 4n
I	*S. platorynchus*	Shovelnose sturgeon	4.7[a]	112[d]	2n / 4n
I	*S. albus*	Pallid sturgeon	Unknown	Unknown	2n /4n ?
I	*S. suttkusi*	Alabama sturgeon	Unknown	Unknown	2n /4n ?
I	*A. oxyrinchus*	Atlantic sturgeon	Unknown	99-112[c]	2n / 4n
I	*A. o. desotoi*	Gulf sturgeon	4.5[a]	Unknown	2n / 4n
II	*A. fulvescens*	Lake sturgeon	8.9[a]	Unknown	4n / 8n
II	*A. transmontanus*	White sturgeon	9.5[a]	271[b]	4n / 8n
II	*A. medirostris*	Green sturgeon	8.8[a]	249[e]	4n / 8n
III	*A. brevirostrum*	Shortnose sturgeon	13.1[a]	Unknown	8n / 12n

Evidence supporting a diploid-tetraploid relationship between the two main chromosome groups in sturgeons was initially provided by researchers using fluorescent in situ hybridization (FISH) techniques to examine genomic organization in Acipenseriformes (Fontana 1994; Fontana et al. 1998; Tagliavini et al. 1999). These FISH-based techniques clearly resolved NOR regions in sturgeon chromosomes allowing more accurate discrimination between chromosomes that were homologous or only superficially similar. Fontana (1994) found that in members of the supposedly tetraploid 120-chromosome group (Birstein and Vasiliev 1987; Birstein et al. 1997), the four NORs detected were located on 2 morphologically different homologous chromosome pairs. A similar pattern of distribution was found for the 240-chromosome group (i.e. eight NORs arranged in 2 morphologically different groups of chromosomes). These results are indicative of a diploid-tetraploid relationship between the two groups of sturgeons (see Fontana et al. 2001 for review).

In North America, karyotypes have only been determined for five species of Acipenseriformes (Table 10-1; see Birstein et al. 1997 for review); these karyotypes have formed the basis for comparative ploidy estimates for other North American sturgeons based on measurements of cellular DNA content (Blacklidge and Bidwell 1993).

Although karotypic studies indicate two main groups of sturgeons based on chromosome counts, studies of DNA content and genome size show a continuum over a significantly larger range of values (Blacklidge and Bidwell 1993). In a cytological study of eight species of North American Acipenseriformes (i.e. all North American taxa excluding *Scaphirhynchus albus*, pallid sturgeon and *S. suttkusi*, Alabama sturgeon), DNA contents (pgDNA per cell) varied from approximately 4.5 to 13.1, and were roughly distributed in a 1:2:3 ratio, which the author inferred was a reflection of ploidy and polyploidation events in the radiation of the group (Blacklidge and Bidwell 1993). Based on their initial assumptions of ploidy from the 4 North American karotypes (i.e. that the chromosome counts reflected a tetraploid-octoploid relationship), Blacklidge and Bidwell (1993) assumed that these ratios reflected a tetraploid:octoploid:dodecaploid (4n:8n:12n to 16n) relationship in ploidy levels. Interestingly, some individual animals in this study exhibited DNA estimates which were very different from conspecific values. For example, Blacklidge and Bidwell (1993) found that two individuals, one *A. fulvescens* and one *A. oxyrinchus desotoi* contained DNA values which were approximately 1.5 times higher than the values for their conspecifics, indicating that these 2 animals were likely triploids (Blacklidge and Bidwell 1993). The relatively small number of animals that they examined (8 *A. fulvescens* and 15 *A. oxyrinchus desotoi*) suggests that triploidy may occur fairly frequently in these fishes.

10.3 Genome Evolution

Several different explanations have been proposed to explain how polyploidy events have shaped the evolution and radiation of the Acipenseriformes (Vasiliev 1999; Ludwig et al. 2001). It is generally accepted that modern Acipensiformes arose from an extinct diploid ancestor with a chromosome number of approximately 60 (Ludwig et al 2001). Some authors have proposed that complex schemes of ancestral hybridization (i.e. reticulate speciation, see Vasiliev 1999), involving unisexual triploid forms are responsible for the modern distribution of ploidy in the Acipenserids. Changes in ploidy can also arise though non-disjunction events during mitosis and meiosis. Non-disjunction has been documented in the lake sturgeon (Pyatskowit et al 2001), and occurs when sister chromatids fail to separate during cell division. This typically results in the duplication or loss of a small number of chromosomes (Appels et al. 1998). Other authors believe that repeated genome duplication (Fontana et al. 2001, Ludwig et al. 2001) have been the major forces driving ploidy diversity in the acipenserid lineages.

To date, only a small number of studies have attempted to explain how ploidy differences have evolved within sturgeons and paddlefish (Blacklidge and Bidwell 1993; Ludwig et al. 2001). Although these studies differed in their interpretations of actual ploidy, both agreed that several independent polyploidization events had occurred during the radiation of the group (Blacklidge and Bidwell 1993; Ludwig et al. 2001). It is unlikely that a general consensus will be reached in the near future concerning the evolution of ploidy levels within acipenserids. Although linkage mapping techniques have been used to characterize polyploid genomes in other fish species (see Sakamoto et al. 2000 and Nichols et al. 2003 for salmonid examples), this type of approach is likely to be difficult to implement in sturgeons due to the extremely high number of chromosomes and the complexity of a potentially multi-duplicated genome. Although the debate concerning ploidy variation in sturgeons will likely continue, the current research in transmission genetics (see below), and the increasing number of karyotypic studies indicates that the three groupings based on chromosome number reflect a diploid-tetraploid-octoploid relationship of ploidy levels.

10.4 Transmission Genetics

The extremely long generation times of most acipenseriform species makes studies of heredity beyond a single generation extremely difficult, and as a result basic research on the transmission genetics of these fishes is lacking. Hypotheses concerning ploidy levels can be tested through transmission genetics and studies of inheritance, as differences in ploidy level are reflected in ratios of genotypes that are transmitted from parents to their progeny (Burnham 1962). This is generally considered to be an important first step in the testing of newly developed marker systems. In North America, only two Acipenserid species (lake sturgeon and white sturgeon) have been subjected to detailed studies of transmission genetics and inheritance (Pyatskowit et al. 2001; McQuown et al. 2002; Rodzen and May 2002). Previous karyotyping efforts in the white sturgeon (Van Eenennaam et al. 1998) and estimates of DNA content in the lake sturgeon (Blacklidge and Bidwell 1993) place both these taxa within the approximately 240 chromosome group of sturgeon species. Although interpretations as to the ploidy level of this group are inconsistent (it is believed to be either tetraploid or octoploid derived; see Fontana et al. 2001 for review), studies of inheritance provide a potentially powerful tool against which various modes of inheritance can be tested.

Inheritance studies in lake sturgeon initially set out to determine the mode of transmission at a number of microsatellite loci (Pyatskowit et al. 2001, McQuown et al. 2002). In both studies, different marker sets were used but the results were quite similar (Pyatskowit et al. 2001, McQuown et al. 2002). No evidence of octosomic inheritance was found in either study, and the microsatellite loci used displayed modes of transmission that could be accounted for by either disomic or tetrasomic models of inheritance (Pyatskowit et al. 2001, McQuown et al. 2002). These results support the hypothesis that lake sturgeon may be a tetraploid-derived species which is currently undergoing a partial or complete return to disomy.

A recent study of inheritance of microsatellite loci in the white sturgeon revealed highly complex modes of inheritance (Rodzen and May 2002) that could not be explained under strictly disomic or tetrasomic models of transmission. Although this study detected some loci that seemed to demonstrate disomic and tetrasomic patterns of inheritance, several of the microsatellite markers examined seem to reflect an even higher level of duplication (Rodzen and May 2002). The authors of this study acknowledge the potential pitfalls of testing such complex systems of inheritance, and caution that microsatellites in sturgeons are best interpreted on a locus-by locus basis. Unfortunately, the experimental design used in this study (i.e. the number of progeny screened and the genotypes of the parents) could not differentiate between octosomic inheritance and the presence of duplicated tetrasomic loci, necessitating follow up studies in the transmission genetics of this species.

Microsatellite-based inheritance studies in acipenserids have been met with limited success, primarily due to the complications involved with implementing codominant marker systems in highly polyploid species. The use of tetrasomic or octosomic codominant microsatellite loci in sturgeons is complicated by several factors that limit their usefulness in the context of transmission genetic research. The electrophoretic banding patterns produced by higher levels of ploidy are extremely complex and difficult to interpret, as are the relative dosages of each allele. The correct interpretation of gene dosage is especially pertinent to testing between different modes of inheritance for a given locus. Researchers are also typically unable to select parent genotypes for appropriate testing of specific multiple hypotheses (i.e. differentiating between duplicated tetrasomic loci and higher levels of octosomic inheritance). For the time being, it seems that the best approach is to use only microsatellites for which disomic inheritance has been established within a given species. Recently, some authors have suggested that some of the problems associated with codominant markers and higher ploidy (i.e. gene dosage and interpretation of inheritance) can be avoided if the codominant nature of the loci is ignored and if individual alleles are each treated as separate dominant markers (see Ludwig et al. 2001). Although this type of approach may appear feasible for superficial ploidy comparisons between species, it is not adequate for more detailed studies of inheritance or population genetics.

10.5 Phylogenetics

To date, all molecular phylogenetic reconstructions within the Acipenseriformes have been based almost exclusively on mtDNA sequences (Brown et al. 1996; Birstein and Bemis 1997; Birstein and DeSalle 1998; Tagliavini et al. 1999; Krieger et al. 2000; Ludwig et al. 2000; Ludwig et al. 2001) and mtDNA RFLP analysis (Brown et al. 1996), with only a single study using nuclear DNA in a RAPD analysis (Comincini et al. 1998). Surprisingly, many of the main conclusions based on the mtDNA data sets are inconsistent between studies even when similar regions in the mtDNA molecule are sequenced. As a result, several key aspects of the evolutionary history of North American Acipenseriformes have been the focus of several recent phylogenetic studies (Birstein and DeSalle 1998; Krieger et al. 2000; Ludwig et al. 2000). These aspects include the identification of sister taxa within the Acipenseriformes clade, the placement of the genus Scaphirhynchus within the Acipenseridae family, differentiation of the 3 species within the Scaphirhynchus genus, and the occurrence of heteroplasmy (heterogeneity within the mtDNA population of an individual) in certain species. The remaining studies are fragmentary from the point of view of the phylogenetics of North American Acipenseriformes, because they focus mainly on European species (Comincini et al. 1998; Tagliavini et al. 1999), or on a small number of North American taxa (Brown et al. 1996).

In the first molecular phylogeny constructed for 27 species of Acipenseriformes (all extant taxa excluding the North American *S. platorhynchus* and *S. suttkusi*), Birstein and DeSalle (1998) used partial sequencing of 3 mtDNA genes (3 regions of cytochrome b gene, 12s and 16s rRNA) to construct a phylogenetic tree illustrating the relationships between North American, European, and Asian sturgeons and paddlefish. From their distance matrices and tree topologies, Birstein and DeSalle (1998) concluded first that the genus Scaphirhynchus was the sister genus to the other Acipenseridae genera, Acipenser and Huso. They also concluded that *A. fulvescens* and *A. brevirostrum* were distantly related, which was contradictory to previous conclusions concerning these two species based on morphological and ecological considerations (Grande and Bemis 1991; Choudhury and Dick 1998). Regarding the sturgeons of the Pacific Coast, evidence of a sister taxa relationship was demonstrated for *A. transmontanus* and *A. medirostris* (Birstein and DeSalle 1998). Strong relationships were also detected between North American and European sturgeons based on geographical distribution (i.e. Pacific North American sturgeons were closely related to Pacific Asian sturgeons, and sturgeons on the Atlantic coast of North

America were closely related to sturgeons on Europe's Atlantic cost (Birstein and DeSalle 1998).

In a subsequent phylogenetic analysis which included all 10 North American taxa (Krieger et al. 2000), nucleotide sequences of four mtDNA genes (12S rRNA, cytochrome c oxidase II, tRNAPhe, and tRNAAsp) revealed a definitive sister-taxa relationship was observed between *A. fulvescens* and *A. brevirostrum*, and between *A. transmontanus* and *A. medirostris*. The placement of the genus Scaphirhynchus was also not clearly demonstrated (Krieger et al. 2000), and the three species of Scaphirhynchus analyzed exhibited identical nucleotides sequences at all 4 mtDNA genes. The most recent phylogeny for sturgeon, based on the entire cytochrome b mtDNA gene (Ludwig et al. 2001) is generally consistent with previous phylogenies and shows a strong Atlantic Pacific subdivision.

10.6 Species Relationships

Genetic studies have also been used to help resolve some taxonomical issues that are especially pertinent to the Scaphirhynchus genus and to the two Acipenser oxyrinchus subspecies. The acipenseriform genus Scaphirhynchus consists of three species; the pallid sturgeon *S. albus*, the shovelnose sturgeon *S. platorhynchus*, and the Alabama sturgeon *S. suttkusi*. All of these fish are distinguished from other Acipenseriformes by a flattened snout and an elongated caudal peduncle (Bemis et al. 1997). The Alabama sturgeon is native to the Mobile River drainage basin, and the shovelnose and pallid sturgeons co-occur together in the Missouri and lower Mississippi Rivers. All three species are restricted to the central United States.

The morphological similarity and sympatric distribution of two of these species (shovelnose and pallid sturgeon) has led some researchers to question their current status (and designation under the endangered species act) as separate species. Early attempts to differentiate between these three species using allozyme variation at 37 loci found no evidence of differentiation between pallid and shovelnose sturgeon (Phelps and Allendorf 1983). To confound matters more, hybridization events among these species have also been documented (Carlson et al. 1985). Recent research using DNA sequencing of 436 bp sequence of the mtDNA control region has generated support for the current taxonomic separation of these three species (Campton et al. 2000). Campton et al. (2000) found no shared haplotypes between pallid and shovelnose sturgeon in the upper Missouri River, and in areas where they were naturally sympatric, consistent differences in haplotype frequencies between the two species were present. Low levels of mtDNA sequence divergence between these two species may be a result of the possible paraphyletic origin of pallid sturgeon, which could have evolved from an isolated population of shovelnose sturgeon (Campton et al. 2000).

There is also some contradictory evidence as to whether or not *A. oxyrinchus desotoi* and *A. o. oxyrinchus* should be given status as separate subspecies (Ong et al. 1996; Artyukhin and Vecsei 1999). Sequencing comparisons between portions of the mtDNA control region between this subspecies and *A.o. oxyrinchus* found 3 fixed nucleotide substitutions that were diagnostic characters in differentiating between these two species (Ong et al. 1996). Morphological evidence suggest that both subspecies of *A. oxyrinchus* and the European species *A. sturio* (European Atlantic sturgeon) likely represent a single species which is distributed along both coasts of the Atlantic Ocean

(Artyukhin and Vecsei 1999). This view is not supported by a microsatellite DNA based investigation (King et al. 2001) that found ample evidence of large genetic distances (comparable to cogeneric status in other orders) between the two subspecies.

10.7 Population Genetics

Investigations of the population genetics of North American Acipenseriformes have been hampered by several factors. A general lack of molecular markers for use in sturgeon combined with the difficulties in collecting tissues for analysis have made research into sturgeon population genetics relatively inaccessible and difficult to accomplish. These fish are relatively difficult and expensive to rear in a long-term hatchery setting, and natural populations are challenging to sample genetically due to the intermittent spawning that is characteristic of all sturgeons and paddlefish.

Much of the existing body of research concerning North American sturgeon population genetics has been directed towards examining patterns of post-glacial dispersal and phylogeography (Ferguson et al. 1993), or addressing practical management and conservation concerns. The main goals of these studies have been to discriminate between reproductively isolated populations within a larger drainage (Brown et al. 1992a, 1992b; Guenette et al. 1993, Grunwald et al. 2002), analyze stock structure (Wirgin et al. 2000, 2002; King et al. 2001; Quattro et al. 2002; Smith et al. 2002; Waldman et al. 2002), quantify the effects of inbreeding or bottlenecking (Brown et al. 1992a), and to determine interspecific differences in areas where sympatric populations overlap (Phelps and Allendorf 1983; Campton et al. 2000). The majority of these studies have centred on mitochondrial DNA (mtDNA) and used techniques based on sequencing (control region and D-loop) and haplotype variation. Some attention has also been given to the feasibility of using amplified fragment length polymorphisms (AFLP) to identify interspecific hybrids (Congiu et al. 2001) and species (Congiu et al. 2002) in unknown samples of sturgeon.

To date, population genetic studies have been completed on the following North American sturgeon species: lake sturgeon (Ferguson et al. 1993; Guenette et al. 1993), white sturgeon (Brown et al. 1992a; Smith et al. 2002), Atlantic sturgeon (Waldman et al. 1996a, 1996b; Wirgin et al. 2000; King et al 2001; Waldman et al. 2002; Wirgin et al. 2002), Gulf sturgeon (Wirgin et al. 2000; King et al. 2001), shortnose sturgeon (Grunwald et al. 2002; Quattro et al. 2002; Waldman et al. 2002), and to a limited extent all 3 species of Scaphirhynchus, pallid, Alabama, and shovelnose sturgeon (Campton et al. 2000). Although it is impossible to quantitatively examine the genetics of every North American sturgeon population, some general conclusions about the likely state of these populations can be gleaned from the few studies that have been completed.

The population dynamics (Bruch 1999), movements (Rusak and Mosindy 1997), and population genetics (Guenette et al. 1993; Ferguson and Duckworth 1997) have been investigated in all of the major drainages inhabited by lake sturgeon, including the Great Lakes (Guenette et al. 1993), Hudson/James Bay (Ferguson et al. 1993) and Mississippi (Ferguson and Duckworth 1997) watersheds. All of these studies were based on RFLP analysis of mtDNA. Mitochondrial DNA based estimates of genetic diversity in lake sturgeon populations are quite low (Guenette et al. 1993; Ferguson et al. 1993; Ferguson and Duckworth 1997), and the extent of population structure detected in these studies has also been low, with little divergence between or within major drainages (Guenette et al. 1993, Ferguson and Duckworth 1997). The lack of divergence detected may be a result of the mtDNA technique used, whereby mtDNA variation accumulates at such a slow pace that the variation detected often reflects historical postglacial events rather than contemporary population structure.

Studies of other acipenseriform species have detected variable levels of genetic diversity accompanied by significant population structure (Ong et al. 1996; Waldman et al. 1996a, 1996b; Wirgin et al. 2000; King et al. 2001). Recent studies (Wirgin et al. 2000, 2002; King et al. 2001) have analyzed coastal Atlantic sturgeon populations through their entire range and reported from five (King et al. 2001) up to seven (Wirgin et al. 2000, 2002) distinct stocks based on microsatellite DNA analysis and mtDNA haplotype groupings, respectively. Measures of genetic diversity in Atlantic sturgeon populations (measured by the number of mtDNA haplotypes) were found to increase in a latitudinal cline within the populations that were sampled from north to south (Waldman et al. 1996a, 1996b; Wirgin et al. 2000). Wirgin et al. (2000, 2002) hypothesized that the northern populations were recolonized after the Pleistocene glaciation while the southern populations persisted in areas which were unglaciated through the Pleistocene (Wirgin et al. 2000). As a result, the reduced levels of haplotypic diversity observed in the northern populations are indicative of a founder effect from the individuals that recolonized northern areas once the Wisconsin glaciers receded. Although the microsatellite analysis found evidence of decreased genetic diversity in Atlantic sturgeon from the St. Lawrence River, an overall reduction in genetic diversity from populations at the northern end of the species distribution was not detected (King et al. 2001).

The single genetic study of Gulf sturgeon populations deserves special mention for it reveals an interesting life history aspect that may or may not be shared with other North American Acipenseriformes. Strong evidence of distinct stock structure detected in southern populations of Gulf sturgeon (Stabile et al. 1996) suggests that this species of sturgeon is characterized by strong homing fidelity to their natal rivers. This tendency may be especially prevalent (or at least detectable) in populations of Gulf sturgeon due to the geographical history of the Gulf of Mexico region. This area was not subjected to glaciation during the Pleistocene and as a result the distribution of local ichthyofauna is ancient compared to cospecific and cogeneric species that are distributed further north.

It is to be expected that when sexually reproducing populations are severely and rapidly reduced in number, loss of genetic diversity and subsequent bottlenecking will occur. Unfortunately genetic bottlenecks can be difficult to detect without comparative samples collected before and after the putative bottlenecking event (Cornuet and Luikhart 1996). To date, strong evidence of genetic bottlenecking has only been detected in white sturgeon populations from the Columbia River (Brown et al. 1992a). RFLP analysis of mtDNA in the Columbia and Fraser populations has uncovered some surprising results. Although the Columbia River population is expected to have provided post-Pleistocene founders to the Fraser River, higher levels of mtDNA genotype diversity were observed in individuals sampled from the Fraser River (Brown et al. 1992a). Other studies have also cited low levels of genetic diversity in some areas as possible evidence of genetic bottlenecking and loss of genetic diversity in populations of lake sturgeon (Guenette et al. 1993) and Atlantic sturgeon (Wirgin et al. 2000). Unfortunately, mtDNA analyses often lack the resolution needed to confirm these suspicions. In the case of the Fraser River populations, a recent study based on mtDNA and microsatellite DNA variation found that diversity patterns in the Fraser River reflected post-glacial dispersal patterns more than recent bottlenecking events (Smith et al. 2002).

MtDNA sequence variation has also recently been examined in shortnose sturgeon stocks along the eastern coast of North America (Grunwald et al. 2002; Quattro et al 2002). In both studies moderate levels of haplotype diversity and population structuring were detected, suggesting that most populations of shortnose sturgeon exhibit high homing fidelity with limited gene flow between rivers (Grunwald et al. 2002). Interestingly, moderate levels of genetic diversity were also detected in populations from Atlantic Canada, the northern range limit of the shortnose sturgeon (Grunwald et al. 2002). It was hypothesized that levels of diversity were maintained in these historically glaciated areas through recolonization from a northern glacial refuge, presumably off the Georges Bank area of the Northwest Atlantic (Grunwald et al. 2002). Although shortnose sturgeon populations have been historically impacted by overfishing and habitat degradation throughout their entire range, this did not seem to have an effect on diversity levels as detected by mtDNA control region sequence variation (Grunwald et al. 2002; Quattro et al 2002). Different explanations have been offered to explain the apparent lack of an anthropogenic impact on mtDNA diversity in shortnose sturgeon populations. Some authors argue that the time scale of the impacts (100 to 200 years) is too recent to have manifested itself in the form of reductions in mtDNA diversity (Grunwald et al. 2002). Other researchers suggest that overlapping age classes in a long-lived iteroparous species like shortnose sturgeon may act to reduce drift and maintain genetic variation (Quattro et al. 2002).

Although traditional mtDNA-based approaches have proven useful in resolving some genetic differentiation among sturgeon stocks, newly developed hypervariable nuclear DNA markers will likely be more successful in determining fine scale relationships between and within acipenserid populations. For reasons mentioned in the Transmission Genetics section, the use of microsatellite DNA markers in sturgeon population genetics studies should, for the time being, be limited to those loci for which disomic patterns of inheritance have been conclusively demonstrated.

10.8 Conclusions and Future Prospects

Several conclusions about the current state of acipenserid population genetics can be drawn from the studies mentioned in this review. First, mtDNA results may not be entirely accurate in their portrayal of the current state of depressed diversity in our sturgeon populations. Lake sturgeon for example, are characterized by a polyandrous mating system in which a single female simultaneously mates with a large number of males. In spring 2000, a single large female lake sturgeon was observed spawning in Rainy River with more than 25 males simultaneously (Robinson, M., personal observation). If this increased maternal transmission is combined with the extremely long generation times for lake sturgeon (females mature at 20-26 years and spawn once every 6 to 12 years), one would expect that the mtDNA genome would be characterized by low genetic diversity and an extremely slow rate of evolution. Thus, every effort should be made to utilize more hypervariable marker systems (such as microsatellites) to address issues pertaining to population genetics and ESU (Evolutionarily Significant

Unit)-specific management. Furthermore, the type of inheritance exhibited by candidate markers should be established on a locus by locus basis within the taxon of interest before they are used for the analysis of fine-scale population structure.

The loss of genetic diversity resulting from one hundred years of overfishing and population fragmentation may not be readily apparent in our sturgeon and paddlefish populations. Sturgeons and paddlefish are extremely susceptible to anthropogenic disturbances, primarily due to several unique aspects of aspects of their life histories including age at maturity, slow growth rates, intermittent spawning, and migratory behaviours (Krieger et al. 2000). Acipenseriformes grow very slowly and the females often require long periods of time to become sexually mature. For example, female Atlantic sturgeon from the Gulf of St. Lawrence, New Brunswick typically mature at 27 to 28 years of age (Smith and Clugston 1997). Once mature, these fish are intermittent spawners reproducing once every 4 to 10 years depending on the taxon (Auer 1996; Birstein et al. 1997). As a result, populations grow very slowly, have low levels of recruitment, and take long periods of time to recover or to benefit from conservation efforts.

Today, all Acipenseriformes in North America are protected by State, Federal, and Provincial law, and commercial, ceremonial, and recreational harvests are limited to only the most robust populations (Auer 1999). Their unique life histories, long generation times, and intermittent spawning means that the types of genetic changes expected to occur (i.e. bottlenecking and inbreeding) will do so slowly over a period of greater than 100 years. Likewise, the genetic effects of rehabilitation and restoration efforts, either positive or negative, will not be apparent for at least a century after their implementation. This necessitates a precautionary approach to management based on sound research. Genetic provisions in the culture of Acipenserids for restocking and enhancement purposes should be mandatory. To this end, Kincaid et al. (1999) have initiated a national fish strain registry for paddlefish and sturgeon that catalogues all species, populations, and strains that are currently being reared in North America. Unfortunately, cultured strains of sturgeon and paddlefish may someday prove to be essential in our efforts to rehabilitate natural populations. If this unfortunate situation arises, as it already has in many European sturgeon rivers, efforts would be assisted if a clear and traceable pedigree were available for cultured strains.

10.9 References

Appels, R., Morris, R., Gill, B.S., and May, C.E. 1998. Chromosome Biology. Kluwer Academic Publishers, Boston. 401 pp.

Artyukhin, E. and Vecsei, P. 1999. On the status of Atlantic sturgeon: conspecificity of European *Acipenser sturio* and North American *Acipenser oxyrinchus*. J. Appl. Ichthyol. 15: 35-37.

Auer, N.A. 1996. Importance of habitat and migration to sturgeons with emphasis on lake sturgeon. Can. J. Fish. Aquat. Sci. 53 (Suppl.1): 152-160.

Auer, N. 1999. Lake sturgeon: a unique and imperiled species in the Great Lakes. In: Great Lakes fishery policy and management: A binational perspective. Taylor, W.W., and Ferreri, P. Editors. Michigan State University Press, East Lansing, Michigan. 551 pp.

Bemis, W.E, Findeis, E.K., and Grande, L. 1997. An overview of Acipenseriformes. Env. Biol. Fish. 48: 1-15.

Birstein, V.J. and Bemis, W.E. 1997. How many species are there within the genus Acipenser? Env. Biol. Fish. 48: 157-153.

Birstein, V.J. and DeSalle, R. 1998. Molecular phylogeny of Acipenserinae. Mol. Phylo.Evol. 9 (1): 141-155.

Birstein, V.J., and Vasiliev V.P. 1987. Tetraploid-octoploid relationships and karyological evolution in the order Acipenseriformes (Pisces): karyotypes, nulceoli, and nucleolus-organizer regions in four acipenserid specimens. Genetica 72: 3-12.

Birstein, V. J., Poletaev, A.I., and Goncharov, B.F. 1993. The DNA content in Eurasian sturgeon species determined by flow cytometry. Cytometry 14 :377-383.

Birstein, V.J., Hanner, R., and DeSalle, R. 1997. Phylogeny of the Acipenseriformes: cytogenetic and molecular approaches. Env. Biol. Fish. 48: 127-155.

Blacklidge, K.H. and Bidwell, C.A. 1993. Three ploidy levels indicated by genome quantification in Acipenseriformes of North America. Journal of Heredity, 84(6): 427-431.

Brown, J.R., Beckenback, A.T., and Smith, M.J. 1992a. Influence of Pleistocene glaciations and human intervention upon mitochondrial DNA diversity in white sturgeon (*Acipenser transmontanus*) populations. Can. J. Fish. Aquat. Sci. 49: 358-367.

Brown, J.R., Beckenback, A.T., and Smith, M.J. 1992b. Mitochondrial DNA length variation and heteroplasmy in populations of white sturgeon (*Acipenser transmontanus*). Genetics 132: 221-228.

Brown, J.R., Beckenback, K., Beckenbach, A.T., and Smith, M.J. 1996. Length variation, heteroplasmy and sequence divergence in the mitochondrial DNA of four species of sturgeon (Acipenser). Genetics. 142: 525-535.

Bruch, R.M. 1999. Management of lake sturgeon on the Winnebago System – long term impacts of harvest and regulations on population structure. J. Appl. Ichthyol. 15: 142-152.

Burnham, C.R. 1962. Discussions in Cytogenetics. Burgess Pub. St. Paul Minnesota.

Campton, D.E., Bass, A.L., Chapman, F.A., and Bowen, B.W. 2000. Genetic distinction of pallid, shovelnose, and Alabama sturgeon: emerging species and the US endangered species act. Conservation Genetics. 1: 17-32.

Carlson, D.M., Pflieger, W.L., Trial, L., and Haverland, P.S. 1985. Distribution, biology, and hybridization of *Scaphirhynchus albus* and *S. platorynchus* in the Missouri and Mississippi Rivers. Envir. Biol. Fish. 14: 51-59.

Choudhury, A., and Dick, T.A. 1998. The historical biogeography of sturgeons (Osteichthyes: Acipenseridae): a synthesis of phylogenetics, palaeontology and palaeogeography. Journal of Biogeography 25: 623-640.

Comincini, S., Lanfredi, M., Rossi, R., and Fontana, F. 1998. Use of RADP markers to determine the genetic relationships among sturgeons (Acipenseridae, Pisces). Fisheries Science 64 (1): 35-38.

Congiu, L., Dupanloup, I., Patarnello, T., Fontana, F., Rossi, R., Arlatis, G., and Zane, L. 2001. Identification of interspecific hybrids by amplified fragment length polymorphism: the case of sturgeon. Mol. Ecol. 10: 2355-2359.

Congiu, L., Fontana, F., Patarnello, T., Rossi, R., and Zane, L. 2002. The use of AFLP in sturgeon identification. J. Appl. Ichthyol. 18: 286-289.

Cornuet, J.M. and Luikhart, G. 1996. Description and power analysis of two tests for detecting recent population bottlenecks from allele frequency data. Genetics 144 (4): 2001-2014.

Dingerkus, G., and Howell, W.H. 1976. Karyotypic analysis and evidence of tetraploidy in the North American paddlefish, *Polyodon spathula*. Science 177: 664-669.

Ferguson, M.M. and Duckworth, G.A. 1997. The status and distribution of lake sturgeon, *Acipenser fulvescens*, in the Canadian provinces of Manitoba, Ontario and Quebec: a genetic perspective. Env. Biol. Fish. 48: 299-309.

Ferguson, M.M., Bernatchez, L., Gatt, M., Konkle, B.R., Lee, S., Malott, M.L., and McKinley, R.S. 1993. Distribution of mitochondrial DNA variation in lake sturgeon (*Acipenser fulvescens*) from the Moose River Basin, Ontario, Canada. J. Fish. Biol. 43 (Suppl. A): 91-101.

Fontana, F. 1994. Chromosomal nucleolar organizer regions in four sturgeon species as markers of karotype evolution in Acipenseriformes (Pisces). Genome 37: 888-892.

Fontana, F. 2002. A cytogenetic approach to the study of taxonomy and evolution in sturgeons. J. Appl. Ichthyol. 18: 226-233. Fontana, F., Tagliavini, J., and Congiu, L. 2001. Sturgeon genetics and cytogenetics: recent advancements and perspectives. Genetica 111: 359-373.

Fontana, F., Lanfredi, M., Chicca, M., Aiello, V. and Rossi, R. 1998. Localization of the repetitive telomeric sequence (TTAGGG) in four sturgeon species. Chromosome Research 6: 303-306.

Fontana, F., Rossi, R., Lanfredi, M., Arlati, G. and Bronzi, P. 1997. Cytogenetic characterization of cell lines from three sturgeon species. Caryologia. 50 (1): 91-95.

Fontana, F., Rossi, R., Lanfredi, M., Arlati, G., and Bronzi, P. 1999. Chromosome banding in sturgeons. J. Appl. Ichthyol. 15: 9-11.

Grande, L., and Bemis, W.E. 1991. Osteology and phylogenetic relationships of fossil and recent paddlefishes (Polyodontidae) with comments on the interrelationships of Acipenseriformes. J. Vert. Paleo. 11 (Suppl.1): 1-121.

Grunwald, C., Stabile, J., Waldman, J., Gross, R., and Wirgin, I. 2002. Population genetics of shortnose sturgeon *Acipenser brevirostrum* based on mitochondrial DNA control region sequences. Mol. Ecol. 11: 1885-1898.

Guenette, S., Fortin, R., and Rassart, E. 1993. Mitochondrial DNA variation in lake sturgeon (*Acipenser fulvescens*) from the St. Lawrence River and James Bay Drainage Basins in Quebec, Canada. Can. J. Fish. Aquat. Sci. 50: 659-664.

Kincaid, H.L., Mengel, L.J., Gray, M.J., and Brimm, S. 1999. National fish strain registry: paddlefish and sturgeon (NFSR-PS) species tables on reported populations and broodstocks. U.S. Geological Survey Research and Development Laboratory, Biological Resources Division, Wellsboro PA. 85 pp.

King, T.L., Lubinski, B.A., and Spidle, A.P. 2001. Microsatellite DNA variation in Atlantic sturgeon (*Acipenser oxyrinchus oxyrinchus*) and cross-species amplification in the Acipenseridae. Journal of Conservation Genetics 2:103-119.

Krieger, J., Fuerst, P.A., and Cavender, T.M. 2000. Phylogenetic relationships of the North American sturgeons (Order Acipenseriformes) based on mitochondrial DNA sequences. Mol. Phylo. Evol. 16 (1): 64-72.

Li, M.F., Marrayatt, V., Annand, C., and Odense, P. 1985. Fish cell culture: two newly developed cell lines from Atlantic sturgeon (*Acipenser oxyrinchus oxyrinchus*) and guppy (*Poecilia reticulata*). Can. J. Zool. 63: 2867-2874.

Ludwig, A., May, B., Debus, L., and Jenneckens, Ingo. 2000. Heteroplasmy in the mtDNA control region of sturgeon (Acipenser, Huso, and Scaphirhynchus). Genetics 156: 1933-1947.

Ludwig, A., Belfiore, N.M., Pitra, C., Svirsky, V., and Jenneckens, I. 2001. Genome duplication events and functional reduction of ploidy levels in sturgeon (Acipenser, Huso, and Scaphirhynchus). Genetics 158: 1203-1215.

McQuown, E.C., Gall, G.A.E., and May, B. 2002. Characterization and inheritance of six microsatellite loci in lake sturgeon. Trans. Amer. Fish. Soc. 131: 299-307.

Nichols, K.M., Young, W.P., Danzmann, R.G., Robison, B.D., Rexroad, C., Noakes, M., Phillips, R.B., Bentzen, P., Spies, I., Knudsen, K., Allendorf, F.W., Cunningham, B.M., Brunelli, J., Zhang, H., Ristow, S., Drew, R., Brown, K.H., Wheeler, P.A., and Thorgaard, G.H. 2003. A consolidated linkage map for rainbow trout (Oncorhynchus mykiss). Animal Genetics 34: 102-115.

Ohno, S., Muramoto, J., Stenius, C., Christian, L. and Kittereli, W.A. 1969. Microchromosomes in holocephalian, chondrostean, and holostean fishes. Chromosoma 226: 35-40.

Ong, T.L., Stabile, J., Wiring, I., and Waldman, J.R. 1996. Genetic divergence between *Acipenser oxyrinchus oxyrinchus* and *A. o. desotoi* as assessed by mitochondrial DNA sequencing analysis. Copeia 2: 464-469.

Phelps, S.R., and Allendorf, F.W. 1983. Genetic identity of pallid and shovelnose sturgeon (*Scaphirhynchus albus* and *S. platoryhchus*). Copeia. 1983: 696-700

Pyatskowit, J.D., Krueger, C.C., Kincaid, H.L., and May, B. 2001. Inheritance of microsatellite loci in the polyploid lake sturgeon (*Acipenser fulvescens*). Genome 44: 185-191.

Quattro, J.M., Greig, T.W., Coykendall, D.K., Bowen, B.W., and Baldwin, J.D. 2002. Genetic issues in aquatic species management: the shortnose sturgeon (*Acipenser brevirostrum*) in the southeastern United States. Cons. Genet. 3: 155-166.

Rodzen, J.A., and May, B. 2002. Inheritance of microsatellite loci in the white sturgeon (*Acipenser transmontanus*). Genome 45: 1064-1076.

Rusak, J.A., and Mosindy, T. 1997. Seasonal movements of lake sturgeon in Lake of the Woods and Rainy River, Ontario. Can. J. Zool. 74: 383-395.

Sakamoto, T., Danzmann, R., Gharbi, K., Howard, P., Ozaki, A., Kean Khoo, S., Woram, R., Okamoto, N., Ferguson, M., Holm, L., Guyomard, R., and Hoyheim, B. 2000. A microsatellite linkage map of rainbow trout characterized by large sex-specific differences in recombination rates. Genetics 155: 1331-1345.

Smith, C.T., Nelson, R.J., Pollard, S., Rubidge, E., McKay, S.J., Rodzen, J., May, B., and Koop, B. 2002. Population genetic analysis of white sturgeon (*Acipenser fulvescens*) in the Fraser River. J. Appl. Ichthyol. 18: 307-312.

Smith, T.I.J., Clugston, J.P. 1997. Status and management of Atlantic sturgeon, *Acipenser oxyrinchus*, in North America. Env. Biol. Fish. 48: 335-346.

Stabile, J., Waldman, J.R., Parauka, F., and Wirgin, I. 1996. Stock structure and homing fidelity of Gulf of Mexico sturgeon (*Acipenser oxyrinchus desotoi*) based on RFLP and sequence analysis of mitochondrial DNA. Genetics 144: 767-775.

Tagliavini, J., Cinterio, F., Gondolfi, G., and Fontana, F. 1999. Mitochondrial DNA sequences of six sturgeon species and phylogenetic relationships within Acipenseridae. J. Appl. Ichthyol. 15: 17-22.

Van Eenennaam, A.L., Murray, J.D., and Medrano, J.F. 1998. Mitotic analysis of the North American white sturgeon, *Acipenser transmontanus* Richardson (Pisces, Acipenseridae), a fish with a very high chromosome number. Genome. 41: 266-271.

Van Eenennaam, A.L., Murray, J.D., and Medrano, J.F. 1999. Karyotype of the American green sturgeon. Trans. Amer. Fish. Soc. 128 (1): 175-177.

Vasiliev, V.P. 1999. Polyploidization by reticular speciation in acipenseriform evolution: a working hypothesis. J. Appl. Ichthyol. 15: 29-31.

Waldman, J.R., Hart, J.T., and Wirgin, I.I. 1996a. Stock composition of the New York Bight Atlantic sturgeon fishery based on analysis of mitochondrial DNA. Trans. Amer. Fish. Soc. 125: 364-371.

Waldman, J.R., Nolan, K., Hart, J.T., and Wirgin, I.I. 1996b. Genetic differentiation of three key anadromous fish populations of the Hudson River. Estuaries 19: 759-768.

Waldman, J.R., Grunwald, C., Stabile, J., and Wirgin, I. 2002. Impacts of life history and biogeography on the genetic stock structure of Atlantic sturgeon *Acipenser oxyrinchus oxyrinchus*, Gulf sturgeon *A. oxyrinchus desotoi*, and shortnose sturgeon *A. brevirostrum*. J. Appl. Ichthyol. 18: 509-518.

Wirgin, I.., Stabile, J.E., and Waldman, J.R. 1997. Molecular analysis in the conservation of sturgeons and paddlefish. Env. Biol. Fish. 48: 385-398.

Wirgin, I., Waldman, J., Stabile, J., Lubinski, B., and King, T. 2002. Comparison of mitochondrial DNA control region sequence and microsatellite DNA analyses in estimating population structure and gene flow rates in Atlantic sturgeon *Acipenser oxyrinchus*. J. Appl. Ichthyol. 18: 313-319.

Wirgin, I., Waldman, J.R., Rosko, J., Gross, R., Collins, M.R., Rogers, S.G., and Stabile, J. 2000. Genetic structure of Atlantic sturgeon populations based on mitichondrial DNA control region sequences. Trans. Amer. Fish. Soc. 129: 476-486.

CHAPTER 11

MULTIJURISDICTIONAL MANAGEMENT OF LAKE STURGEON IN THE GREAT LAKES AND ST. LAWRENCE RIVER

NANCY J. LÉONARD[1], WILLIAM W. TAYLOR[1], AND CHRISTOPHER GODDARD[2]

[1]*Michigan State University, Department of Fisheries & Wildlife, 13 Natural Resource Building, East Lansing, Michigan 48824.*

[2]*Great Lakes Fishery Commission, 2100 Commonwealth Blvd, suite 100, Ann Arbor, Michigan 48105.*

11.1 Introduction

The presence of multiple entities with management authority of a shared natural resource, such as the lake sturgeon, can easily result in confusing and often opposing management strategies. Jurisdictional stress can also occur on a resource when there are numerous and different pressures being applied to the shared resource, resulting in the demands exceeding the capacity of the resource (Ferreri et al. 1999). These problems, however, can be minimized through integrated multijurisdictional management of the resource, which requires that the jurisdictions, whether within a country or between countries, work together to manage the resource. When a successful cooperative management of a shared resource is attained, this can improve the management of a shared fishery resource as well as achieving the most efficient and effective use of the limited fiscal and personnel resources available to each jurisdiction.

As a result of the extensive distributional range of lake sturgeon in the Great Lakes-St. Lawrence River system, numerous federal, provincial, state, and aboriginal jurisdictions have fisheries management authority for the species in the basin. Each of these jurisdictions has its own regulations and management plans for lake sturgeon, and this has resulted in differing management actions to protect the species. As a result of the multitude of management plans, lake sturgeon populations in the Great Lakes-St. Lawrence River currently support a variety of recreational fishing, aboriginal subsistence fishing, and a limited lake sturgeon commercial harvest (Todd 1999) in which both non-aboriginal and aboriginal fishers participate while in the other areas harvest is prohibited (e.g., Lloyd Mohr, Assessment Biologist Upper Great Lakes Management Unit, Ontario Ministry of Natural Resources, personal communication).

Reflecting the myriad of management plans, the opinions of fisheries managers in the Great Lakes-St. Lawrence River system are divided on whether or not the lake sturgeon populations are sufficiently large to support the current recreational fishing

G.T.O. LeBreton et al. (eds.), Sturgeons and Paddlefish of North America, 231–251.
© 2004 *Kluwer Academic Publishers. Printed in the Netherlands.*

and commercial fishing harvest levels. Nonetheless, the vast majority of managers in the Great Lakes basin do appear to share the concern that the future of lake sturgeon may be threatened by the continuation of the current harvest level. Furthermore, there is significant worry about a potential increase in harvest in response to the international demand for sturgeon products (Hoover 1999). This chapter will examine some of the regulatory features that enhanced the survival of lake sturgeon in the Great Lakes-St. Lawrence River ecosystem, the growing interest in managing lake sturgeon using an integrated, consensual, multijurisdictional approach, and the impact of the international market on this fishery and rehabilitation initiatives.

11.2 Lake Sturgeon in the Great Lakes-St. Lawrence River

The abundance of lake sturgeon in the Great Lakes-St. Lawrence River was severely depressed as a result of intense harvesting during the mid-1800s for their caviar and flesh (Scott and Crossman 1998; Ferguson and Duckworth 1997; see chapters by Saffron and Auer), and by human induced changes within the Great Lakes-St. Lawrence watershed that negatively impacted lake sturgeon populations (Scott and Crossman 1998; Stegemann 1994; Moisan and Laflamme 1999; see chapter by Auer). With the drastic decrease in lake sturgeon abundance, most fisheries managers and ecologists believed in the early to mid 1900s that lake sturgeon would eventually disappear as a result of compounding negative pressures (Smith 1968). However, lake sturgeon in the Great Lakes proved to be more resilient than previously assumed, and despite the polluted waters, loss of habitat, and overharvesting pressures; populations of lake sturgeon persist throughout the Great Lakes and St. Lawrence River region (e.g. Houston 1987; Fortin et al 1993; Thomas and Haas 2002).

In addition to the work undertaken by fisheries management agencies, the passage of regulations likely played an important role in enabling lake sturgeon populations to persist. These include: the enactment of federal, provincial, and state laws aimed at protecting endangered species, and improvement of water quality through the International Joint Commission.

11.2.1 INTERNATIONAL JOINT COMMISSION

The International Joint Commission (IJC) was established by the *Boundary Waters Treaty* in 1909, to provide the principles and mechanisms to resolve and prevent disputes about water quantity and quality along the boundary between Canada and the United States, including in the Great Lakes Region (IJC 1989). The role of the IJC in restoring and maintaining the chemical, physical, and biological integrity of the waters of the Great Lakes basin ecosystem was increased significantly in 1972 and again in 1978, with the signing of the *Great Lakes Water Quality Agreement* (Beeton et al. 1999; IJC 1989).

The 1978 *Great Lakes Water Quality Agreement* (GLWQA) proposed to maintain the upper Great Lakes' higher water quality standards and to restore and enhance the water quality in the lower Great Lakes. To achieve these goals the GLWQA focused on diminishing anthropogenic eutrophication and the concentration level of toxic substances in the Great Lakes. To reduce anthropogenic eutrophication, phosphorous inputs were successfully lowered by building and improving wastewater treatment plants in Canada and United States and by reducing non-point source pollution (Beeton et al. 1999). To decrease the levels of persistent organic chemicals and heavy metals in the Great Lakes, these types of chemicals were banned from the region (Environmental Canada 1991; DeVault et al. 1994). Subsequent to the bans, the concentrations of these contaminants in the Great Lakes' biota, including fish, decreased rapidly and markedly (DeVault et al. 1994).

Similarly to other Great Lakes biota, lake sturgeon were impacted by the degradation of the Great Lakes habitat and water quality (Smith 1972). Lake sturgeon foraging and spawning sites most likely were negatively affected by logging, urbanization and agricultural activities that increased sediment and nutrient load input into the lakes' tributaries and into the Great Lakes. This reduction in water quality combined with the increase in sediments altered the benthic community, covered fish spawning area with silt, and led to the depletion of dissolved oxygen (Beeton et al. 1999). In addition, lake sturgeon, being a long-lived benthivore with high lipid content, is susceptible to the bioaccumulation of contaminants (Moisan and LaFlamme 1999). Thus, the presence of biologically toxic contaminants in the water and sediments of the Great Lakes, which cause reproductive failure and negatively affect the survival and reproductive success of fish (IJC 1991; USEPA 1992), could have impaired lake sturgeon. Indeed, Doyon et al. (1999) found that the percentage of lake sturgeon physical deformities was higher in lake sturgeon residing in polluted waterways than in non-polluted system, and that these deformities affected the offspring of adult sturgeons living in these polluted areas. Thus, the water quality improvements likely benefited populations of the lake sturgeon in the Great Lakes-St. Lawrence River basin, by improving the quality of habitat used for spawning and foraging in addition to decreasing the level of biologically toxic chemicals and their impacts on the lake sturgeon.

11.2.2 PROTECTION OF ENDANGERED/THREATENED SPECIES

11.2.2.1 The Endangered Species Act of the United States

The United States passed the *Endangered Species Act* (ESA) in 1973 (USFWS 1996; Czech and Krausman 2001), which aims to "provide a means whereby the ecosystems upon which endangered species and threatened species depend may be conserved, [and] to provide a program for the conservation of such endangered species and threatened species" (USFWS 1988). In this act, conservation refers to "the use of all methods and procedures which are necessary to bring any endangered species or threatened species to the point at which the measures provided pursuant to this ESA are no longer necessary" (USFWS 1996).

Presently, the ESA may be the most comprehensive and powerful species protection act worldwide (Reffalt 1991; Stanford Environmental Law Society 2001). One of the strengths of this act is that it requires all federal agencies to actively work towards the conservation of listed species, to develop and implement conservation programs for endangered and threatened species (Stanford Environmental Law Society 2001), and to provide federal funds to States, under a cooperative agreement with the Secretary of the Interior, to establish and maintain active programs for the conservation of endangered species and threatened species (Kohm 1991; USFWS 1996). Although, the ESA currently addresses the conservation of species, many managers see the need to further increase the protection of genetically unique fish strains within a species (e.g. Lowie 2000).

In 1982, lake sturgeon were being considered as potential candidates for protection under the ESA. As such, lake sturgeon were listed in the United States Federal Notices of Review Register as a Category 2 (C2) species (Klar and Schleen 2001). This category is assigned to a species that is being considered for potential listing as either a threatened or endangered, but for which more information is needed about the species status prior to making the decision of whether or not to include the species on the list. The C2 listing of lakes sturgeon was maintained through 1994 as indicated in the 1994 Federal Notice of Review Register (vol. 59, no. 219, Dept. of Interior F.W.S. 50-CFR-part 17). but in 1995 lake sturgeon were listed as a species at risk (Klar and Schleen 2001), thus the species does not receive protection under the ESA. The decision not to list lake sturgeon was based on the assessment that there was a minimal threat to the species survival at that time (Charles Wooley, Assistant Regional Director, U.S. Fish and Wildlife Service, personal communication). Although lake sturgeon are not listed under the ESA, the inclusion of lake sturgeon in the federal notice for twelve years as a C2 species, and the resulting possibility that lake sturgeon would be listed, most likely provided the incentive for all agencies to more closely monitor the health and status of their lake sturgeon populations, and to provide appropriate management plans for their restoration. Currently, several states list lake sturgeon as being either threatened or endangered within their jurisdiction, based on lake sturgeons' population decline and current abundance in their waters (Table 11-1).

*Table 11-1. Lake sturgeon status by jurisdiction. *Species of Concern - an informal term indicating that the United States Fish and Wildlife Service has some degree of concern for the future well-being of the taxon, but the taxon does not receive any protection under the Endangered Species Act.*
***Appendix II-lists species that could become rare or endangered if the international trade is not regulated.*

United States		Canada		Global	
Jurisdiction	**Status**	**Jurisdiction**	**Status**	**Jurisdiction**	**Status**
Federal level (USFWS)	Species of concern*	Federal level (COSEWIC)	Not at risk	CITES	Listed in Appendix II**
Pennsylvania	Endangered	Ontario	Not listed		
New York State	Threatened	Québec	Not listed		
Michigan	Threatened				
Illinois	Threatened				
Indiana	Endangered				
Ohio	Endangered				
Wisconsin	Not listed				
Minnesota	Special concern				

11.2.2.2 Canada's COSEWIC and SARA

In Canada, the provincial and federal governments both have responsibility for the fish and the fisheries within all inland water bodies such as the Great Lakes-St. Lawrence River (Thompson 1974). The provincial governments of Ontario and Quebec have the authority to regulate the licensing of fishing within their provinces (Thompson 1974), as outlined by the Ontario Fish Licensing Regulations (Fish and Wildlife Conservation Act, 1997, Ontario regulation 664/98, amended to o. reg. 546/00, fish licensing), and the Québec Fishing Activities Regulation (An Act respecting the Conservation and Development of Wildlife, R.S.Q. C-61.1; Fishing Activities Regulation, R.R.Q. C-61.1, r.0.00001.1). The Canadian federal government regulates the fishery for its conservation and protection through the Federal Fisheries Act. Generally, however, the provinces of Ontario and Quebec, on behalf of the federal government, will enforce the provisions of the Fisheries Act within their territory. Nonetheless, even though the provincial and federal governments share the responsibility for inland fish and fisheries, the federal law can override the provincial law if the federal government believes that the health of the Great Lakes-St. Lawrence River fishery is threatened by the existing provincial law (Dochoda 1999).

In 1977, the federal government established the Committee on the Status of Endangered Wildlife (COSEWIC) in Canada to aid in the protection of the fauna and flora (Rishikof 1997; COSEWIC 2002a, b; Egan 2002). This committee has the responsibility of assessing "the status of wild Canadian species, subspecies, and separate populations suspected of being at risk...[based on] the best up-to-date scientific information and Aboriginal traditional knowledge available" (COSEWIC 2002a), and lists species that it assesses as being at risk in terms of its survival within Canada (Freedman et al. 2001). Listing of species by COSEWIC is based on several criteria including that the species must be native to Canada, its abundance, and how dependent the species' survival is on habitat located within Canada (COSEWIC 2003). In addition, the Canadian Wildlife Service (CWS), to assist with the efforts of protecting the species listed by COSEWIC, created an endangered species unit in 1978 (Bocking 2001), and then enacted the Recovery of Nationally Endangered Wildlife (RENEW) program in 1988 (CWS n.d.; Freedman et al. 2001). The RENEW program was implemented to coordinate the recovery efforts between federal and provincial government agencies and non-governmental organizations, for the species listed by COSEWIC (Bocking 2001).

More recently, after many failed attempts, the federal Bill C-5, named the *Species at Risk Act* (SARA), passed through the Canadian House of Commons in 2002 (HC 2002), with two-thirds of SARA being in effect by 5 June 2003 and with a projected date of 1 June 2004 for the entirety of SARA to come in effect (http://www.sararegistry.gc.ca/gen_info/HTML/approach_e.cfm, Environment Canada 2002). SARA is an act "respecting the protection of wildlife species at risk in Canada" (HC 2002) and "to protect wildlife at risk from becoming extinct or lost from the wild, with the ultimate objective of helping their numbers to recover" (CWS 2002). The endangered species to be listed under SARA consist of species recommended by the COSEWIC, but will not necessarily include all species listed by COSEWIC prior to the enactment of SARA. Once a species is listed under SARA, a scientific assessment of the species is mandatory, including the identification and protection of the species' habitat, developing recovery strategies, and management plans.

On a provincial level, the COSEWIC list does not have legal standing (Freedman et al. 2001; Rideout and Ritter 2002) such that provincial governments are not obligated to recognize the listing (Lindgren 2001; COSEWIC 2002a,b). Instead of adopting the COSEWIC list, most provinces have developed their own lists (Lindgren 2001; Freedman et al. 2001), and have enacted their own laws for endangered species. In the case of Ontario and Québec, these provinces have each enacted an act that protects both the critical habitat and the listed species, which are, respectively, the 1971 *Ontario Endangered Species Act* and the 1989 Québec *Act Respecting Threatened or Vulnerable Species*. In addition to these acts, both provinces have other rules and programs that could also apply to endangered species (Gauthier and Wiken 2001).

Currently, lake sturgeon populations within the Canadian jurisdiction of the Great Lakes-St. Lawrence River are not on the federal COSEWIC list, because COSEWIC does not believe the species to be at risk. Similarly, neither Ontario nor Quebec has assigned a special status to lake sturgeon under their provincial legislation. Although not included on the federal and provincial lists, the lake sturgeon does fall under the Canadian Federal Fisheries Act and Provincial Fishery Regulations. Thus this species receives some protection through these fisheries acts, which manages and protects fish species through the use of fishing permits, harvest quotas and fishing season (Houston 1987).

The possibility exists that lake sturgeon residing within the Great Lakes-St. Lawrence River may eventually be assigned a status in these two provinces. Québec is currently considering listing lake sturgeon as either a threatened or vulnerable species (Moisan and Laflamme 1999; SFPQ 2001). Whereas, in the Ontario waters of Lake Huron some sections are considered by fisheries managers to have rare or threatened lake sturgeon populations, but these populations not currently assigned an official status (Fielder et al. 2003). If lake sturgeon were to be included under the current Ontario and Québec lists pertaining to endangered species, this would allow for: 1) the protection, restoration, or creation of new habitat for the species 2) enforcement of the act in order to protect the species and its habitat by the respective ministry, and additionally, under the Quebec Act: 3) allow for agreements to be made between the provincial government and either individuals or other government, in order to attain the objective of protecting the listed species, 4) allow for research to be conducted on the species, and 5) provide for mitigation of damage to the species and its habitat. Although the provinces are presently investing some effort in protecting and managing the lake sturgeon, officially listing the lake sturgeon may augment the provinces' effort, such as by modifying harvest as needed, increasing the provinces' incentives to protect or expand the amount of habitat available, and increasing our knowledge of the species and its ecological needs.

11.3 Successes of Cooperative Management for Lake Sturgeon

When a resource is shared between multiple jurisdictions, conflicting viewpoints and resource uses may arise. Under integrated multijurisdictional management, these types of counterproductive actions can be eliminated or at least modified to assure that the overall goals are met. An example of an ongoing cooperative venture is the work being done to address the negative impacts of dams on migrating sturgeons. For example the USFWS and its partners are developing sturgeon fishways to bypass dams; the National Fish Passage Program is being implemented to remove unnecessary dams and/or installing fishways to decrease the impact of existing dams on migrating fish species; and the GLFC and its partners aim to reduce the impact of the sea lamprey control program on sturgeon.

11.3.1 BARRIERS AND FISHWAYS

Multiple stakeholders, such as wildlife populations, humans, and industries often share streams, rivers and lakes, thus resulting in the overlap of jurisdictions between agencies and organizations that are trying to meet these various needs. An example of conflicting use between lake sturgeon and other stakeholders is the construction of dams for reservoirs, flood control, water retention, hydroelectric power, and recreational purposes. These dams, in addition to affecting water levels and quality, generally pose formidable obstacles to migratory aquatic species, and in the case of the lake sturgeon, block migrating lake sturgeon from reaching upstream spawning sites. A wide variety of fish ladders or fishways have been installed to allow fish migration over these obstacles (Scheidegger 2002). These fishways, however, have been primarily designed to accommodate jumping species such as salmon (Gowans et al. 1999, Laine et al. 2002; Bunt et al. 1999, Bunt et al. 2000). As such, fishways have not been very effective in facilitating upstream migration of lake sturgeon above dams (Great Lakes Basin Ecosystem Team Lakes Sturgeon Committee n.d.; USFWS-Green Bay n.d.). However, research is being conducted to investigate how fishway designs can be modified to better accommodate lake sturgeon (Peake et al. 1997). A spiral fishway design for lake sturgeon is being tested at the S. O. Conte Anadromous Fish Research Centre-USGS (Kynard et al. 2003), and the USFWS hopes to try out a prototype of this fishway in 2005 at the White Rapids Hydroelectric Project on the Menominee River (Janet Smith, Field Supervisor, USFWS Green Bay ES Field Office, personal communication). Through cooperation, the impact of dams may be reduced or eliminated for some lake sturgeon populations and thus be less of an impediment to the lake sturgeon rehabilitation and management effort

11.3.2 SEA LAMPREY CONTROL PROGRAM

The implementation of the sea lamprey control program by the GLFC and its partners, while positive in regards to reducing the impact of sea lamprey induced mortality, can have the potential to negatively impact lake sturgeon. For instance, the sea lamprey control program uses low-head barriers to prevent migrating sea lamprey from reaching spawning grounds, the release of sterilized males to decrease the reproductive success, and the application of the specific chemical lampricide 3-trifluoromethyl-4-nitrophenol (TFM) in Great Lakes tributaries to kill sea lamprey larvae (Weisser 2000, GLFC 2002). Some of these techniques, while effective in reducing sea lamprey abundance, can negatively affect the spawning success of lake sturgeon by blocking migrating lake sturgeon from reaching upstream spawning grounds, and by killing larval lake sturgeon through the application of TFM. The GLFC and its partners have worked hard at ensuring that these potential impacts be minimized or eliminated. As such the GLFC and its partners have expended considerable effort to investigate ways to minimize the impact of sea lamprey control, such as ensuring improved fish passage over the lamprey barriers (GLFC 2000a), and implementing TFM treatment protocols that contain less chemical and are not applied during lake sturgeon spawning migration and incubation times (Hay-Chmielewski and Whelan 1997; Weisser 2000; Auer 2002). Therefore, the sea lamprey management program is a good example of how two potentially opposing management actions, reducing sea lamprey populations yet protecting and enhancing lake sturgeon populations, can be fulfilled through cooperation among all parties and applying holistic fishery management.

11.4 Current Management of Lake Sturgeon

Jurisdictions with management responsibilities for lake sturgeon in the Great Lakes-St. Lawrence River are multifold, including two countries, Canada and the United States, eight states, two provinces, and two aboriginal authorities (Table 11-2). Lake sturgeon are known to move across jurisdictional boundaries (Thomas and Haas 2002; Fielder et al. 2003), and although some of the agencies are beginning to cooperate under the auspice of the Lake Committees of the Great Lakes Fishery Commission, the multiple entities vested in managing lake sturgeon are largely, still making their final management decisions independently from the other entities. This jurisdictional network has lead to the development of numerous plans, and as shown in Table 11-3, regulations with seemingly opposing objectives that may prevent the fulfilment of basin-wide lake sturgeon management goals.

One of the potential outcomes of having each jurisdiction managing lake sturgeon in isolation of the other jurisdiction is that the eight states and the two provinces may differ in their decisions, for example there are differences in the status of lake sturgeon assigned by the province of Ontario and the eight states, even for lake sturgeon populations residing within the same lake. A possible explanation for this discrepancy may be that some lake sturgeon stocks of the Great Lakes may have access to more and healthier habitat than others and thus do not require as stringent regulations to maintain the population. Conversely, these discrepancies may arise due lack of information on the population that prevents managers from estimating the population size, differences in opinion as to what population size requires protective action, or may be related to the differing cultural and historic significance of lake sturgeon among the inhabitants of the basin. For instance, lake sturgeon are listed in seven of the eight Great Lakes states as being endangered, threatened, or of special concern; whereas lake sturgeon are not listed by the provinces as being at risk (Table 11-1). In addition, commercial fishing of lake sturgeon is not allowed in the United States, whereas in Canada, a limited commercial fishing is allowed within sections of the Ontario and Québec waters (Table 11-3). There is also a discrepancy between the state and provincial recreational fishing regulations for lake sturgeon. In provincial waters of the Great Lakes, recreational fishing of lake sturgeon is *only prohibited* in Lake Ontario; in the State waters of the Great Lakes recreational fishing is *only allowed* in Lake St. Clair (Table 11-3, tributaries of the Great Lakes may differ in their regulations). Even in lakes where recreational fishing is allowed by both the states and the provinces the quotas differ greatly, with the states having a highly restricted fishing limit, with one sturgeon per license quota, and the provinces currently having a one sturgeon per day quota.

Table 11-2. Federal, state, provincial, and aboriginal entity with authority over lake sturgeon in the Great Lakes-St. Lawrence River basin.

Jurisdiction		**Entity**
Federal	U.S.A.	United States Fish and Wildlife Service United States Geological Service – Biological Resource Division
	Canada	Department of Fisheries and Ocean
States	Illinois	Illinois Dept. of Natural Resources
	Indiana	Indiana Dept. of Natural Resources
	Michigan	Michigan Dept. of Natural Resources
	Minnesota	Minnesota Dept. of Natural Resources
	New York	NY Dept. Environmental Conservation
	Ohio	Ohio Dept. of Natural Resources – Lake Erie Unit
	Pennsylvania	Fish and Boat Commission
	Wisconsin	Wisconsin Depart. Natural Resources
Provinces	Ontario	Ontario Ministry of Natural Resource
	Québec	Société de la faune et parcs du Québec
Aboriginal	United States	CORA and member tribes which are the Bay Mills Indian Community, Sault Ste. Marie Tribe of Chippewa Indians, Grand Traverse Band of Ottawa and Chippewa Indians, Little River Band of Ottawa Indians and Little Traverse Bay Bands of Odawa Indians. GLIFWC and member tribes which are Bay Mills Indian Community, Keweenaw Bay Indian Community, Lac Vieux Desert band, Bad River, Lac Courte Oreilles, Lac du Flambeau, Mole Lake/Sokaogon, Red Cliff, St. Croix bands, Fond du Lac and Mille Lacs bands.
	Canada	Various First Nation tribes in the Great Lakes-St. Lawrence River basin but involvement is in the early stages as aboriginal rights are being defined in courts.

Further, in 2003 the United States stated that it would not engage in the commercial harvest or export of lake sturgeon from the Great Lakes and St. Lawrence River drainage, whereas Canada stated its expectation of commercial catch and export of lake sturgeon meat to be over 100t from the Great Lakes-St. Lawrence River drainage (CITES 2003a). The United States is managing most lake sturgeon populations as threatened, endangered or a species of concern, whereas Canada assumes that most populations are viable for a limited commercial harvest. These opposing views and actions will likely decrease the effectiveness of the actions taken by the United States to protect the species, and potentially lead to conflict between Great Lakes fisheries agencies in the future. However, with increasing knowledge of lake sturgeon populations, and sharing of data between jurisdictions such as recapture data on tagged sturgeon individuals, which is used for population estimates, these discrepancies in status and management plans may diminish or be resolved.

11.5 The Potential for Multijurisdictional Management of Lake Sturgeon

The majority of lake sturgeon fisheries managers appear to believe that, given the habitat requirements, movements and life history of lake sturgeon, this species may be best managed using an ecosystem or basin-wide approach rather than the jurisdictional approach. The implementation of this approach would result in decisions being based not on individual jurisdictional opinions but on what all involved parties believe is best for the shared resource (Ferreri et al. 1999). The growing interest among lake sturgeon managers in the Great Lakes-St. Lawrence River to expand cooperation throughout the basin, have led, in recent years, to an increase in the number of cooperative efforts being formed, such as the Central Great Lakes Bi-National Lake Sturgeon Group (McClain 1997; Hill and McClain 2002; USFWS 2003), the partnerships instigated by the United States Fish and Wildlife Service (USFWS n.d.), the Great Lakes lake sturgeon website (http://midwest.fws.gov/sturgeon/default2.htm), and the National Paddlefish and Sturgeon Committee (1993; Holey et al. 2000; Fielder et al. 2003). In addition, this effort has been reflected in the agenda of the GLFC through the Lake Committees by the inclusion of lake sturgeon within the Lake Committees' Fish Community Objectives.

The GLFC facilitates the implementation of *A Joint Strategic Plan for the Management of Great Lakes Fisheries* (JSP) through the Lake Committees, to which the eight Great Lakes States, the Province of Ontario, two aboriginal fisheries agencies and the federal governments of Canada and the United States are signatory parties (GLFC 1997). The JSP is generally recognized as one of the most effective vehicles for cooperative, international fisheries management in the world (GLFC 2000b). The JSP provides the framework for fisheries management decision-making based on sound science and consensus (GLFC 2000b), and is based on the philosophy that each of the Great lakes fisheries management agencies has a share in the Great Lakes ecosystem and that in the interest of the "common good," fisheries management authority must be cooperative, rather than unilateral. The strength of the JSP lies in its implementation through the Lake Committees. These committees have been established for each of the Great Lakes, including the upper portion of the St. Lawrence River, and consist of representatives from each of the state, provincial, and aboriginal agencies with management authority on each lake. Within the Lake Committees, fisheries management decisions, including the desired fish community structures, stocking strategies, law enforcement activities, and harvest rates, are determined collectively by consensus. Each management agency then reports to the Committee for implementing the shared decisions within its own jurisdiction. To help the Lake Committees in reaching a decision based on a common scientific knowledge of the status of the fisheries resources, the Lake Committees established Technical Sub-Committees, which adopted common fisheries assessment and analysis protocols, share fisheries assessment, research information, and databases. Each Lake Committee reports to the Council of Lake Committees, which is comprised of representatives from each of the Lake Committees, to resolve basin-wide Great Lakes fisheries issues.

The JSP is beginning to be applied to rehabilitate and manage lake sturgeon within the Great Lakes-St. Lawrence River, as such the lake sturgeon is now included in the Lake Committees' Fish Community Objectives, and depending on the Lake Committee, these Committees are at various stages of implementing joint management decisions for lake sturgeon. For instance, as of 2003, the Lake Michigan Committee has established a lake sturgeon task force (LMC 2003); the Lake Huron Committee is in the process of developing a lake sturgeon management plan (LHC 2003); and the Lake Superior Committee and its Technical Committee formed the Lake Sturgeon Subcommittee in 1994 to evaluate the status of lake sturgeon, and in 2002 developed a rehabilitation plan for lake sturgeon in Lake Superior (Auer 2002).

Thus, if the current endeavours to initiate joint management of lake sturgeon prove successful and favourable to all the Parties of the JSP, then the Lake Committees and the Council for Lake Committees, under the umbrella of the GLFC, may become the forum for basin-wide, cooperative management of lake sturgeon within the Great Lakes. However, ideally, to encompass the entire range of lake sturgeon in the Great Lakes-St. Lawrence River, an agreement with the Province of Québec would need to be signed to include the province within the cooperative effort. The inclusion of the Province of Québec would allow management of lake sturgeon migrating between the Québec section of the St. Lawrence River and the Ontario-New York section of the river.

11.6 International Perspective on Lake Sturgeon

The survival and health of lake sturgeon in North America is no longer simply affected by the local or national demand for caviar and flesh, but also by a global demand. This demand for lake sturgeon products is increasing, as other sturgeon stocks in the northern hemisphere, such as the sturgeon species from the Caspian Sea, become depleted and lake sturgeon products are viewed as an acceptable alternative source. Currently, Russia and Iran are the main exporters of caviar from the Caspian Sea area, with nearly 80% of all caviar exports being supplied by these two countries (WWF 2002). The sturgeon stocks found within these countries, including the stellate sturgeon, Russian sturgeon, and *Huso huso*, have been decreasing. As in the Great Lakes, the decline in sturgeon populations from the Caspian Sea area has been blamed on a combination of habitat destruction, water pollution, and overfishing, the latter having been exacerbated by intensified poaching activities on these stocks for the caviar trade (CITES 2001). With the disintegration of the Union of Soviet Socialist Republics, the level of poaching in the former Soviet republics has increased to 10 to 12 times the amount of the current legal catch (CITES 2001). Should the Caspian Sea caviar export supply collapse, a gap between demand and supply would be created, which would likely increase the pressure for North American sturgeon and paddlefish products (Graham and Rasmussen 1997).

Table 11-3. Recreational fishing and commercial fishing regulations for each state and province located along the Great Lakes-St. Lawrence River.

			Recreational fishing			Commercial fishing		
State/Province	**Status**	**Water body**	**Open season**	**Catch limit**	**Size regulations**	**Open season**	**Catch limit**	**Size regulations**
Minnesota	Special concern	Lake Superior	None	N/A	N/A	None	N/A	N/A
Wisconsin	None	Lake Superior	None	N/A	N/A	None	N/A	N/A
		Lake Michigan	None	N/A	N/A	None	N/A	N/A
Illinois	Threatened	Lake Michigan	None	N/A	N/A	None	N/A	N/A
Indiana	Endangered	Lake Michigan	None	N/A	N/A	None	N/A	N/A
Ohio	Endangered	Lake Erie	None	N/A	N/A	None	N/A	N/A
Pennsylvania	Endangered	Lake Erie	None	N/A	N/A	None	N/A	N/A
New York	Threatened	Lake Erie	None	N/A	N/A	None	N/A	N/A
		Lake Ontario	None	N/A	N/A	None	N/A	N/A
Michigan	Threatened	Lake Superior	July 16 to Nov 30	Catch & release	N/A	None	N/A	N/A
		Lake Michigan	July 16 to Nov 30	Catch & release	N/A	None	N/A	N/A
		Lake Erie	July 16 to Nov 30	Catch & release	N/A	None	N/A	N/A
		Lake Huron	July 16 to Nov 30	Catch & release	N/A	None	N/A	N/A
	Threatened	Lake St. Clair	July 16 to Sep 30	1/license yr	>42”-< 50”	None	N/A	N/A

Table 11-3. (Continued)

State/Province	Status	Water body	Open season	Catch limit	Size regulations	Open season	Catch limit	Size regulations
Ontario[2]	None	Lake Superior	All year	1 per day	114 cm	None	N/A	N/A
		Lake Erie	All year	1 per day	114 cm	None	N/A	N/A
		Lake Huron	All year	1 per day	114 cm	Year round[3]	5661 kg from lake allowed	58cm dressed length
		Lake Ontario	None	N/A	N/A	None	N/A	N/A
		Lake St. Clair	All year	1 per day	114 cm	None[4]	N/A	N/A
		Upper St. Lawrence River	None	N/A	N/A	None	N/A	N/A
Québec	None	St. Lawrence River	Depends on zone	1 per day	>45cm	May 1 to Oct 15 Depends on zone	Varies e.g. 12173 sturgeon taken in 2002	>45cm

Aboriginal fishing regulations in Canada and United States sections of the Great Lakes may differ from State fishing regulations[1].

[1]Aboriginal (also called Tribal or First Nations) fishing regulations can differ between Canada and the United states as well as among tribes. As example, the CORA does not allow any harvest of lake sturgeon in the Great Lakes; whereas subsistence, ceremonial and commercial lake sturgeon fishing may be allowed by other tribes.

[2]The province of Ontario is considering changing the lake sturgeon recreational fishing quota to one fish per year for 2004 (David M. Reid, Lake Management Supervisor - Lake Huron, Upper Great Lakes Management Unit, Ontario Ministry of Natural Resources, personal communication).

[3]In the Province of Ontario all fishery can be closed by the middle of December, but generally are kept open year round.

[4]In the province of Ontario Lake St. Clair has one remaining commercial fisher, fishing with 2 licenses that will be appropriated with the fisher's retirement from the industry. These 2 licenses permit the fisher to use setlines (baited hooks) to harvest 2200 lb annually with no size restriction. Approximately 35-50 fish per year are harvested.

11.6.1 MANAGING A THREATENED AND ENDANGERED SPECIES GLOBALLY

The *Convention on International Trade of Endangered Species* (CITES), is based on the voluntary participation of numerous countries, and is currently the best option for the protection of species on a global level. The CITES was implemented in 1975 (Wijnstekers 2001), and as of 2003 there are 162 member countries (CITES 2003b). This convention was enacted to respond to the threat imposed by international trade on threatened or endangered species (Stanford Environmental Law Society 2001, Wijnstekers 2001). CITES lists species into three Appendices. Species listed in Appendix I are considered to be threatened with extinction; Appendix II species are not threatened with extinction but uncontrolled trade might threaten their existence; and species in Appendix III are currently being protected by at least one country which has requested the aid of CITES Parties in controlling the species trade (CITES 2003c).

CITES monitors the trade of species listed in the appendices through a system of export and import permits (Wijnstekers 2001). The use of these permits allows CITES to monitor the trade demand, to detect illegal shipments, and make adjustments to better protect the species if necessary, such as changing the trade quota for the species which are usually set by individual Parties or by the Conference of Parties (Wijnstekers 2001). The enforcement of CITES permits and quotas, due to the nature of the agreement, lies on the signatory countries (list of signatory countries available on the CITES website (www.cites.org; Wijnstekers 2001; Le Prestre and Stoett 2001). In the United States, CITES is implemented through the *Endangered Species Act* and the United States Fish and Wildlife Service is the designated management authority for CITES in the United States (Stanford Environmental Law Society 2001). In Canada, CITES is implemented through *The Wild Animal and Plant Protection and Regulation of International and Interprovincial Trade Act* (WAPPRIITA). The Canadian Wildlife Service, Department of Environment is the management and scientific Authority for CITES in Canada (CITES Secretariat 2003).

In April of 1998, all sturgeon and paddlefish species (Acipenseriformes) were listed in Appendix I or II of the CITES (Wijnstekers 2001; Raymakers and Hoover 2002). In listing all Acipenseriformes under Appendix I or II, this required that participating countries enforce the use of permits and certifications for all international trade involving these species. In addition to the usual CITES permits, the Parties agreed to several other aspects that they believed necessary for the conservation of Acipenseriformes, including the Conservation of Sturgeons, Resolution Conf. 10.12 (Wijnstekers 2001). This later agreement states that the fishery and other management initiatives for protection of Acipenseriformes should include the necessary national legislative improvements, the need for regional agreements, the development of a universal marking system for all fish parts to aid in tracking the origin of the products, the role of aquaculture, and the need to focus on decreasing/eradicating illegal trade of Acipenseriformes (Wijnstekers 2001). CITES also stipulated in decision 11.58, that to help protect Acipenseriformes, countries which share a common stock, such as the lake sturgeon in the Great Lakes, need to consult and agree on the annual catch and export quota for international trade for that shared stock (Meffe and Carroll 1994, Wijnstekers 2001).

Although CITES is an agreement among signatory countries that does not have enforcement authority, the signatory countries can cooperate together to employ tactics such as economic incentives, to pressure a member country that is not properly managing a listed species, to address the conservation issues occurring within that country's jurisdiction. An example of this is when the members of the CITES agreement became concerned about the dwindling Caspian Sea sturgeon stocks in June 2001. In an attempt to force the countries surrounding the Caspian Sea to address the declining stock abundance, the CITES Standing Committee recommended that imports of sturgeon products from these stocks be suspended in 2002, unless these countries took action in accord with the approved resolution (CITES 2003d). This tactic appeared to be successful for the Caspian region as the Russian, Azerbaijan and Kazakhstan governments announced on 21 July 2001 a temporary ban on commercial fishing of sturgeon in the Caspian Sea in response to international pressure (Anonymous 2001). This type of tactic, however, only works well if the main importing countries are CITES members, and if the loss of their market demand results in a significant decrease in demand for those products or alternative products are available.

11.7 Future Outlook

Lake sturgeon is a truly unique and important fish in the Great Lakes ecosystem. While its numbers have dwindled due to decades of pollution, habitat alterations, and overharvesting, the renewed commitment of fishery managers, aboriginal tribes, and other stakeholders to its restoration in the Great Lakes-St. Lawrence River makes the outlook for lake sturgeon rehabilitation very positive. With an increase in our understanding of the ecology of this species, along with habitat improvements, rehabilitative stocking, and strict harvesting regulations; these populations should rebound from their reduced numbers basin-wide. The Winnebago Lake system lake sturgeon population in Wisconsin, USA is an example of a lake sturgeon population that has responded well to targeted management actions, including limited fishing season, habitat management, increased enforcement, particularly during the spawning season, and increased public awareness, while continuing to maintain a recreational winter spear fishery (Bruch 1999a).

Currently, in the Great Lakes-St. Lawrence River system, there is a move toward appreciation of the need for ecosystem management and co-management activities. Even so, some of the jurisdictions in the Great Lakes basin that govern lake sturgeon currently allow for the commercial trade of parts or live fish, and for the recreational fishing harvest of lake sturgeon, while other jurisdictions do not. The lack of a coordinated management approach for lake sturgeon throughout the Great Lakes-St. Lawrence River makes it easier for illegally harvested sturgeon to be sold into the legal market (Bruch 1999b). The Law Enforcement Committee under GLFC has recognized the potential ease of infractions as a result of the mosaic of rules, and the committee is investigating possible recommendations to prevent and reduce infractions between the jurisdictions. One method being employed to address this threat is the use of combined enforcement teams (CET) in the Lake St Clair and the St Mary's River, which have already proven successful in other parts of the Great Lakes. CETs involve the participation of conservation officers from multiple state, provincial, federal, tribal agencies, from the United States and Canada, with law enforcement authority in the Great Lakes to enforce Great Lakes fishery regulations. The success of these teams arises from having representatives of each jurisdiction present during the apprehension of an individual committing an offence outside of his home jurisdiction (Kirshman and Léonard 2003). Nonetheless, the disparate status and management actions being taken by the various jurisdictions need to be addressed if effective coordinated management of the lake sturgeon is to be achieved.

Presently, in addition to the need for coordinated management and regulations in the Great Lakes, there is an urgent need to further increase understanding of the life history, migratory patterns, and population status of unique genetic lake sturgeon stocks in the Great Lakes-St. Lawrence River. Until the status and ecological needs of the lake sturgeon in the Great Lakes-St. Lawrence River is better understood, the fisheries management agencies should err on the side of caution, using the precautionary principle, so as to not lose this opportunity to rehabilitate this important and unique species. The first step that must be addressed is the need for a common management goal amongst fisheries authorities, which is reflected in the fishing regulations for lake sturgeon throughout the entire Great Lakes-St. Lawrence River. This will require the establishment of an official cooperative effort, such as the JSP and its implementation through the Lake Committees structure and the Council for Lake Committee under GLFC. This effort will need support and representation by Canadian and American entities with management authority on lake sturgeon, assured funding for programs, and accountability by the members for implementing joint decisions. Additionally, this cooperative effort should include the Province of Québec to ensure the protection of lake sturgeon throughout the Great Lakes-St. Lawrence basin. Secondly, the cooperative effort should have as mandate the joint cooperative management of populations within the Great Lakes-St. Lawrence River for the preservation, restoration, and enhancement of lake sturgeon. Data and information sharing will be needed for partners to agree on the status and best management for lake sturgeon throughout the basin on a stock/population level versus a jurisdictional management. Thirdly, the cooperative effort should investigate methods for protecting wild lake sturgeon from the sturgeon flesh and caviar market demand, and encourage research and application of alternative ways of providing for the demand of sturgeon meat and caviar. Lastly, effective enforcement of lake sturgeon regulations on the national, international, and global level, such as CITES, is imperative to the success of any management plan.

The future of lake sturgeon in the Great Lakes-St. Lawrence River will depend greatly on the development of a cooperative relationship between jurisdictions to manage this shared fish resource. The extensive geographic and jurisdictional area inhabited by lake sturgeon further validates the need to have a coordinated management plan among agencies sharing a common stock, to ensure that the financial and human resources will be utilized effectively. This coordination across agencies will facilitate management planning and resource allocations, and fulfil the basin-wide/ecosystem level common goal of achieving healthy lake sturgeon stocks. The JSP currently brings agencies together to engage in this type of cooperation within the Great Lakes basin, and with the Lake Committees already listing lake sturgeon within their Fish Committee Objectives for each Great Lake, the Lake Committees and the Council for Lake Committees under the umbrella of GLFC would seem to provide the ideal forum under which the lake sturgeon management jurisdictions could meet. Having such a forum will provide in the future an integrated, multijurisdictional management program for lake sturgeon populations, which will enhance the chance for a viable, genetically diverse, sustainable commercial and recreational fishery of lake sturgeon in the Great Lakes-St. Lawrence River.

However, the growing demand for lake sturgeon on an international level and the impact of globalization (Lechner and Boli 1999; Friedman 2000; Raymakers and Hoover 2002) will require that the cooperative group keep abreast of the global market demand and global agreements that may affect lake sturgeon populations in North America. This rising demand and trade potential will increasingly require an active monitoring of acipenseriform stocks worldwide, and as a result, it will be progressively more important to assure strict enforcement of existing federal, provincial, state, and tribal regulations and CITES to assure the conservation of our Great Lakes lake sturgeon.

Acknowledgements

We would like to acknowledge Tracy Hill from the USFWS-Alpena Fishery Resource Office for his assistance, and Gary Whelan, John Robertson, Lloyd Mohr and David Reid for reviewing our chapter.

11.8 References

Anonymous. 2001. Russia: Russia temporary halts sturgeon fishing. Marketnews, 21 July 2001, Fishround Inc. Retrieved April 30, 2003 from https://www.fishround.com/fishround/JapaneseVersion/gate/news/mnews/Russia/7_21_1.html

Auer, N.A (ed). 2002. A lake sturgeon rehabilitation plan for Lake Superior. Great Lakes Fishery Commission Miscellaneous Publication. Great Lakes Fishery Commission, Ann Arbor, Michigan.

Beeton, A. M., Sellinger, C.E. and Reid, D. E. 1999. An introduction to the Laurentian Great Lakes ecosystem. In Great Lakes policy and management. Edited by W. W. Taylor and C. P. Ferreri. pp. 3-54. Michigan State University, East Lansing, Michigan.

Bocking, S. 2001. The politics of endangered species: a historical perspective. In Politics of the Wild: Canada and endangered species. Edited by K. Beazley and R. Boardman. pp. 119-136. Oxford University Press, Ontario.

Bruch, R. M. 1999a. Management of lake sturgeon on the Winnebago System – long term impacts of harvest and regulations on population structure. J. Appl. Icthyol. 15(4-5): 142-152.

Bruch. R. M. 1999b. Management and trade of lake sturgeon in North America. In Proceedings of the Symposium on the Harvest, Trade and Conservation of North American Paddlefish and Sturgeon, May 7-8, 1998, Chattanooga, TN. Edited by D. F. Williamson, G. W. Benz and C. M. Hoover. TRAFFIC North America/World Wildlife Fund, Washington, DC, USA. pp. 106-120.

Bunt, C. M., Katopodis, C. and McKinley, R.S. 1999. Attraction and passage efficiency of white suckers and smallmouth bass by two Denil fishways. N. Am. J. Fish. Manag. 19(3): 793-803.

Bunt, C. M., Cooke, S. J., and McKinley R. S. 2000. Assessment of the Dunnville Fishway for passage of walleyes from Lake Erie to the Grand River, Ontario. J. Gt. Lakes Res. 26 (4): 482-488.

CITES. 2001. Press Release: Governments to consider trade restrictions on Caspian Sea caviar. Retrieved March 13, 2003 from http://www.CITES.org/eng/news/press/SC45_caviar1.shtml

CITES. 2003a. Catch and export quotas for specimens of acipenseriformes species included in Appendix II for 2003. Retrieved March 13, 2003 from http://www.CITES.org/common/quotas/2003/2003sturgeon.pdf

CITES. 2003b. Member countries. Retrieved June 13, 2003 from http://www.CITES.org/eng/parties/index.shtml

CITES. 2003c. How CITES works? Retrieved June 13, 2003 from http://www.CITES.org/eng/disc/how.shtml

CITES. 2003d. A brief history of sturgeon & CITES. Retrieved March 13, 2003 from http://www.CITES.org/eng/prog/sturgeon.shtml

CITES Secretariat. 2003. National contacts: Canada. Retrieved March 13, 2003 from http://www.CITES.org/common/directy/e_directy.html

COSEWIC. 2002a. About us: operating procedures. Retrieved March 18, 2003 from http://www.COSEWIC.gc.ca/eng/sct6/sct6_5_e.cfm

COSEWIC. 2002b. About us: celebrating 25 years of science and knowledge for species at risk in Canada. Retrieved March 18, 2003 from http://www.COSEWIC.gc.ca/eng/sct6/sct6_3_e.cfm#hist

COSEWIC. 2003. COSEWIC's Assessment Process and Criteria. Retrieved on June 12, 2003 from http://www.COSEWIC.gc.ca/pdf/English/Assessment_process_e.pdf

CWS (Canadian Wildlife Service). n.d.. Backgrounder: recovery of nationally endangered wildlife. Ref: 830-728/04A4. Retrieved March 18, 2003 from http://www.scics.gc.ca/cinfo01/83072804a4_e.html

CWS (Canadian Wildlife Service). 2002. News release: Species at Risk. Retrieved March 11, 2003 from www.speciesatrisk.gc.ca/species/media/back1_e.cfm

Czech, B. and Krausman, P. 2001. The endangered species act: history, conservation, biology, and public policy. The John Hopkins University Press, Maryland. p.213.

DeVault, D., Bertram, P., Whittle, D. and Rang, S. 1994. Toxic contaminants in the Great Lakes. State of the Lakes Conference Working Paper presented at the State of the Lakes Conference (SOLEC), October 26-28, 1994, Dearborn, Michigan. Environmental Canada and United States Environmental Protection Agency.

Dochoda, M. R. 1999. Authorities, responsibilities, and arrangements for managing fish and fisheries in the Great Lakes ecosystem. In Great Lakes fisheries policy and management: a binational perspective. Edited by W. W. Taylor and C. P. Ferreri. Michigan State University Press, East Lansing, Michigan. pp. 93-110.

Doyon, C., Fortin, R., and Spear, P.A.1999. Retinoic acid hydroxylation and teratogenesis in lake sturgeon (*Acipenser fulvescens*) from the St. Lawrence River and Abitibi region, Québec. Can. J. Fish. Aquat. Sci. 56: 1428-1436.

Egan, M. 2002. Hinterland Who's Who fact sheet: endangered species in Canada. Canadian Wildlife Service, Environment Canada. Retrieved March 18, 2003 from http://www.cws-scf.ec.gc.ca/hww-fap/hww-fap.cfm?ID_species=84&lang=e

Environmental Canada. 1991. Toxic chemicals in the Great Lakes and associated effects-synopsis. Department of Fisheries and Oceans, Environmental Canada.

Environmental Canada. 2002. News release: Species at risk act given royal assent. The Green Lane, Environmental Canada's World Wide Web Site. Retrieved March 11 2003 from http://www.ec.gc.ca/press/2002/021212_n_e.htm

Ferguson, M. M. and Duckworth, G. A. 1997. The status and distribution of lake sturgeon, *Acipenser fulvescens,* in the Canadian Province of Manitoba, Ontario and Quebec: a genetic perspective. Environmental Biology of Fishes 48: 299-309.

Ferreri, C. P., Taylor, W. W. and Robertson, J. M. 1999. Great Lakes fisheries futures: balancing the demands of a multijurisdictional resource. In Great Lakes fisheries policy and management: a binational perspective. Edited by W. W. Taylor and C. P. Ferreri. Michigan State University Press, East Lansing, Michigan. pp. 515-538.

Fielder, D.G., Baker, J.P., Mohr, L.C., Liskaukas, A., McClain, J.R. 2003. Status of the near-shore fish community. In The State of Lake Huron in 1999. Edited by M.P. Ebener. Great Lakes Fishery Commission Special Publication 03-draft. Great Lakes Fishery Commission, Ann Arbor, Michigan.

Fortin, R., Mongeau, J-R, Desjardins, G., and Dumont, P. 1993. Movements and biological statistics of lake sturgeon (*Acipenser fulvescens*) populations from the St. Lawrence and Ottawa River system, Quebec. Can. J. Zool. 71: 638-650.

Freedman, B., Rodger, L., Ewins, P., and Green, D. M. 2001. Species at risk in Canada. In Politics of the Wild: Canada and endangered species. Edited by K. Beazley and R. Boardman. Oxford University Press, Ontario. pp. 26-48.

Friedman, T. L. 2000. The lexus and the olive tree. New York: Anchor Books. 490p.

Gauthier, D. and E. Wiken. 2001. Avoiding the endangerment of species: the importance of habitats and ecosystems. In Politics of the Wild: Canada and endangered species. Edited by K. Beazley and R. Boardman. Oxford University Press, Ontario. pp. 49-74.

Gowans, A.R.D., Armstrong, J. D. and Priede, I. G. 1999. Movements of adult Atlantic salmon in relation to a hydroelectric dam and fish ladder. J. Fish Biol. 54(4) 713-726.

Graham, L. K. and Rasmussen, J.L. 1997. A MICRA perspective on closing paddlefish and sturgeon commercial fisheries. In Proceedings of the Symposium on the Harvest, Trade and Conservation of North American Paddlefish and Sturgeon, May 7-8, 1998, Chattanooga, TN. Edited by D. F. Williamson, G. W. Benz and C. M. Hoover. TRAFFIC North America/World Wildlife Fund, Washington, DC, USA. pp. 130-142.

Great Lakes Basin Ecosystem Team Lake Sturgeon Committee. n.d. Priority research need: passage. Retrieved March 13, 2003 from http://midweSt.fws.gov/sturgeon/pass2.htm

GLFC (Great Lakes Fishery Commission). 1997. A Joint Strategic Plan for Management of Great Lakes Fisheries. Ann Arbor, MI.

GLFC (Great Lakes Fishery Commission). 2000a. Factsheet 5: Sea lamprey barriers new technologies help solve and old problem. Great Lakes Fishery Commission. Ann Arbor, Michigan.
GLFC (Great Lakes Fishery Commission). 2000b. Factsheet 10: A Joint Strategic Plan for Management of Great Lakes Fisheries. Great Lakes Fishery Commission. Ann Arbor, Michigan.
GLFC (Great Lakes Fishery Commission). 2002. Brochure: sea lamprey the battle continues to protect our Great Lakes fishery. Great Lakes Fishery Commission. Ann Arbor, Michigan.
Hay-Chmielewski, E. M., and Whelan, G. E. (Editors). 1997. Lake sturgeon rehabilitation strategy. Michigan Department of Natural Resources, Fisheries Division, Special Report No. 18.
HC (House of Commons of Canada). 2002. Bill C-5: an act respecting the protection of wildlife species at risk in Canada. First session, thirty-seventh Parliament, 49-50-51 Elizabeth II, 2001-2002. Retrieved March 18, 2003 from http://www.parl.gc.ca/PDF/37/1/parlbus/chambus/house/bills/government/C-5_3.pdf
Hill, T. D. and McClain, J.R. (Editors). 2002. 2000 activities of the Central Great Lakes Bi-National Lake Sturgeon Group. U.S. Fish and Wildlife Service – Alpena Fisheries Resources Office, Michigan.
Holey, M. E., Baker, E. A., Thuemler, T. F., and Elliott, R. F. 2000. Research and assessment needs to restore lake sturgeon in the Great Lakes. June 27-28, 2000 Workshop in Muskegon, Michigan. Great Lakes Fishery Trust, Lansing Michigan.
Hoover, C. 1999. Import and export of sturgeon and paddlefish in the United States. In Proceedings of the Symposium on the Harvest, Trade, and Conservation of North American Paddlefish and Sturgeon, May 7-8, 1998, Chattanooga, TN. Edited by D.F. Williamson, G.W. Benz, and C. M. Hoover. Traffic North American/World Wildlife Fund, Washington, D.C., USA. Pp. 192-214.
Houston, J. J. 1987. Status of Lake Sturgeon, *Acipenser fulvescens*, in Canada. Canadian Field-Naturalist 101(2): 171-185.
IJC (International Joint Commission). 1989. Revised Great Lakes Water Quality Agreement of 1978, as amended by protocol signed November, 18 1987. Published by the International Joint Commission. Pp 84.
IJC (International Joint Commission). 1991. Persistent toxic substances: virtually eliminating inputs to the Great Lakes. Interim Report of the Virtual Elimination Task Force. International Joint Commission, Windsor, Ontario.
Kirshman, H., and Léonard, N. 2003. Operation Kingfisher. In FORUM: special law enforcement edition. Edited by J. Finster and M. Gaden. Great Lakes Fishery Commission, Ann Arbor, Michigan. pp. 4-5.
Klar, G. T., and Schleen, L. P. 2001. Integrated management of sea lamprey in the Great Lakes 2000. Annual Report to Great Lakes Fishery Commission. Ann Arbor, Michigan. 65p.
Kohm, K. A. 1991. The act's history and framework? In Balancing on the brink of extinction the endangered species act and lessons for the future. Edited by K. A. Kohm. Island Press, Washington D. C. pp. 10-22.
Kynard, B, Pugh, D, and Parker, T. 2003. Development of fish passage for lake sturgeon. Final report to the Great Lakes Fishery Trust, Lansing, Michigan.
Laine A, Jokivirta, T., and Katopodis, C. 2002. Atlantic salmon, *Salmo salar* L., and sea trout, *Salmo trutta* L., passage in a regulated northern river - fishway efficiency, fish entrance and environmental factors. FME 9(2): 65-77.
Lechner, F.J., and Boli, J. (Editors). 1999. The globalization reader. New York: Blackwell.
Le Prestre, P., and Stoett, P. 2001. International initiatives, commitments, and disappointments: Canada, CITES and the CBD. In Politics of the Wild: Canada and endangered species. Edited by K. Beazley and R. Boardman. Oxford University Press, Ontario. pp. 190-234.
LHC (Lake Huron Committee). 2003. Lake Huron Committee Executive Summary. Retrieved June 26, 2003 from http://www.glfc.org/lcexsum/lhc303.htm
Lindgren, R. D. 2001. The species at risk act: an overview. Report No. 408. Canadian Environmental Law Association, Toronto, Ontario. Retrieved March 13, 2003 from http://www.cela.ca/408sara.pdf
LMC (Lake Michigan Committee). 2003. Lake Michigan Committee Executive Summary. Retrieved June 26, 2003 from http://www.glfc.org/lcexsum/lmc303.htm
Lowie, C. (Editor). 2000. Great Lakes sturgeon genetics: status, needs, and standardization workshop proceedings, Holiday Inn – Chicago, Illinois, December 8-9, 1999. USFWS Administrative Report 2000-01. Amherst, New York. Downloaded August 4, 2003 from http://midwest.fws.gov/sturgeon/ Genworkpro.htm#Why%20Genetic%20Information%20is
McClain, J. (Editor). 1997. Summary of the 1996 activities of the interbasin lake sturgeon work group in waters of Lake Huron, St. Clair, and Erie. U.S. Fish and Wildlife Service-Alpena Fisheries Resources Office, Michigan. Retrieved February 8, 2003 from http://midweSt.fws.gov/alpena/rpt-stnib96.html
Meffe, G. K., and Carroll, C. R. 1994. The species in conservation. In Principles of Conservation Biology. Edited by G. K. Meffe and C. R. Carroll. Sinauer Associated, Inc Publishers, Massachusettes. pp. 50-77.
Moisan, M., and Laflamme, H. 1999. Rapport sur la situation de l'esturgeon jaune (*Acipenser fulvescens*) au Québec. Direction de la faune et des habitats, Québec. p. 68.

National Paddlefish and Sturgeon Steering Committee. 1993. Framework for the management and conservation of paddlefish and sturgeon species in the United States. U.S. Fish and Wildlife Service, Washington D.C. p. 41.

Peake, S., Beamish, F.W.H., McKinley, R.S., Scruton, D.A., and Katopodis, C. 1997. Relating swimming performance of lake sturgeon, *Acipenser fulvescens*, to fishway design. Can. J. Fish. Aquat. Sci. 54: 1361-1366.

Raymakers, C., and Hoover, C. 2002. Acipenseriformes: CITES implementation from range States to consumer countries. J. Appl. Icthyo. 18: 629-638.

Reffalt, W. 1991. The endangered species lists: chronicles of extinction? In Balancing on the brink of extinction the endangered species act and lessons for the future. Edited by K. A. Kohm. Island Press, Washington D. C. pp. 77-85.

Rideout, S. G. and Ritter, J. A. 2002. Canadian and U.S. Atlantic salmon institutions and politics: where are the fish and who cares? In: Sustaining North American salmon: perspectives across regions and disciplines. Edited by K. D. Lynch, M. L. Jones and W. W. Taylor. American Fisheries Society, Bethesda, Maryland. pp. 93-117.

Rishikof, R. 1997. Endangered species protection in Canada: an overview. Environment Probe on the web June 1997. Retrieved March 13, 2003 from www.nextcity.com/EnvironmentProbe/pubs/ev539.htm

Scheidegger, K. 2002. Fish gotta swim. Wisconsin Natural Resources Magazine on the web April 2002. Retrieved March 22, 2003 from www.wnrmag.com/stories/2002/apr02/fishlad.htm

Scott, W. B. and Crossman, E. J. 1998. Freshwater fishes of Canada. Ontario: Galt House Publications. p. 69-76.

Smith, S. H. 1968.Species succession and fishery exploitation in the Great Lakes. J. Fish. Res. Board Can. 25: 667-693.

Smith, S. H. 1972. Factors of ecological succession in oligotrophic fish communities of the Laurentian Great Lakes. J. Fish. Res. Board Can. 29: 717-730.

SFPQ (Société de la Faune et des Parcs Québec). 2001. Espèces fauniques menacées ou vulnérables au Québec: Esturgeon jaune. Retrieved March 20, 2003 from http://www.fapaq.gouv.qc.ca/fr/etu_rec/esp_mena_vuln/esp/esturgeon_jaune.htm

Stanford Environmental Law Society. 2001. The endangered species act. Stanford University Press, Stanford, California. p.296.

Stegemann, E. C. 1994. Sturgeon – the king of freshwater fishes. Conservationist 49(1): 18-23.

Thomas, M. V., and Haas, R. C. 2002. Abundance, age structure, and spatial distribution of lake sturgeon, *Acipenser fulvescens*, in the St Clair System. J. Appl. Icthyol. 18(4-6): 495-501.

Thompson, P.C. 1974. Institutional constraints in fisheries management. J. Fish. Res. Board Can. 31: 1965-1981.

Todd, R. M. 1999. Sturgeon and paddlefish commercial fishing in North America. In Proceedings of the Symposium on the Harvest, Trade and Conservation of North American Paddlefish and Sturgeon, May 7-8, 1998, Chattanooga, TN. Edited by In D. F. Williamson, G. W. Benz and C. M. Hoover. TRAFFIC North America/World Wildlife Fund, Washington, DC, USA. p. 42-50.

USEPA (U.S. Environmental Protection Agency). 1992. Exposure and effects of airborne contamination. Great Waters Program Report, Chapter 2.2. pp. 201.

USFWS (U.S. Fish and Wildlife Service). n.d.. Lake sturgeon: giant of the Great Lakes. Brochure published by U. S. Fish and Wildlife Service Great Lakes Basin Ecosystem Team Lake Sturgeon Committee. Retrieved March 18, 2003 from http://midweSt.fws.gov/sturgeon/lakesturgeon.pdf

USFWS (U.S. Fish and Wildlife Service). 1988. Endangered Species Act of 1973, as amended through the 100th congress. U.S. Department of the Interior, Washington, D. C. pp. 66-68.

USFWS (U.S. Fish and Wildlife Service). 1996. The endangered species act of 1973. Retrieved March 18, 2003 from http://endangered.fws.gov/esa.html

USFWS (U.S. Fish and Wildlife Service). 2003. Lake sturgeon population investigations. U.S. Fish and Wildlife Service-Alpena Fishery Resources Office, Michigan. Retrieved January 30 2003 from http://midweSt.fws.gov/alpena/lakesturgeon.htm

USFWS-Green Bay (Wisconsin Ecological Services). n.d. Lake sturgeon restoration activities: lake sturgeon upstream and downstream passage initiative on the Menominee River. Retrieved April 1, 2003 from http://midweSt.fws.gov/sturgeon/green_bay_ES_jim.htm

Weisser, J. 2000. Sea lamprey management program and lake sturgeon. In Update on lake sturgeon in New York State Waters: abstracts and meeting summary. Edited by T. E. Brooking. Lake Sturgeon Research Meeting, January 27, 2000, Syracuse, NY. Retrieved March 18, 2003 from www.dnr.cornell.edu/fieldst/cbfs.htm

Wijnstekers, W. 2001. The evolution of CITES: a reference to the convention on international trade in endangered species of wild fauna and flora, 6th edition. CITES Secretariat, Switzerland. pp.492.

WWF (World Wildlife Fund). 2002. WWF fact sheet: 12th meeting of the conference of the parties to CITES, Santiago, 3-15 November 2002. Retrieved March 18, 2003 from http://www.worldwildlife.org/CITES/pdfs/sturgeon_factsheet.pdf

CHAPTER 12

CONSERVATION

NANCY A. AUER
Michigan Technological University, Department of Biological Sciences, 1400 Townsend Drive, Houghton, Michigan 49931

12.1 Introduction

The total volume of water on the earth is estimated to be 1.4 billion km^3 however only 2.5% or 35 million km^3 occurs as freshwater (Gleick 2000) which is circulated around the planet by the hydrological cycle. Lakes, marshes, wetlands and rivers contain 105,000 km^3 or 30% of all the freshwater needed to sustain humans and freshwater aquatic organisms. These sources of freshwater have been important to humankind for centuries for domestic, industrial and agricultural purposes, and they have also sustained sturgeon and paddlefish species that require freshwater for successful spawning and nursery habitats since the Pleistocene. Of the ten longest and largest rivers in the world, five are located in the Northern Hemisphere and are important habitat for at least ten of the 27 known species of sturgeon and paddlefish on earth (Table 12-1).

Distribution of members of the Acipenseriformes is restricted to the Northern Hemisphere. Six of the ten members of this ancient family of fishes are known from the North American continent. The Atlantic, lake, white, pallid, and shovelnose sturgeons, and the paddlefish, occur in its five longest rivers (Table 12-2). The other four species from North America include the shortnose sturgeon, which inhabits rivers on the east coast from the Saint John River in New Brunswick to the St. Johns River in Florida; the green sturgeon, which frequents the rivers of the west coast from Skeena River, British Columbia to the Sacramento River, California (Scott and Crossman 1973), and two subspecies, the Gulf sturgeon, (Wooley 1985) and the Alabama sturgeon, *S. platorynchus* subsp. (Kincaid et al. 1999) or *S. suttkusi* (Mayden and Kuhajda 1996), which occur in the Gulf of Mexico and Alabama River drainages, respectively.

12.2 Status of Acipenseriformes Species

The conservation status of the eight species and two subspecies of Acipenseriformes in North America vary by country, province, state, or listing organization. Listing status as considered by International, North American, and province/state categories are shown in Table 12-3. The first listing category is the International listing and is provided by the United Nations Environment Programme

G.T.O. LeBreton et al. (eds.), Sturgeons and Paddlefish of North America, 252–276.
© 2004 *Kluwer Academic Publishers. Printed in the Netherlands.*

and the World Conservation Monitoring Centre (Appendix A). This organization has included all North American sturgeon and paddlefish on the IUCN (International Union for Conservation of Nature and Natural Resources) Red List of Threatened Animals for 1996. Their listing criteria can be found through the source listed in Appendix A or in Haywood (1998).

Table 12-1. Ten longest rivers in the world and species of Acipenseriformes occurring in Northern Hemisphere rivers.

River	Hemisphere	Species	River length (km)
Nile	Southern		6,650
Amazon-Ucayali-Apurimac	Southern		6,400
Yangtze	Northern	*A. dabrynchus, Psephurus gladius*	6,300
Mississippi-Missouri-Red Rock	Northern	*A. fulvescens, Scaphirhynchus albus, S. platorynchus, Polyodon spathula*	5,970
Yenisey-Baikal-Selenga	Northern	*A. baerii, A. b. baicalensis*	5,540
Huang Ho (Yellow)	Southern		5,460
Ob-Irtysh	Northern	*A. baerii*	5,410
Parana	Southern		4,880
Congo	Southern		4,700
Amur-Argun	Northern	*A.schrenkii, Huso dauricus*	4,440

Table 12-2. Five longest North American Rivers and known Acipenseriformes in each system

River	Species	River length (km)
Mississippi – Missouri	*A. fulvescens, Scaphirhynchus albus, S. platorynchus, Polyodon spathula*	5,970
St. Lawrence	*A. fulvescens, A. oxyrinchus*	4,000
Columbia – Snake	*A. transmontanus*	3,670
Yukon		3,010
Mackenzie	*A. transmontanus*	4,240

The second listing category in Table 12-3 focuses on the status of a species within the countries of the United States or Canada by such organizations as the United States Fish and Wildlife Service (USFWS) or the Committee on the Status of Endangered Wildlife in Canada (COSEWIC) respectively. The USFWS lists organisms as Threatened (TH), Endangered (EN), or Candidates for listing (C) while the COSEWIC lists organisms as of special concern (SPC). The third listing category in Table 12-3 summarizes individual province or state status. Johnson (1987) and Williams et al. (1989) summarized status of fishes in individual Canadian provinces and states of the United States as receiving legal protection or of special concern. These listing characteristics have been summarized in Table 12-3 and illustrate how most sturgeon and paddlefish species have been extirpated or show declines in abundance and require some protected status in historic ranges in North America.

12.3 Reasons for the Initial Decline of the Species in North America

Initial declines in paddlefish and sturgeon populations worldwide were primarily due to three factors: overharvest, barriers to migration, and habitat loss (Rochard et al. 1990). For many species in North America these three influences to recruitment and survival occurred simultaneously during the early 1900s. Almost all North American fish species experienced strong fishing pressure between 1880 and 1930 (Bogue 2000). This fishing pressure was reported for sturgeon species in the United States (Tower 1908), Atlantic sturgeon in Delaware Bay (Secor and Waldman 1999), lake sturgeon in the Great Lakes (Baldwin et al. 1979), white sturgeon in the Fraser River (Semakula and Larkin 1968) and paddlefish (Russell 1986). Small hydroelectric facility construction on many North American rivers blossomed between 1880 and the early 1900s. Electricity was first generated with flowing water using a water wheel on the Fox River, Wisconsin in 1882 (Appendix A), a river critical to lake sturgeon. The small hydroelectric facility construction boom in North America was followed by a post World War II era emphasis on large dams (Bureau of Reclamation – Appendix A and Pringle et al. 2000). Large and small hydroelectric facilities have affected most of the rivers used by species of Acipenseriformes in North America and around the world. Not only do these structures block migration corridors but flow regime manipulations can impact river access and spawning site scouring.

Log drives and pulp and paper mill industries of the early 1900s also destroyed and smothered spawning sites and nursery areas for lake sturgeon (Brousseau and Goodchild 1989). Mining activities, especially in the western states and provinces, have also contributed to significant habitat impacts on white and green sturgeon (Graham 1981, USFWS 1999). Flood control and channelization of many rivers has resulted in lost habitat for several Acipenseriformes species (Carlson et al. 1985, USFWS-GSMFC 1995). Either singly or in combination, these numerous influences on survival of all life stages of sturgeon and paddlefish lead to the initial decline in populations and contributed to the current status of Acipenseriformes in North America (Table 12-3).

12.4 Persistent Threats to Recovery of North American Species

Sturgeon and paddlefish have unique life history strategies when compared to other fish species. Most Acipenseriformes reach sexual maturity at a great age, often 15 to 20 years, recruitment to a spawning population is therefore slow, and once reproductive age is reached spawning occurs often intermittently. These life history traits, coupled with low abundance resulting from historic over-harvest, barriers to migration, and habitat loss, have resulted in finding only small, remnant populations or stocks in many regions during assessment efforts. Most harvest of these species has been terminated or strongly regulated and habitat restoration has been attempted in some areas, but seeing results of these efforts will take time.

Table 12-3. Sturgeon and Paddlefish status listings worldwide, country, and state or province. In the UNEP-WCMC listing organisms are listed as Extinct (EX), Critically Endangered (CR), Endangered (EN), Vulnerable (VU) or at Lower Risk (LR). In the province or state list species are considered either legally protected or of special concern unless otherwise noted in parentheses as threatened (th) or endangered (en)..

Species	UNEP-WCMC Listing	Status in Canada and USA	Status in Province or State
A. brevirostrum	VU – CA and USA	EN – USA – CT, DE, FL, GA, MA, MD, ME, NJ, NY, NC, RI, SC, VA, SPC – Canada	CT, DE, FL,GA, MA, MD, ME, NC, NH, NJ, NY, PA, RI, SC, VA; N.B.
A. fulvescens	VU – CA and USA		AL, AR, GA, IA, IL, IN, KS, KY, MI(th), MN, MO (en), MS, NC, NE, NY, OH, PA, SD, TN, VT, WI, WV
A. medirostris	VU – CA and USA	SPC – Canada	
A.o. oxyrhhynchus	LR – CA and USA		C – GA and Canada, CT (th), DE, FL, GA, MA, ME, MS, NC, NH, NJ, NY, PA, PQ, RI, SC, VA
A o. desotoi	VU – USA	TH – USA – AL, FL, LA, MS	AL, FL, GA, LA, MS
A. transmontanus	LR – CA and USA	EN – USA - ID, MT, SPC – Canada	ID, MT, B.C.
A. transmontanus Kootenai River	EN – USA	EN – USA – ID, MT	ID
Polyodon spathula	VU – USA, EX – CA		EX – MI and Canada; AL, AR, GA, IA, IL, KS, KY, LA, MD, MN(th), MO, MS, MT,NC,ND, NE, NY, OH, OK, PA, SD,TN, TX, VA, WI, WV
Scaphirhynchus albus	EN – USA	EN – USA – AR, IL, IA, KS, KY, LA, MS, MO, MT,NE, ND, SD, TN,	AR, IA, IL, KS, KY, LA, MO, MS, MT, ND, NE, SD, TN
Scaphirhynchus platorynchus	VU – USA		KY, MN, ND, TX, WV, WY
Scaphirhynchus p. suttkusi	CR – USA	EN – USA – AL, MS	AL, MS

Population responses to protection and rehabilitation measures have been slow. In Michigan, a hydroelectric facility, operating in a peaking mode since 1931 and impacting successful reproduction of lake sturgeon, was relicensed in 1990 to operate near run-of-the-river (Auer 1996a). Only as we near 2010 will success of change in flow regime and its benefit to the sturgeon population be observed in a hoped for increase in recruitment to the spawning population. The following section will discuss the detrimental factors that can be manipulated or mitigated by humans to decrease negative impacts of human activity on sturgeon and paddlefish populations.

12.4.1 COMMERCIAL HARVEST AND BY-CATCH

Today commercial and sport harvest is strictly limited or discontinued for many sturgeon and paddlefish populations in North America. Sport fishing regulations change almost yearly in some areas for sturgeon and paddlefish; a recent summary of allowable sport fishing harvest by region is given in Mosher (1998). The Alabama, Gulf, pallid, shortnose and Kootenai River white sturgeon are protected throughout their ranges in the United States and experience no commercial or recreational harvest. These species are subject to incidental catch and by-catch. Commercial fishing gears, used to target other ocean species on the East and Gulf coasts, have produced by-catches of primarily juvenile Atlantic sturgeon and adult shortnose sturgeon (Collins et al. 1988).

Populations of green, lake, shovelnose, white sturgeon and paddlefish remain open to sport, commercial and subsistence harvest for flesh and roe in the USA and Canada (Table 12-4). Harvest of these five species is closely monitored in many remaining fisheries but their often large geographic range (Auer 1996b) and movement between jurisdictions continues to complicate management and recovery programs. Between 10,000 and 80,000 kg/yr of green sturgeon were harvested in Washington and Oregon between 1990 and 1995 (Todd 1998). Commercial harvest of white sturgeon during those years in those states ranged from 89,000 to 155,000 kg/yr with a maximum size limit of 122 cm for white sturgeon (Todd 1998). There is a sport fishery for lake sturgeon in Canadian and United States waters, and a commercial harvest in Ontario (2,000kg/yr), Quebec (100,000 to 200,000 kg/yr) and Saskatchewan (4,000 kg/yr) (Todd 1998). Commercial fisheries exist in eight states for shovelnose sturgeon (850 to 20,000 kg/yr) and in seven states for paddlefish (4,000-30,000 kg/yr) (Todd 1998). An example of states having different management policies is shown in the moratorium on paddlefish harvest initiated in the Alabama waters of the Tennessee River in 1988. Continued commercial harvest of these fish in adjacent states has kept abundance low in Alabama (Hoxmeier and DeVries 1996). Commercial fisheries for North American Acipenseriformes have and do provide eggs for caviar industries worldwide (Hoover 1998). Currently, the depletion of Russian and Iranian stocks of sturgeon, which historically sustained major caviar industries, has forced caviar industries to look for new stocks or supplies of Acipenseriformes. Increased poaching and fishing pressure on remnant populations and on those only recently protected is a concern as this search for new caviar supplies reaches North America. Development of aquaculture practices which will provide relief from overharvest of natural stocks to supply caviar (Waldman and Secor 1998) may reduce these pressures, as will new research which reduces adult mortality during roe removal. In response to market demand for caviar both as a current (Russia and Iran) and future threat (North America) CITES (Convention on International Trade in Endangered Species of Wild Fauna and Flora) (Gnam 1998) listed all sturgeon and paddlefish species in an attempt to protect them from over exploitation due to international trade.

Table 12-4. Locations with allowable sport, commercial or tribal harvest. (1) Waldman and Wirgin (1998). (2) Keenlyne (1997). (3) www.dnr.state.mn.us

Species	Area of Harvest – USA	Area of Harvest – Canada
Atlantic (1)	None	St. Lawrence, St. John, and Shubenacadie Rivers
Lake	MI, WI	St. Lawrence, ON,
Shovelnose (2) (3)	AR, IA, IL, IN, KS, KY, MO, MT, NE, WI, WY	None
White	WA, OR	B.C.
Paddlefish	AR, , IL, IN, IO, KS, NE, KY, MO, MT, ND, OK, TN	None

12.4.2 BARRIERS TO MIGRATION AND WATER LEVEL MANIPULATIONS

Most sturgeon and paddlefish migrate within or into freshwater river systems to spawn each spring over a clean, rubble-gravel substrate in flowing water. Large-bodied sturgeon species tend to migrate long distances to spawn (Auer 1996b). For example, white sturgeon move as much as 1,300 km during spawning runs and spring spawning movements of 160 to 320 km for paddlefish are common. Paddlefish were also found to move 1,920 km down the Missouri River (Russell 1986). Currently most North American river systems have some barrier structure preventing unrestricted movement of these fishes. Hydroelectric dams which span rivers have segmented sturgeon and paddlefish populations, prevented access to spawning grounds and resulted in out-migration mortality for juveniles and adults passing over dams and through turbines. Isolated stocks remaining in reservoirs above the dams show reduced recruitment as newly-hatched larvae and juveniles are lost from the stocks when they drift or drop downstream, and some adults are killed as they try to pass turbines or barrier spillways (Pitman and Parks 1994, G. Kornely, Wisconsin Department of Natural Resources, personal communication).

Barriers to migration also appear to influence growth and feeding of fish isolated in small sections of river systems. Galbreath (1985) found white sturgeon isolated in reservoirs to experience lower and more irregular growth patterns than white sturgeon having access to the estuary and sea. McKinley et al. (1993) found that a variable flow regime below a hydroelectric facility produced altered nutritional status of lake sturgeon in the Mattagami River, Ontario Canada. In addition to these effects, river water flows and temperature, manipulated or changed from patterns occurring naturally by hydroelectric facility operation and water storage, can delay or retard spawning (Auer 1996b, Pringle et al. 2000). This delay is seen in reproductively mature adult Acipenseriformes halted in migration below barriers, who do not release eggs because environmental cues in water flow and temperature needed to initiate spawning are absent.

Adult sturgeon and paddlefish blocked from spawning in extensive historic rapids above barriers may spawn in unusually large and concentrated groups below barriers (Kynard 1997). These fish may deposit eggs in mats and clumps (Kempinger 1988) thereby reducing egg survival and hatch success. Hybridization may occur where there is unusual co-mingling of spawning ready species (Carlson et al. 1985). In the Great Lakes lake sturgeon have been observed congregating over, and may spawn on, alternative substrates such as coal cinders or clinkers, dumped in some areas years ago by steamboats, because natural substrate material has been buried or covered by invasive species or algal growth. Spawning over coal cinders is suspected to have occurred in the Wolf River, Wisconsin (Kempinger 1988) and the lower St. Clair River, Michigan (S. Jerrine Nichols, USGS-BRD, Ann Arbor, MI). Unfortunately these substrates may not support good egg development or survival of newly-hatched larvae, thereby contributing to a further decline in abundance.

12.4.3 HABITAT LOSS

Industry, logging, mining and hydroelectric dams began to change rivers in North America by the mid-1800s (Brousseau and Goodchild 1989, Chiasson et al. 1997, Collins 1988, Graham 1981, Kelso et al. 1996). Habitat is lost when industrial and mining operations discharge toxic substances into the water or when these discharges change the property of the water, such as the pH or salinity. The use of rivers for moving logs from the forest into bays for transport to mills resulted in streambed scouring and widening and erosion of stream banks which increased sedimentation, burying spawning substrate.

Hydroelectric facilities were often placed in rivers at locations of natural falls or areas of sharp loss in gradient to take advantage of natural water forces. These sites are suspected of being natural spawning sites for many fish species. Establishment of dams, spillways, powerhouses and reservoirs at such locations has reduced or eliminated the fast-flowing water, clean rubble habitat needed by sturgeons and paddlefishes for spawning. As mentioned earlier, hydroelectric dams block migration routes and can prevent adults from moving onto historic spawning grounds. They produce reservoirs which change the flow characteristics and water temperature of the river above the dam and increase sedimentation in sections of the river. Hydroelectric dams often need to manipulate flows to meet electric demand which can impact fish preparing to spawn. Rivers segmented by dams and reduced in flow by reservoir water storage also impact nursery and juvenile habitat. Settling of sediment and debris in slow flowing reservoirs buries benthic organisms and habitat, reducing food sources for sturgeon (Chiasson, et al. 1997). For Scapirhynchus species in the Missouri and Mississippi Rivers, dredging and channelization changed habitat and river flow characteristics (Carlson et al. 1985). Growth of the white sturgeon in the Columbia River was greater for stocks with access to the natural river estuary and ocean than it was for stocks sandwiched between dams higher up in the system (Galbreath 1985).

Documenting lost habitat for sturgeon and paddlefish is difficult because of the lack of records and little early quantitative assessment of fish stocks. For most rivers, knowledge of historic use by Acipenseriformes has been lost. Only in a few rivers in North America do sturgeon and paddlefish have access to historic and unaffected natural habitats for spawning and nursery areas. It is these areas which need intensive study to describe habitat requirements for the various life stages of these species and serve as examples for habitat restoration.

12.4.4 CONTAMINANTS

Acipenseriformes are long-lived, have a high lipid content and feed, especially at adult life stages, on organisms and organic material on the bottom of rivers and lakes. Due to this strategy they may accumulate contaminants in their oil rich flesh and eggs that have concentrated in benthic organisms or settled in the substrate. Because sturgeon and paddlefish are legally protected and low in abundance in most areas and since sampling sufficient tissue results in fish mortality, few studies of contaminant burden have been conducted. Tissue samples are usually collected from poached or incidental catches, however, sample size for analysis remains low for most species.

In locations where sufficient sample material has been obtained analysis confirms that sturgeon and paddlefish do accumulate PCBs, heavy metals and other contaminants, and that these burdens increase with age. For lake sturgeon, polychlorinated biphenyls (PCBs) were found to be above U.S. Food and Drug Administration (FDA) levels set for human consumption in Lake Wisconsin, Green Bay, and the Pestigo River, Wisconsin, while below advisory levels for some lake sturgeon collected in Lake Superior and the Menominee River, Wisconsin (Larson 1988, and T. Thuemler, WDNR personal communication). Lake sturgeon in the St. Lawrence River, showing some fin and craniofacial malformations, were found to have elevated levels of PCBs in liver samples (Doyon et al. 1999). Tissue samples from 12 Gulf sturgeon revealed contaminant concentrations at levels high enough to warrant concern (USFWS-GSMFC 1995) and juvenile Gulf sturgeon from the Suwannee River, Florida, were found to have accumulated high levels of some heavy metals such as iron, lead, nickel and zinc (Alam et al. 2000). Samples of white sturgeon flesh from six fish taken from the upper and lower Fraser River, British Columbia revealed that the fish taken from a section of the upper river, which had been the site of effluent from bleached-kraft pulp mills and a wood preservation facility, contained concentrations of heavy metals and organochlorines above Health Canada working guidelines for human health (MacDonald et al. 1997). Elevated concentrations of contaminants have also been found in a small sampling of pallid sturgeon from the Missouri River (Dryer and Sandvol 1993). Three pallid sturgeons from Nebraska and North Dakota had PCB levels in ovary, muscle and liver tissue above the FDA action level of 2mg/kg wet weight (Ruelle and Keenlyne 1993). Working with tissue samples from 15 pallid sturgeon from across the United States, Ruelle and Henry (1994) found strong correlations between age, weight and percent lipid of fish and mg/kg dry weight p,p'-DDE in liver tissue. Contaminants are considered to be a problem in paddlefish from the Ohio River basin, and after a ban on Chlordane use in 1994, Chlordane levels dropped and are now not a problem in Missouri paddlefish populations (K. Graham, 1997).

Understanding the diet of sturgeon and paddlefish at all life stages, evaluating the contaminant pathways, and determining affects of contaminants on reproductive health of these long lived fishes will be important to future management and survival of these species. Toxic substances and heavy metals discharged into North American waters prior to the Clean Water Act have remained in the sediments and continue to affect fishes such as sturgeon and paddlefish which grow to great size and age, contain relatively high concentrations of lipids, and feed on benthic organic material.

12.4.5 ILLEGAL HARVEST AND OTHER HUMAN RELATED FISH MORTALITY

Poaching is, and probably will remain, a continuous threat to all paddlefish and sturgeon populations. All Acipenseriformes are valued for their flesh, both fresh and smoked, and for their eggs for caviar. Both flesh and eggs bring good prices in legal and black markets, and caviar from some species has been purposely been mislabelled to represent caviar from more highly esteemed species (Birstein et al. 1998). Pre-spawning sturgeons often begin to gather in the fall near river mouths making them more vulnerable to capture. In the spring they are even more vulnerable to poaching as they are more exposed when moving into river systems and onto spawning grounds. Spawning migrations often cover great distances, usually in remote locations, and protection and surveillance of fishes is difficult and costly for management units. Organized poaching of paddlefish occurred in Table Rock Lake and Lake of the Ozarks, Missouri, in 1986. One poacher took over 5000 pounds of eggs for caviar, selling the eggs for over $187,000 in New York City (K. Graham, personal communication). Poaching for lake sturgeon on the Menominee River, Wisconsin, also occurred in 1986 by individuals from the Ukraine (S. Daye, personal communication) and at least three lake sturgeon were poached from the Sturgeon River, Michigan in 2000 (Daily Mining Gazette, June 12, 2000). Long-term poaching operations were exposed in 1990 near Olympia, Washington which included the taking of white sturgeon (Detroit News, April 1991 and Boyle 1994).

Aside from the illegal harvesting of individuals, sturgeon and paddlefish are also subject to mortality during legal sport fishing activities. In the United States limited spearing seasons are managed in the inland waters of Michigan and Wisconsin for lake sturgeon. Anxious fishermen, hoping to spear a large sturgeon, can accidentally spear an undersized fish or spear a large fish which escapes landing but sustains a mortal injury (Richey 1987). Snagging for paddlefish can produce injury to a fish that escapes before being landed and in some cases paddlefish are mortally handled when high-grading occurs; the exchanging of a large fish caught later in the day for a small one caught earlier (Scarnecchia et al. 1996). Sturgeon can also be mortally wounded in late summer, when boaters often hit fish laying just below the water surface, a phenomenon often observed but little understood.

12.4.6 LACK OF BASIC KNOWLEDGE OF THE SPECIES AND OF PUBLIC APPRECIATION

The lack of knowledge and appreciation for these unique fishes by the public continues to threaten the recovery of most sturgeon and paddlefish species. Public support is essential to recovery efforts since some agency financial resources are directed to management decisions based upon public interest. The overall low abundance of most populations and the lack of visibility of these species have prevented the public from developing an appreciation for sturgeon and paddlefish. Until recently maintaining and feeding large sturgeon and paddlefish in aquaria was difficult and few zoos or aquariums showcased these fishes. Most of the fisheries for these species have been closed since the 1950s and sturgeon and paddlefish are now considered nongame species in many locations. Nongame species management did not become popular until many of the "Fish Community Objectives" were developed in the late 1980s for Great Lakes fisheries. These reports focused on predators, forage species, sea lamprey and 'other species'; recognizing in 'other species' the need to re-establish depleted stocks of native species such as sturgeon, brook trout, and walleye (Busiahn 1990).

12.5 Recent Advances in Recovery of Acipenseriformes in North America

12.5.1 MANAGEMENT AND REHABILITATION PLANS

The initial drastic decline in abundance of many sturgeon and paddlefish caused populations levels to drop so low that management agencies and scientists believed the fish to be nearly exterminated, so little attention was focused on them (Tower 1908). By the mid 1960s it was generally believed that "The lake sturgeon undoubtedly will continue to live in the Great Lakes at low abundance until influenced adversely by the changing environment, or by a fishery for another species that raises the incidental catch above the limit of its biological yield" (Smith 1968). However, protecting the environment came to the forefront of political agendas in the 1970s and brought about the safeguarding of wetland and aquatic habitat as well as efforts to reduce the pollution of all water through the Clean water Act of 1972. A growing awareness of the significance of habitat loss and affects of fluctuating water regimes on fisheries and other aquatic organisms resulted in requests for many hydroelectric facilities to alter operations to maintain more natural river flow regimes. This became possible as many hydroelectric facility licenses also became due for renewal in the 1990s. These flow changes, usually proposed for salmonids, also benefited other aquatic riverine species.

Within the past decade, many natural resource agencies began focusing on sturgeon and paddlefish as viable members of historic fish communities, and have developed protection and rehabilitation plans for stocks of these fishes in North America. Over 24 management and rehabilitation plans have been drafted for the majority of species on either a state, federal, provincial, local or watershed level since about 1990 (Table 12-5). These plans are important guidelines for cooperative management and protection of these unique species that are low in abundance or extirpated from some portion of their historic range. Examining this list reveals a lack of status evaluations or recovery plans for green, shortnose and shovelnose sturgeons. Within the last 10 years only a few reports were published on work devoted to understanding green sturgeon. One recent report describes a tagging study initiated in 1992 on spawning movements of 25 green sturgeon in the Klamath River (NTIS 1998). Several groups have developed to increase our understanding and protection of paddlefish and sturgeon such as the Mississippi Interstate Cooperative Resource Association (MICRA). This group was formed in 1989 due to concern for the status of the paddlefish in the Mississippi River drainage, and interstate management (Rasmussen 1998). Since its formation, the group has included the three sturgeon species also found in the drainage within its scope of concern (Graham and Rasmussen 1998). In 1997 the MICRA recommended that all commercial fishing for sturgeon be banned and a closure for paddlefish be considered (Graham and Rasmussen 1998).

The majority of the rehabilitation and management plans for sturgeon and paddlefish (Table 12-5) include summaries of the history of the fishery, current impediments to re-establishing populations, and management and research needs. Most plans agree that overharvest, barriers to migration and habitat loss impacted the fisheries or specific populations. The majority of the plans developed within the last decade report continued low abundance of populations and habitat loss as two important current impediments to recovery.

Table 12-5. Status and Rehabilitation Plans for Paddlefish and Sturgeon in North America

Species	Status and Restoration Plans/Reports	Year	Author/ Organization
A. brevirostrum	Status review of shortnose sturgeon in the Androscoggin and Kennebec Rivers	1996	NMFS
	Life history, latitudinal patterns, and status of the shortnose sturgeon, *Acipenser brevirostrum*.	1997	Kynard
A. fulvescens	Species recovery plan for lake sturgeon in the Lower Saskatchewan River	1991	Wallace
	A plan for recovery of the lake sturgeon in Missouri	1992	Graham
	DRAFT Management strategy for lake sturgeon in Ontario edition 6.	1992	OMNR
	A study of the feasibility of restoring lake sturgeon to Lake Champlain	1994	Moreau and Parrish
	Ohio's lake sturgeon management and recovery program for Lake Erie	1994	Cavender
	Fisheries Long Range Plan – Sturgeon and Paddlefish	1995	Minnesota DNR
	DRAFT Lake sturgeon rehabilitation within the Bay of Quinte	1996	Mathers
	The status and distribution of lake sturgeon, *Acipenser fulvescens*, in the Canadian Provinces of Manitoba, Ontario and Quebec: a genetic perspective	1997	Ferguson and Duckworth
	Status of lake sturgeon in Lake Superior	1997	Slade and Auer
	Lake Sturgeon Rehabilitation Strategy – Michigan Department of Natural Resources	1997	Hay-Chmielewski and Whelan
	Sturgeon restoration plan for the Upper St. Louis River, MN	1999	Fond Du Lac Natural Resources
	DRAFT Lake sturgeon plan for the Green Bay basin	1999	Thuemler et al.
	Plan de gestion de L'esturgeon jaune du fleuve Saint-Laurent	2000	Societe de la faune et des parcs du Quebec
	Alberta's lake sturgeon management plan	2000	Alberta Environment
	DRAFT A recovery plan for the lake sturgeon (*Acipenser fulvescens*) in N. Y.	2000	Carlson and Bouton
A. fulvescens	DRAFT Wisconsin Lake Sturgeon Management Plan	2000	Scheidegger
	A Lake Sturgeon Rehabilitation Plan for Lake Superior	2003	Auer
A. medirostris	Status of the green sturgeon, *Acipenser medirostris*, in Canada	1988	Houston
A.o. oxyrinchus	Status and Restoration Options for Atlantic Sturgeon in North America	1998	Waldman and Wirgin

Table 12-5 (Continued)

Species	Status and Restoration Plans/Reports	Year	Author/ Organization
A. o. desotoi	Gulf Sturgeon Recovery/Management Plan	1995	USFWS-GSMFC
	Status of white sturgeon, *Acipenser transmontanus*, in Idaho.	1985.	Cochnauer et al.
	State of Florida Conservation Plan for Gulf Sturgeon.	2001	Wakeford
A. transmontanus	Status of the White Sturgeon *Acipenser transmontanus*, in Canada	1991	Lane
	Status of the White Sturgeon Resource in the mainstem Columbia River	1993	Tracy et al
A.transmontanus Kootenai River	Recovery Plan for the Kootenai River Population of the white sturgeon	1999	U.S.F.W.S.
Polyodon spathula	The Paddlefish: Status, Management and Propagation	1986	Dillard et al.
	Status of the Paddlefish, *Polyodon spathula*, in Canada	1988	Parker
	Contemporary status of the North American paddlefish, *Polyodon spathula*.	1997	Graham
Scaphirhynchus albus	Status of the Pallid Sturgeon	1983	Kallemeyn
	Recovery Plan for the Pallid Sturgeon	1993	Dryer and Sandoval
Scaphirhynchus platorynchus	Life history and status of the shovelnose sturgeon, *Scaphirhynchus platorynchus*.	1997	Keenlyne
Scaphirhynchus p. suttkusi	Systematics, Taxonomy, and Conservation Status of the Endangered Alabama Sturgeon, *Scaphirhynchus suttkusi* Williams and Clemmer (Actinopterygii, Acipenseridae)	1996	Mayden and Kuhajda
Acipenseriformes	Framework for the Management and Conservation of Paddlefish and Sturgeon Species in the United States	1993	Booker et al.

The plans (Table 12-6) generally identify the most significant management and research needs to include five topics. These are population and habitat assessment throughout North America, closer oversight and management of current commercial and sport-harvest, implementation of stocking programs and genetic studies, greater public education, and increased multi-jurisdictional efforts where appropriate (Table 12-6).

12.5.2 CHANGES IN FISHING REGULATIONS

For the Alabama, Atlantic, Gulf, Pallid, shortnose and Kootenai River white sturgeon commercial and sport fisheries are now closed while populations are protected and allowed to increase in abundance.

Table 12-6. Summary of principal threats and management needs from status and rehabilitation plans

Species	Report	Current Impediments	Management Needs
A. brevirostrum	NMFS 1996	Overharvest/bycatch, habitat loss	Population assessment, Genetic studies
	Kynard 1997	By-catch, Barriers to fish passage	Completion of recovery plan, artificial spawning substrate
A. fulvescens	Wallace 1991	Habitat loss, loss of mature fish	Implement plan, inform public, habitat improvement
	Graham 1992	Low abundance	Stock assessment, regulatory framework,
	OMNR 1992	Habitat degradation, commercial exploitation	Habitat protection and restoration, stock/harvest management,
	Moreau and Parrish 1994	Low abundance	Stocking and assessment
	Cavender 1994	Low abundance, multi-jurisdictional mgmt	Protection, public education and assessment
	MNDNR 1995		
	Mathers 1996	Habitat degradation, barriers and flow change	Population assessment, habitat protection and creation
	Ferguson and Duckworth 1997	Variable harvest, fishery compliance	Habitat protection, refined genetic techniques to identify stocks
	Hay-Chmielewski and Whelan 1997	Habitat loss	Conserve and rehabilitate self-sustaining stocks, assessment
	Fond Du Lac 1999	Low abundance, dams, no passage	Stocking, re-establishment
	Societe de la faune et des parcs du Quebec 2000	Over exploitation	Reduction of fishing season Restrict fishing quota
	Alberta Environment 2000	Low abundance of older, large fish, increased fishing pressure	Diligent regulation
	Carlson and Bouton 2000	Extirpated or low abundance, multi-jurisdictional mgmt	Population assessment, stocking
	WIDNR 2000	Habitat loss, low abundance	Life history and genetic information, harvest mgmt. Public education
	Auer, 2001	Low abundance, habitat loss	Population and habitat assessment

Table 12-6. (Continued)

Species	Report	Current Impediments	Management Needs
A. medirostris	Houston, 1988	Estuary pollution, habitat alteration, harvest	Population assessment, Life history
A. o. oxyrinchus	Waldman and Wirgin 1998	Commercial harvest and by-catch	Set restoration goals, Genetics
	Smith and Clugston, 1997	Harvest regulations (CA), habitat loss	Life history, assessment, culture techniques
A. o. desotoi	USFWS-GSRMP 1995	Barriers, habitat loss, Water quality, harvest	Population assessment, Regulatory framework, reduce mortality
A. transmontanus	Cochnauer 1985	Barriers, change in river water flow, land practices	Catch and release fishery
	Lane 1991	Environmental change, commercial fishery	Life history data
	Tracy et al. 1993	Lack of life history data	Reduce harvest and habitat degradation in reservoirs.
A. transmontanus Kootenai River	USFWS 1999	Flow regulation, lost recruitment	US and Canada enact recovery plan
Polyodon spathula	Dillard et al. 1986	Inadequate management, Habitat loss	Regional approach to management, assessment
	Graham 1997	Population assessment, Habitat degradation	Need for more life history information Mitigate for habitat losses
Scaphirhynchus albus	USFWS 1993 Kallemeyn 1983	Habitat loss, alteration Habitat destruction	Restore and protect habitat Assess habitat, Coop effort
Scaphirhynchus platorynchus	Keenlyne 1997	Habitat alteration, pollution	
Scaphirhynchus p. suttkusi	Mayden and Kuhajda 1996	Low abundance and recruitment, lack of life history data	Effective recovery plan, Protection of species

On the West Coast, commercial and sport harvest occurs for white and green sturgeon (Schaffter and Kohlhorst 1999). In the mid-west, spearing, hook and line, and some commercial fisheries remain for lake sturgeon (Bruch, 1999, Baker and Borgeson 1999, Ferguson and Duckworth 1997) and shovelnose sturgeon and paddlefish can be taken in commercial and sport fisheries including snagging for paddlefish (Graham, 1997, Keenlyne 1997, Scarnecchia et al. 1996). Where sport fishing for lake sturgeon is permitted, fishing intensity, as evidenced by growing numbers of permit applications, has increased, and gears and techniques have become more effective. Regulations in some regions are changed on a yearly basis (R. Bruch Wisconsin Department of Natural Resources and E. Baker, Michigan Department of Natural Resources, personal communication). Remaining sport fisheries for Acipenseriformes are closely monitored, and regulations can include a predetermined total harvest for a particular area, total length limits, and season limits. Unregulated and unstudied fisheries for some species remain a concern (Beamesderfer and Farr 1997). Catch and release programs are being considered to help preserve these unique species and reduce fishing mortality as sport fishery interests grow (Scarnecchia and Stewart 1997, Thuemler 1997, Cochnauer et al. 1985).

12.5.3 RELICENSING OF DAMS AND FISH PASSAGE

Many dams built in the early 1900s operated as 'peaking facilities' producing electricity during peak hours of demand, usually 0800 to 1500h. At peaking facilities water from a reservoir is allowed to pass through turbines and is released downstream during hours of peak electrical demand. Generation is terminated when demand decreases or when reservoir water levels drop below established levels. During the hours between peaking, usually from 1500 to 0800h, reservoirs are allowed to fill and little or no water is released for stream biota. Many old hydroelectric facilities have been relicensed in the late 1980s and '90s and others are due for relicensing in the early 2000s. Instream needs of the diverse flora and fauna of some systems are being considered in these relicensing processes, and changes in hydroelectric facility operation from 'peaking' to 'run-of-the-river', ROR or near-ROR are being implemented (Auer 1996a). ROR operation requires that river flows coming into a reservoir must be matched by flows released by the hydroelectric facility, thereby stabilizing river flow downstream.

For some river systems where populations of species at risk are being rehabilitated, the removal of small hydroelectric dams is being suggested (G. Whelan, Michigan Department of Natural Resources, personal communication). For other systems barrier industries and hydroelectric facilities are looking to install fish passage facilities (T. Thuemler, Wisconsin Department of Natural Resources, personal communication). Sturgeon and paddlefish migrate along the river bottom and can be three to five times larger than walleye and salmonids so fishways designed for species which jump, such as salmon, are not effective in passing Acipenseriformes. No fish passage facility to-date has been designed specifically to meet the needs of paddlefish or sturgeon but some groups are working on such designs (Peake et al. 1997, B. Kynard, Conte Fish Laboratory, MA, personal communication).

12.5.4 HABITAT PROTECTION

Barrier structures and water level manipulations have not been the only factors to continue to influence habitat in river systems important to paddlefish and sturgeon. Human development of and encroachment upon wetlands, dredging and the straightening of rivers for boat passage, and draining of backwaters accounts for a tremendous loss of habitat for these species. These organically rich areas are believed to be important nursery and forage areas during early life. Where self-sustaining stocks of sturgeon persist, wetlands and backwater slough habitat are present.

To sustain wetlands the North American Wetlands Conservation Act of 1989 encourages partnerships to conserve North American wetland systems (Appendix A). Other agencies like Environment Canada, the U.S. Environmental Protection Agency, and the U.S. Army Corp of Engineers are concerned with protecting wetlands (Appendix A). Public groups and societies have also formed river alliances and watchdog groups internationally like the International Rivers Network and in North America the American Rivers Program, and the Canadian Heritage Rivers System and American Heritage Rivers as well as the Society for Ecological Restoration (Appendix A). These efforts to protect and maintain entire river ecosystems and habitat can benefit sturgeon and paddlefish.

Awareness of the importance of conserving and enhancing aquatic habitat has increased among management agencies, industries and the general public but human economic concerns and global population growth require an ever-vigilant guard of our freshwater resources. The important connection between habitat protection and restoration and preservation of self-sustaining populations of organisms is getting attention from scientists. A workshop on the 'Science and Management for Habitat Conservation and Restoration Strategies (HabCARES) in the Great Lakes' was held in Barrie, Ontario in 1994 (Kelso and Wooley 1996). This workshop sought to synthesize knowledge on habitat, production, and structure of aquatic and wetland communities, and provide recommendations for effective conservation, restoration and enhancement of aquatic habitat in North America.

12.5.5 GENETICS, AQUACULTURE, STOCKING AND CRYOPRESERVATION

Recent restoration plans for Acipenseriformes (Table 12-5) include the need to re-establish populations of these species in regions where they have been extirpated. Intentional re-establishment of historic populations requires knowledge of available habitat within the system to be stocked, species genetics, availability of brood stock, efficient rearing and hatchery techniques, and hatchery facilities for each species. Research to gain an understanding of the genetic diversity within and between such species as lake, white, Atlantic and Kootenai white sturgeon is underway (Brown et al. 1992, Ferguson et al. 1993, Ferguson and Duckworth 1997, Guenette et al. 1993, Lowie and Krise 1995, Lowie 2000, Porter et al. 1995, Pyatskowit et al 1999, Wirgin et al. 1997). One study has identified a group of genetic markers in eight species of North American sturgeon which can be obtained from non-lethal sampling (May et al. 1997). Another study conducted on lake sturgeon has reveal distinct stock identities, even in rivers separated by only 200 km, such as the Bad River, Wisconsin, and the Sturgeon River, Michigan, in Lake Superior (K. Scribner, Assistant Professor, Michigan State University, personal communication). This finding underscores the importance of identifying brood stocks within regions where re-establishment of stocks is being considered. Little genetic work has been completed for the pallid, shovelnose, shortnose, Alabama, or Gulf sturgeon, or the paddlefish. Without a priori identification of genetic diversity, re-establishing stocks of these species in areas where they have been extirpated from within their historic range will compromise the success of the re-establishment.

Currently there have been some advances made to protect genetically different stocks. A registry has been created to facilitate sharing of knowledge of new stocks as they are being identified. Broodstock strains of Atlantic, Gulf, lake, pallid, shortnose, shovelnose, and white sturgeon and paddlefish are identified in the National Fish Strain Registry-Paddlefish and Sturgeon (Kincaid et al. 1999). This registry can then be used to identify and create clean and appropriate individuals for stocking. Hatchery and rearing techniques have been developed for lake, white, and shovelnose sturgeon and for paddlefish (Anderson 1984, Czeskleba et al. 1985, Conte et al. 1988, Graham et al. 1986, Mims and Shelton 1998). Rearing and enhancement of populations of such species as pallid and Alabama sturgeon, species believed to be close to extinction, will need to be addressed in the near future by management agencies. Mobile and instream hatchery techniques are being proposed to address the potential problem of imprinting at early life stages. Another technique being developed for the preservation of genetic information and endangered species is the creation of a storehouse of frozen gametes. For some species of sturgeon and paddlefish, especially those found in Asia and Europe, techniques to freeze sperm and eggs to aid in preserving individual stocks has already been undertaken. In North America these techniques are being incorporated into aquaculture practices (Mims and Shelton 1998) but may need to be utilized to save endangered or threatened species in captive propagation programs (Andreasen 1998).

12.5.6 EXOTIC SPECIES CONTROL

There is little information available on exotic species and their competition with or predation on or by Acipenseriformes. It is suspected that the exotics such as the round goby, Eurasian carp, and rusty crayfish may prey upon eggs of spawning sturgeon in the Great Lakes. Zebra mussels and invertebrate introductions may impact water quality and indirectly affect spawning areas or food webs critical to sturgeon and paddlefish. Control of exotics using chemicals or barriers also impacts Acipenseriformes in North America. Chemical treatments for sea lamprey are now conducted in late summer in the Great Lakes region and concentrations of 3-trifluoromethyl-4-nitrophenol TFM and the river water pH are closely monitored to insure proper effectiveness of treatment and to reduce mortality of early life stages of lake sturgeon (Whelan, G., Michigan Department of Natural Resources, personal communication). Barriers proposed to keep exotics from moving into river systems will also prevent sturgeon and paddlefish from moving on spawning grounds. More research is needed on methods to prevent the spread of exotics yet to preserve migration corridors for native fishes. There is a tremendous lack of information on interactions between sturgeon and paddlefish and exotic species. This information would be helpful in adding to our ability to manage and restore sturgeon and paddlefish populations.

12.6 The Future for Sturgeon and Paddlefish in North America – Walking a Tightrope

The future of sturgeon and paddlefish species can be assured only through total ecosystem consideration and balanced, long-term commitment to recovery. Conservation efforts, to more effectively protect the earth's biodiversity, have shifted from developing management plans to protect individual endangered and threatened species in limited areas to keeping entire portions of landscapes or ecosystems intact for the benefit of many organisms (Poiani et al. 2000). Conservation management, focused on a landscape or ecosystem scale, often assures the protection of the target species of concern (Poiani et al. 2000). Sturgeon and paddlefish, which can range within and beyond entire watersheds (Auer 1996b) have had their ranges disrupted by barriers. These barriers have isolated populations, degraded water quality and changed flow and thermal regimes. Conservation biologists are convinced that management and rehabilitation decisions must be based on biological data that is collected from remaining natural environments (Angermeier 2000, Povilitis 2001). At present there is no standard definition of a natural environment but they are described as systems that are self-maintaining and renewing (Odum 1997), systems that are not made or influenced by humans or technology (Angermeier 2000) and systems with balance and increasing diversity of forms (McCarty 2001). Fishery biologists and managers working with sturgeon and paddlefish populations are relying on their ability to re-introduce these fishes to areas where the species was extirpated from historic ranges. Continued changes in the environment may prove that "It is no longer safe to assume that all of a species' historic range remains suitable" (McCarty 2001). Conservation efforts for Acipenseriformes should be concentrated on populations persisting in natural systems or near-natural systems and on improving habitat connectivity for remnant populations isolated by barriers. The importance of habitat connectivity has been shown to be an important conservation tool (Beier and Noss 1998). As hydropower development increases there is concern that disruptions in river connectivity will interfere with the ability of aquatic biota to adapt to other human induced changes in environmental condition, such as global warming (Pringle et al. 2000). Conservation biologists are being urged to view climate change as a current threat to species, not simply some future worry (McCarty2001).

Sturgeon and paddlefish are unique fishes and they have an ancestral record dating back 100 million years (Harkness and Dymond 1961). Their unusual life history allows them to attain ages of 100 years or more and their tendency to spawn intermittently, combined with their behaviour of staying near the bottom and in the darkest regions of rivers and lakes, has contributed to their persisting through times of drought and other natural and human induced perturbations. The lengthy period as juveniles and slow recruitment to the adult population has had a significant influence on recovery of Acipenseriformes from such factors as overharvest (Boreman 1997) which is not seen for other commercially sought species. Reduced abundance, combined with these life history characteristics, impedes thorough investigations of basic biology for many sturgeon and paddlefish species.

Perhaps one aspect in the life of Acipenseriformes for which there is the least information concerns impacts of predation on all life stages. It is believed that all species are at their most vulnerable from the fertilized egg, through newly-hatched larvae, and into the early juvenile stage. Usually, when the development of sharp scutes occurs in sturgeons at about 25 cm or when fully functional gill rakers are developed for paddlefish at about 40 cm (Russell 1986), the young fish disperse and are physically less vulnerable. Some early life history strategies such as drifting downstream at night, light avoidance, and finding refuge in gravel substrate during the day help protect sturgeon and paddlefish young from predation. However, most sturgeon spawn in the spring when other species such as suckers and redhorse are also in rivers for spawning. Lake sturgeon in the act of spawning are often closely followed by white suckers in the Sturgeon River, Michigan (Auer, personal observation) but predation on spawned eggs has not been verified. Dadswell et al. (1984) found young-of-the-year, 5 cm FL, shortnose sturgeon in the stomachs of yellow perch in the Androscoggin River, Maine. Non-native salmonids, often stocked in rivers for sport fishing opportunities, may feed on drifting and newly-hatched sturgeons. Other organisms may prey on sturgeon and paddlefish as well. Hay-Chmeilewski (Michigan Department of Natural Resources, personal communication) found crayfish to consume young lake sturgeon stocked into a rearing pond. More research is needed on the early life history of all Acipenseriformes and especially on mortality of and predation on these valuable fishes.

Research and management plans need to consider and address these issues of gaps in our knowledge of basic life history and habitat preferences. Beamesderfer and Farr (1997) illustrated the conflict between protecting stocks, enhancing habitat and allowing for natural recovery and the tendency of biologists and managers to turn toward harvest restrictions and aquaculture programs as the sole method to prevent stocks of Acipenseriformes from losing ground. Stock assessment and general status of at least four North American species of sturgeon remains poorly understood (Table 12-5). There are few status reports or cooperative management programs developed for the Alabama, green, shovelnose, or shortnose sturgeon in North America and there is little basic life history information available for Alabama or green sturgeon. Only two area-specific status reports are available for shortnose sturgeon (Table 12-5), while one 13 year old general review was found for green sturgeon (Houston 1988). A summary work on paddlefish is now 15 years old (Dillard 1986) and for such wide-ranging, commercially valuable species as shovelnose sturgeon and paddlefish there have only recently been published status reports by Keenlyne (1997) and Graham (1997) respectively. Fishery researchers can contribute much to the science of Acipenseriformes if valuable information reported in proceedings of workshops and agency reports were to be published and more readily available to the scientific and management community, and management plans were regularly up-dated.

In efforts to recover stocks of sturgeon and paddlefish, a great deal of attention has been focused on the old threats of barriers to migration and lost habitat. However, new threats to population recovery are on the horizon. Exotic species in the Great Lakes, such as the round *Neogobius melanostomus* and tubenose *Proterorhinus marmoratus* gobies have been observed near spawning lake sturgeon and are suspected of consuming spawned eggs and larvae, yet no information has been published. The influence of exotic species can be subtle and impacts on aquatic food webs remain undocumented. This is seen in the extirpation of a native diatom *Stephanodiscus niagarae* (Julius et al. 1998) from Lake Ontario which was closely followed by the establishment of *Thalassiosira baltica*, a diatom originally described from the Baltic Sea (Edlund et al. 2000). Exotic species invasions can also be extremely obvious, such as the dramatic changes in water clarity produced by the spread of the zebra mussel *Dreissena polymorpha* in the Great Lakes (Mills et al. 1993). How changes in food webs and water quality brought about by exotics has affected native organisms remains poorly understood. Exotics are becoming established throughout North America and some may influence sturgeon and paddlefish at one particular life stage. *Bythotrephes cederstroemi,* the spiny water flea, may reduce the feeding efficiency of young filter-feeding paddlefish, while other exotics may become food sources for or prey upon some life stage of Acipenseriformes. The presence of established exotic species may determine the success of recovery efforts where re-establishment of sturgeon and paddlefish stocks is a priority and needs further research.

The demands on freshwater supplies resulting from human population growth, global climate change, and increased hydroelectric power development clash with efforts to protect natural continuous river ecosystems used by Acipenseriformes. This chapter began with a look at the importance of freshwater to sturgeon, paddlefish, other flora and fauna, and man. As human populations continue to increase, sources of freshwater are and will be dammed to control flooding, to provide hydropower and to create reservoirs of freshwater, placing the future of sturgeon and paddlefish in jeopardy. The future of the single remaining, of three, native species of the genus of shovelnose sturgeon Pseudoscaphirhynchus found in the Aral Sea region, is threatened by the unprecedented withdrawal of water for irrigation of cotton and other crops in Asia (Zholdasova 1997). This scenario can be and is being repeated as demands for water and power increase around the world. The draw-down of reservoirs in Washington and Oregon to generate hydroelectric power demands set in California during the winter of 2000-2001 has decreased the ability of these power facilities to provide the much needed spring flood conditions for spawning salmon and white sturgeon in the Northwest. Nilsson and Berggren (2000) believe that the question of how to protect river ecosystems and human needs "remains one of the most important questions of our time." It will be one of the most important questions for those scientists working with sturgeon and paddlefish.

Those concerned and responsible for the future of Acipenseriformes walk a fine line as they try to balance the life history needs of these fish with the growing human demand for water, energy and land. Government agencies have new directives to protect and restore sturgeon and paddlefish as evidenced by the growing number of management and rehabilitation plans specific to these species. Whether there are sufficient resources to enact the plans remains a concern. Research on a diverse array of habitat, life history, and management questions with regard to these species has increased dramatically over the last 15 years and hopefully will continue. Public interest is on the increase. A number of sizable populations of some species remain and given time and protection they may be able to rebuild to self-sustaining levels. In other areas populations, especially those impacted by human activities or by new threats resulting from the spread of exotic species, appear to have dropped to questionable levels of restoration. The challenge lies in continuing efforts to educate, study, and champion a place for these fishes in the 21st century and beyond.

12.7 References

Alam, S.K., Brim, M.S., Carmody, G.A., and Parauka, F.M. 2000. Concentrations of heavy and trace metals in muscle and blood of juvenile Gulf Sturgeon (*Acipenser oxyrinchus desotoi*) from the Suwannee River, Florida. J. Envir. Sci. Health Pt A, 35: 645-660.

Alberta Environment. 2000. Alberta's lake sturgeon management plan. Fisheries and Wildlife Division, Alberta Environment. 3 pp. www.gov.ab.ca/env/fw/fishing/strgnpln.html.

Anderson, E. R. 1984. Artificial propagation of lake sturgeon *Acipenser fulvescens* (Rafinesque), under hatchery conditions in Michigan. Michigan Department of Natural Resources, Fisheries Division, Report No. 1898. 32 pp.

Andreasen, L. 1998. Captive propagation as a recovery tool for North American sturgeon. *In* Proceedings of the Symposium on the Harvest, Trade and Conservation of North American Paddlefish and Sturgeon, May 7-8, Chattanooga, TN. *Edited by* D.F. Williamson, G.W. Benz, and C.M. Hoover TRAFFIC, North America/World Wildlife Fund, Washington, DC, USA. pp.121-129.

Angermeier, P.L. 2000. The natural imperative for biological conservation. Conserv. Biol. 14:373-381.

Auer, N.A. 1996a. Response of spawning lake sturgeons to change in hydroelectric facility operation. Trans. Amer. Fish. Soc.125: 66-77.

Auer, N.A. 1996b. Importance of habitat and migration to sturgeons with emphasis on lake sturgeon. Can. J. Fish. Aquat. Sci. 53(Suppl.1): 152-160.

Auer, N.A. (*Editor*). 2003. A lake sturgeon rehabilitation plan for Lake Superior. Great Lakes Fishery Commission, Misc. Publ. 2003-02.

Baker, E.A. and Borgeson. D.J. 1999. Lake sturgeon abundance and harvest in Black Lake, Michigan, 1975-1999. N. Amer. J. Fish. Mgmt. 19: 1080-1088.

Baldwin, N.S., Saalfeld, R.W., Ross, M.A. and Buettner, H.J. 1979. Commercial fish production in the Great Lakes 1867-1977. Great Lakes Fishery Commission Technical Report No.3, Ann Arbor, MI.

Beamesderfer, R.C.P. and Farr, R.A. 1997. Alternatives for the protection and restoration of sturgeons and their habitat. Envir. Biol. Fishes 48: 407-417.

Beier, P. and Noss, R.F. 1998. Do habitat corridors provide connectivity? Conserv. Biol. 12(6): 1241-1252.

Birstein, V.J., Doukakis, P., Sorkin, B. and DeSalle, R. 1998. Population aggregation analysis of three caviar-producing species of sturgeons and implications for the species identification of black caviar. Conserv. Biol. 12(4): 766-775.

Bogue, M.B. 2000. Fishing the Great Lakes: An environmental history, 1783-1933. The University of Wisconsin Press. U.S.A.

Booker, W.V., Chapman, F.A., Filipek, S.P., Graham, L.K., Rakes, A.A., . Rasmussen, J.L., Semmens, K.J., St. Pierre, R.A., and Smith, T.I.J. 1993. Framework for the management and conservation of paddlefish and sturgeon species in the United States. U.S. Fish and Wildlife Service. U.S.A.

Boreman, J. 1997. Sensitivity of North American sturgeons and paddlefish to fishing mortality. Environ. Biol. Fishes 48: 399-405.

Boyle, R.H. 1994. The cost of caviar. Amicus Jour. 16: 22-28.

Brousseau, C.S. and Goodchild, G.A. 1989. Fisheries and yields in the Moose River Basin, Ontario. *In* Proceedings of the International Large River Symposium. Vol. 106 *Edited by* D.P. Dodge. Spec. Publ. Can. J. Fish. Aquat. Sci. pp. 145-158.

Brown, J.R., Beckenbach, A.T. and Smith, M.J. 1992. Influence of Pleistocene glaciations and human intervention upon mitrochondrial DNA diversity in white sturgeon (*Acipenser transmontanus*) populations. Can. J. Fish. Aquat. Sci. 49: 358-367.

Bruch, R.M. 1999. Management of lake sturgeon on the Winnebago System – long term impacts of harvest and regulations on population structure. J. Appl. Ichthyol. 15: 142-152.

Busiahn, T.R. (ed.) 1990. Fish Community Objectives for Lake Superior. Great Lakes Fishery Commission Special Publication 90-1.

Carlson, D.M., Pflieger, W.L., Trail, L. and Haverland, P.S. 1985. Distribution, biology and hybridization of *Scaphirhynchus albus* and *S. platorynchus* in the Missouri and Mississippi rivers. *In*. North American Sturgeons, *Edited by* F.P. Binkowski and S.I. Doroshov. Dr. W. Junk Publishers, Dordrecht, Netherlands. pp. 51-59

Carlson, D. and Bouton, D. 2000. DRAFT A recovery plan for the lake sturgeon (*Acipenser fulvescens*) in New York. N.Y. Dept. Envir. Conserv.

Cavender, T.M. 1994. Ohio's lake sturgeon management and recovery program for Lake Erie. Ohio State University.

Chiasson, W.B., Noakes, D.L.G. and Beamish, F.W.H. 1997. Habitat, benthic prey, and distribution of juvenile lake sturgeon *(Acipenser fulvescens)* in northern Ontario rivers. Can. J. Fish. Aquat. Sci. 54: 2866-2871.

Cochnauer, T.G., Lukens, J.R. and Partridge, F.E. 1985. Status of the white sturgeon, *Acipenser transmonuanus*, in Idaho. *In*. North American Sturgeons. *Edited by* F.P. Binkowski and S.I. Doroshov Dr. W. Junk Publishers, Dordrecht, Netherlands. pp.127-133.

Collins, J.J. 1988. Changes in the North Channel fish community, with emphasis on pink salmon (*Oncorhynchus gorbuscha* Walbaum). Hydrobiologia 163: 195-213.

Conte, F.S., Doroshov, S.I., Lutes, P.B. and Strange, E.M. 1988. Hatchery manual for the white sturgeon. Coop. Exten. Univ. Calif. Publ. 3322.

Czeskleba, D.G., AveLallemant, S. and Thuemeler, T.F. 1985. Artificial spawning and rearing of the lake sturgeon, *Acipenser fulvescens*, in Wild Rose State Fish Hatchery, Wisconsin, 1982-1983. *In*. North American Sturgeons. *Edited by* F.P. Binkowski and S.I. Doroshov Dr. W. Junk Publishers, Dordrecht, Netherlands. pp.79-85.

Dadswell, M.J., Taubert, B.D., Squires, T.S., Marchette, D. and Buckley, J. 1984. Synopsis of biological data on shortnose sturgeon, *Acipenser brevirostrum* LeSeuer 1818. FAO Synopsis 140. NMFS Tech. Rept. 14.

Dillard, J.G., Graham, L.K., Russell, T.R., Bassett, B.K. and Ross, D. 1986. The paddlefish: status, management and propagation. North Central Div., Amer. Fish. Soc., Columbia, MO.

Doyon, C, Fortin, R. and Spear, P.A. 1999. Retinoic acid hydroxylation and teratogenesis in lake sturgeon (*Acipenser fulvescens*) from the St. Lawrence River and Abitibi region, Quebec. Can. J. Fish. Aquat. Scis. 56(8):1428-1436.

Dryer, M.P. and Sandvol, A.J. 1993. Recovery Plan for the Pallid Sturgeon (*Scaphirhynchus albus*). Dept. Interior, USFWS.

Edlund, M.B., Taylor, C.M., Schelske, C.L.and Stoermer, E.F. 2000. *Thalassiosira baltica*, (Ballariophyta), a new exotic species in the Great Lakes. Can. J. Fish. Aquat. Sci. 54: 610-615.

Ferguson, M.M., Bernatchez, L., Gatt, M., Konkle, B.R., Lee, S., Malott, M.L. and McKinley, R.S. 1993. Distribution of mitrochondrial DNA variation in lake sturgeon (*Acipenser fulvescens*) from the Moose River basin, Ontario, Canada. J. Fish Biol. 43 (Suppl. A): 91-101.

Ferguson, M.M. and Duckworth, G.A. 1997. The status and distribution of lake sturgeon, *Acipenser fulvescens*, in the Canadian provinces of Manitoba, Ontario, and Quebec: a genetic perspective. Envir. Biol. Fishes 48: 299-309.

Fond Du Lac Natural Resources Program. 1999. Sturgeon Restoration Plan for the Upper St. Louis River. Cloquet, MN.

Galbreath, J.L. 1985. Status, life history, and management of Columbia River white sturgeon, *Acipenser transmontanus*. *In*. North American Sturgeons. *Edited by* F.P. Binkowski and S.I. Doroshov Dr. W. Junk Publishers, Dordrecht, Netherlands. pp.119-125.

Gleick, P.H. 2000. The World's Water 2000-2001. Island Press. Washington, D.C.

Gnam, R. 1998. Implementation of 1997 CITES listing made effective April 1, 1998. *In* Proceedings of the Symposium on the Harvest, Trade and Conservation of North American Paddlefish and Sturgeon, May 7-8, Chattanooga, TN. *Edited by* D.F. Williamson, G.W. Benz, and C.M. Hoover TRAFFIC, North America/World Wildlife Fund, Washington, DC, USA. pp.192-214.

Graham, L.K. 1992. A plan for recovery of the lake sturgeon in Missouri. Missouri Dept Conserv., Jefferson City, MO.

Graham, L.K. 1997. Contemporary status of the North American paddlefish, *Polyodon spathula*. Environ. Biol. Fishes 48: 279-289.

Graham, L. K. and Rasmussen, J.L. 1998. A MICRA perspective on closing paddlefish and sturgeon commercial fisheries. *In* Proceedings of the Symposium on the Harvest, Trade and Conservation of North American Paddlefish and Sturgeon, May 7-8, Chattanooga, TN. *Edited by* D.F. Williamson, G.W. Benz, and C.M. Hoover TRAFFIC, North America/World Wildlife Fund, Washington, DC, USA. pp.130-142.

Graham, L.K.,Hamilton, E.J., Russell, T.R., and Hicks, C.E. 1986. The culture of paddlefish – A review of methods. *In* The paddlefish: Status, management and propagation. *Edited by* J.G. Dillard, L.K. Graham, T.R. Russell, B.K. Bassett and D. Ross. North Central Div., Amer.Fish. Soc., Columbia, MO. pp 78-94.

Graham, P.J. 1981. Status of white sturgeon in the Kootenai River. Montana Dept Fish, Wildl. and Parks. Kalispell, MT.

Guenette, S., Fortin, R and Rassat, E. 1993. Mitrochondrial DNA variation in lake sturgeon (*Acipenser fulvescens*) from the St. Lawrence River and James Bay drainage basins in Quebec, Canada. Can. J. Fish. Aquat. Sci. 50(30): 659-664.

Harkness, W.J.K. and Dymond., J.R. 1961. The lake sturgeon. Ontario Department of Lands and Forests.

Hay-Chmielewski, E.M. and Whelan, G.E. 1997. Lake sturgeon rehabilitation strategy. Michigan Dept. Nat. Res., Fish. Div., Spec. Rept. Number 18.

Haywood, M. 1998. Sturgeon and paddlefish and their current conservation status. *In* Proceedings of the Symposium on the Harvest, Trade and Conservation of North American Paddlefish and Sturgeon, May 7-8, Chattanooga, TN. *Edited by* D.F. Williamson, G.W. Benz, and C.M. Hoover. TRAFFIC, North America/World Wildlife Fund, Washington, DC, USA. pp. 8-37.

Hoover, C. 1998. Import and export of sturgeon and paddlefish in the United States. *In* Proceedings of the Symposium on the Harvest, Trade and Conservation of North American Paddlefish and Sturgeon, May 7-8, Chattanooga, TN. *Edited by* D.F. Williamson, G.W. Benz, and C.M. Hoover TRAFFIC, North America/World Wildlife Fund, Washington, DC, USA. pp.162- 171.

Houston, J.J. 1987. Status of the lake sturgeon, *Acipenser fulvescens*, in Canada. Can. Field-Nat. 10(2): 171-185.

Houston, J.J. 1988. Status of the green sturgeon, *Acipenser medirostris*, in Canada. Can. Field-Nat. 102(2): 286-290.

Hoxmeier, R.J.H., and DeVries, D.R. 1996. Status of paddlefish in the Alabama waters of the Tennessee River. N. Amer. J. Fish. Mgmt. 16: 935-938.

Johnson, J. J. 1987. Protected Fishes of the United States and Canada. Amer.Fish. Soc., Special Publ.

Julius, M.L., Stoermer, E.F., Taylor, C.M. and Schelske, C.L. 1998. Local extirpation *of Stephanodiscus niagarae* Ehrenb. (Bacillariophyta) in the recent limnological record of Lake Ontario. J. Phycol. 34: 766-771.

Kallemeyn, L. 1983. Status of the pallid sturgeon. Fisheries 8(1): 3-9.

Keenlyne, K. D. 1997. Life history and status of the shovelnose sturgeon, *Scaphirhynchus platorynchus*. Envir. Biol. Fishes 48: 291-298.

Kelso, J.R.M. and Wooley, C. 1996. Introduction to the International Workshop on the Science and Management for Habitat Conservation and Restoration Strategies (HabCARES). Can. J. Fish. Aquat. Sci. 53(Suppl.1): 1-2.

Kelso, J.R.M., Steedman, R.J. and Stoddart, S. 1996. Historical causes of change in Great Lakes fish stocks and the implications for ecosystem rehabilitation. Can. J. Fish. Aquat. Sci. 53(Suppl.1): 10-19.

Kempinger, J.J. 1988. Spawning and early life history of lake sturgeon in the Lake Winnebago System, Wisconsin. Amer. Fish. Soc. Symp.5: 110-122.

Kincaid, H.L., Mengel,L.J., Gray, M.J. and Brimm, S. 1999. National Fish Strain Registry – Paddlefish and Sturgeon (NFSR-PS). USGS, Wellsboro, PA.
Kynard, B. 1997. Life history, latitudinal patterns, and status of shortnose sturgeon *Acipenser brevirostrum*. Envir. Biol. Fishes. 48: 319-334.
Lane, E.D. 1991. Status of the white sturgeon, *Acipenser transmontanus*, in Canada. Can. Field-Nat. 105(2): 161-168
Larson, T. 1988. The lake sturgeon fishery of Lake Wisconsin, 1978-1985. Wisconsin Department of Natural Resources, Fish Management Report 136. 34 pp.
Lowie, C. 2000. Workshop proceedings: Great Lakes lake sturgeon genetics: status, needs and standardization. USFWS administrative report 2000-01, Amherst, N.Y.
Lowie, C.E. and Krise, W.F. 1995. Proceedings on the workshop for the conservation and management of the lake sturgeon in the Great Lakes, Niagara Falls, N.Y., October 26-27, 1994. Administrative report 95-01, U.S.F.W.S., Lower Great Lakes Fishery Resources Office, Amherst, NY.
MacDonald, D.D., Ikonomou, M.G., Rantalaine, A.L., Rogers, I.H., Sutherland, D. and Van Oostdam, J. 1997. Contaminants in white sturgeon (*Acipenser transmontanus*) from the upper Fraser River, British Columbia, Canada. Enviro. Tox. Chem. 16(3): 479-490.
Mathers, A. 1996. DRAFT Lake sturgeon rehabilitation within the Bay of Quinte. Ontario Ministry of Natural Resources.
May, B., Krueger, C.C. and Kincaid, H.L. 1997. Genetic variation at microsatellite loci in sturgeon: primer sequence homology in *Acipenser* and *Scaphirhynchus*. Can. J. Fish. Aquat. Sci. 54: 1542-1547.
Mayden, R.L. and Kuhajda, B.R. 1996. Systematics, taxonomy, and conservation status of the endangered Alabama sturgeon, *Scaphirhynchus suttkusi*, Williams and Clemmer (Actinopterygii, Acipenseridae). Copeia (2): 241-273.
McCarty, J.P. 2001. Ecological consequences of recent climate change. Conserv. Biol. 15(2): 320-331.
McKinley, R.S., Singer, T.D., Ballantyne, J.S., Power, G. 1993. Seasonal variation in plasma nonesterified fatty acids of lake sturgeon (*Acipenser fulvescens*) in the vicinity of hydroelectric facilities. Can. J. Fish. Aquat. Sci. 50: 2440-2447.
Mills, E.L., Leach, J.H., Carlton, J.T. and Secor, C.L. 1993. Exotic species in the Great Lakes: a history of biotic crises and anthropogenic introductions. J. Great Lakes Res. 19(1): 1-54
Mims, S.D. and Shelton, W.L. 1998. Monosex culture of paddlefish and shovelnose sturgeon. *In* Proceedings of the Symposium on the Harvest, Trade and Conservation of North American Paddlefish and Sturgeon, May 7-8, Chattanooga, TN. *Edited by* D.F. Williamson, G.W. Benz, and C.M. Hoover TRAFFIC, North America/World Wildlife Fund, Washington, DC, USA. pp. 51-66
Minnesota Department of Natural Resources. 1995. Fisheries Long Range Plan – Sturgeon and Paddlefish.
Moreau, D.A. and Parrish, D.L. 1994. A study of the feasibility of restoring lake sturgeon to Lake Champlain. Lake Champlain Basin Program Technical Report No. 9.
Mosher, T.D. 1998. Sturgeon and paddlefish sport fishing in North America. *In* Proceedings of the Symposium on the Harvest, Trade and Conservation of North American Paddlefish and Sturgeon, May 7-8, Chattanooga, TN. *Edited by* D.F. Williamson, G.W. Benz, and C.M. Hoover TRAFFIC, North America/World Wildlife Fund, Washington, DC, USA. pp. 51-66.
National Marine Fisheries Service. 1996. Status review of shortnose sturgeon in the Androscoggin and Kennebec Rivers. Gloucester, MA.
Nilsson, C. and Berggren, K. 2000. Alterations of riparian ecosystems caused by river regulation. Bioscience. 50(9): 783-792.
NTIS. 1998. Tagging of green sturgeon for estimation of population size and range of migrations. Hoopa Valley Tribal Council, Fish. Dept., Fish Wildl. Rest. Prog.,National Tech. Infor. Serv., Springfield, VA
Odum, E.P. 1997. Ecology: a bridge between science and society. Sinauer Associates, Sunderland, Mass.
Ontario Ministry of Natural Resources. 1992. A Draft management strategy for lake sturgeon in Ontario. OMNR, Edition 6.
Parker, B.J. 1988. Status of the Paddlefish, *Polyodon spathula*, in Canada. Can. Field-Nat. 102(2): 291-295.
Peake, S., Beamish,F.W.H., McKinley, R.S., Scruton, D.A. and Katopodis, C. 1997. Relating swimming performance of lake sturgeon, *Acipenser fulvescens*, to fishway design. Can. J. Aquat. Sci. 54: 1361-1366.
Pitman, V.M. and Parks, J.O. 1994. Habitat use and movement of young paddlefish (*Polyodon spathula*). J. Freshwater Ecol. 9(3): 181-189.
Poiani, K.A., Richter, B.D., Anderson, M.G. and Richter, H.E. Biodiversity conservation at multiple scales: Functional sites, landscapes, and networks. Bioscience 50(2): 133-146.
Porter, B., Cavender, T., Fuerst, P. and Nickell, T. 1995. The genetic status of lake sturgeon in Lake Erie and other populations from the Laurentian Great Lakes. Ohio State University, Ohio Dept. Nat. Res., Columbus, Ohio.
Povilitis, T. 2001. Toward a robust natural imperative for conservation. Conserv. Biol. 15(2): 533-535.

Pringle, C. M., Freeman, M.C. and Freeman, B.J.. 2000. Regional effects of hydrologic alterations on riverine macrobiota in the new world: Tropical-Temperate comparisons. Bioscience 50(9): 807-823.

Pyatskowit, J.D., Kruger, C.C., Kincaid, H.L. and May, B. 2001. Inheritance of microsatellite loci in the polyploidy lake sturgeon (*Acipenser fulvescens*). Genome. 44(2):185-191.

Rasmussen, J.L. 1998. What is MICRA and why does it work? *In* Proceedings of the Symposium on the Harvest, Trade and Conservation of North American Paddlefish and Sturgeon, May 7-8, Chattanooga, TN. *Edited by* D.F. Williamson, G.W. Benz, and C.M. Hoover TRAFFIC, North America/World Wildlife Fund, Washington, DC, USA. pp. 218-229.

Richey, D. 1987. Lake Huron Spotlight: Sturgeon Spearing. Great Lakes Fisherman 17(2): 61-62.

Rochard, E., Castelnaud, G. and LePage, M. 1990. Sturgeons (Pisces:Acipenseridae); threats and prospects. J. Fish Biol. 37(Suppl. A): 123-132.

Ruelle, R. and Henry, C. 1994. Life history observations and contaminant evaluation of pallid sturgeon. U.S. Fish and Wildlife Service Contaminants Program, Pierre, S.D.

Ruelle, R. and Keenlyne, K.D. 1993. Contaminants in Missouri River Pallid sturgeon. Bull. Environ. Contam. Toxicol. 50: 898-906.

Russell, T.R. 1986. Biology and Life History of the Paddlefish – A Review. P. 2-19 *In:* Dillard, J.G., L.K. Graham and T. R. Russell (eds.) The Paddlefish: Status, management and Propagation. American Fisheries Society.

Scarnecchia, D.L. and Stewart, P.A. 1997. Implementation and evaluation of a catch-and-release fishery for paddlefish. N. Amer. J. Fish. Mgmt. 17: 795-799.

Scarnecchia, D.L., Stewart, P.A and Lim, Y. 1996. Profile of recreational paddlefish snaggers on the lower Yellowstone River, Montana. N. Amer. J. Fish. Mgmt. 16: 872-879.

Scott, W.B. and Crossman, E.J. 1973. Freshwater Fishes of Canada. Fisheries Research Board of Canada, Ottawa.

Schaffter, R.G. and Kohlhorst, D.W. 1999. Status of white sturgeon in the Sacramento-San Joaquin Estuary. Calif. Fish. Game 85(1): 37-41.

Scheidegger, K. 2000. DRAFT Wisconsin Lake sturgeon management plan. Wisconsin Department of Natural Resources. www.dnr.state.wi.us/org/water/ftp/fish/sturgeon/Index.htm.

Secor, D.H. and Waldman, J.R. 1999. Historical abundance of Delaware Bay Atlantic sturgeon and potential rate of recovery. American Fish. Soc. Symp. 23: 203-216.

Semakula, S.N. and Larkin, P.A. 1968. Age, growth, food, and yield of the white sturgeon (*Acipenser transmontanus*) of the Fraser River, British Columbia. J. Fish. Res. Bd. Can. 25(12): 2589-2602

Slade, J.W. and Auer, N.A. 1997. Status of lake sturgeon in Lake Superior. Report to the Lake Superior Technical Committee.

Smith, S.H. 1968. Species succession and fishery exploitation in the Great Lakes. J. Fish. Res. Bd. Can. 25(4): 667-693.

Smith, T.I.J., and Clugston, J.P. 1997. Status and management of Atlantic sturgeon, *Acipenser oxyrinchus*, in North America. Envir. Biol. Fishes 48: 335-346.

Societe de la faune et des parcs du Quebec. 2000. Plan de gestion de l'esturgeon jaune du fleuve Saint-Laurent 2000-2002. FAPAQ, Directions de l'amenagement de la faune du Centre-du-Quebec, de Lanaudiere, de la Monteregie et de Montreal.

Thuemler, T.F. 1988. Movements of young lake sturgeons stocked in the Menominee River, Wisconsin. American Fisheries Society Symposium 5: 104-109.

Thuemler, T.F. 1997. Lake sturgeon management in the Menominee river, a Wisconsin-Michigan boundary water. Envir. Biol. Fishes 48: 311-317.

Thuemler, T., Baker, E.A. and Elliott, R. 1999. DRAFT Lake sturgeon plan for the Green Bay basin. Wis. Dept Nat. Res.

Todd, R.M. 1998. Sturgeon and paddlefish commercial fishery in North America. *In* Proceedings of the Symposium on the Harvest, Trade and Conservation of North American Paddlefish and Sturgeon, May 7-8, Chattanooga, TN. *Edited by* D.F. Williamson, G.W. Benz, and C.M. Hoover TRAFFIC, North America/World Wildlife Fund, Washington, DC, USA. pp.42-50.

Tower, W.S. 1908. The passing of the sturgeon: A case of the unparalleled extermination of a species. The Popular Science Monthly. 73: 361-371.

Tracy, C.A. 1993. Status of the white sturgeon resource in mainstem Columbia River. Washington Department of Fisheries, Battle Ground, WA.

U.S. Fish and Wildlife Service and Gulf States Marine Fisheries Commission. 1995. Gulf Sturgeon Recovery Plan. Atlanta, Georgia.

U.S. Fish and Wildlife Service. 1999. Recovery plan for the white sturgeon (*Acipenser transmontanus*):Kootenai river population. U.S. Fish and Wildlife Service, Portland, OR 96 pp.

Wakeford, A. 2001. State of Florida Conservation Plan for Gulf Sturgeon (*Acipenser oxyrinchus desotoi*). Florida Fish and Wildlife Conservation Commission FMRI, St. Petersburg, Florida, Technical Report TR-8.
Waldman, J.R. and Secor, D.H. 1998. Caviar trade in North America: An Historical Perspective. *In* Proceedings of the Symposium on the Harvest, Trade and Conservation of North American Paddlefish and Sturgeon, May 7-8, Chattanooga, TN. *Edited by* D.F. Williamson, G.W. Benz, and C.M. Hoover TRAFFIC, North America/World Wildlife Fund, Washington, DC, USA. pp.77-91.
Waldman, J.R. and Wirgin, I.I. 1998. Status and Restoration Options for Atlantic Sturgeon in North America. Conserv.Biol. 12(3): 631-638.
Wallace, R.G. 1991. Species recovery plan for lake sturgeon in the lower Saskatchewan River. Saskatchewan Parks and Renewable Resources, Fisheries Technical report 91-3, Saskatoon, CA.
Williams, J.E., Johnson, J.E., Hendrickson, D.A., Contreras-Balderas, S., Williams, J.D., Navarro-Mendoza, M., McAllister, D.E.,and J.E. Decon. 1989. Fishes of North America endangered, threatened, or of special concern. Fisheries. 14(6): 2-20.
Wooley, C.M. 1985. Evaluation of morphometric characters used in taxonomic separation of Gulf of Mexico sturgeon, *Acipenser oxyrinchus desotoi*. *In*. North American Sturgeons. *Edited by* F.P. Binkowski and S.I. Doroshov Dr. W. Junk Publishers, Dordrecht, Netherlands. pp.97-103.
Wirgin, I.I., Stabile, J.E. and Waldman, J.R. 1997. Molecular analysis in the conservation of sturgeons and paddlefish. Envir. Biol. Fishes 48: 385-398
Zholdasova, I. 1997. Sturgeons and the Aral Sea ecological catastrophe. Envir. Biol. Fish. 48: 373-380.

Appendix A. Internet addresses for organizations cited in this chapter.

American Heritage Rivers: www.epa.gov/rivers/initiative.html
American Rivers: www.amrivers.org
Bureau of Reclamation: www.usbr.gov/power/edu/history.htm
Canadian Heritage Rivers System: www.chrs.ca/Main_e.html
Committee on the Status of Endangered Wildlife in Canada (COSEWIC): www.speciesatrisk.gc.ca/Species
International Rivers Network: www.irn.org
IUCN, International Union for Conservation of Nature and Natural Resources, Red List of Threatened Animals for 1996: www.iucn.org/themes/ssc/redlists/criteria.htm
North American Wetlands Conservation Act of 1989: http://northamerican.fws.gov/NAWCA/ Environment Canada: www.on.ec.gc.ca/wildlife/conservation/wetland/wetland.html
The Society for Ecological Restoration: www.serontario.org
United Nations Environment Programme and the World Conservation Monitoring Centre: www.unep-wcmc.org
United States Fish and Wildlife Service: ecos.fws.gov/webpage/webpage_usa_lists.html 2/6/2001
United States Army Corp of Engineers: www.wes.army.mil/el/wetlands
United States Environmental Protection Agency: www.epa.gov/owow/wetlands
Water Wheel on the Fox River, Wisconsin in 1882: www.iclei.org

CHAPTER 13

AQUACULTURE

JOEL P. VAN EENENNAAM[1], FRANK A. CHAPMAN[2], AND PETER L. JARVIS[3]
[1]*University of California, Department of Animal Science, One Shields Avenue, Davis, California 95616 U.S.A. (email: jpvaneenennaam@ucdavis.edu).*
[2]*University of Florida, Department of Fisheries and Aquatic Sciences, 7922 N.W. 71st Street, Gainesville, Florida 32653 U.S.A. (email:* fchapman@ufl.edu*).*
[3]*Great Lakes Laboratory for Fisheries & Aquatic Sciences Fisheries and Oceans Canada, 867 Lakeshore Road, Burlington, Ontario, Canada L7R 4A6 (email: JarvisP@dfo-mpo.gc.ca).*

13.1 Introduction

Historically, North America was an important source of sturgeon flesh and caviar. During 1880-1900 the Atlantic sturgeon was considered one of the most commercially important fisheries in the U.S. and Canada, with a peak harvest of 3,350 metric tons (t) during 1890 (Smith and Clugston 1997), lake sturgeon landings peaked around 1895 with 3,000t harvested (Harkness and Dymond 1961), the white sturgeon catch on the Columbia River peaked at 2,494t during 1892 (Pruter 1972), and Canada was contributing 75% of the world's supply of caviar during the late 1800s, with an estimated production of 86t during 1896 (Prince 1905). However, overfishing and habitat degradation (e.g., damming of spawning rivers) rapidly decimated sturgeon populations at which point the fishery collapsed and the caviar trade shifted to the Caspian Sea region. The decline in North American sturgeon fisheries during the late 1800s and early 1900s prompted the first attempts at artificial reproduction, to enhance and rehabilitate the fishery (Post 1890; Ryder 1890; Stone 1900; Carter 1904; Meehan 1909; Harkness and Dymond 1961). These attempts were only moderately successful due to difficulties in procuring broodstock and incubating adhesive eggs. Sturgeon hatchery research and development that was continuing in the former USSR (Barannikova 1987) provided crucial information which was needed when interest in sturgeon culture was rekindled. The next significant period of sturgeon culture in North America was from 1960 to the mid-1980s when paddlefish and Atlantic, shortnose, white and lake sturgeon were first spawned and cultured in hatcheries and laboratories (Purkett 1963; Smith et al. 1980, 1981; Buckley and Kynard 1981; Doroshov et al. 1983; Ćzeskleba et al. 1985). Since then, all North American species have been successfully reproduced in captivity, including gulf sturgeon (Parauka et al. 1991), shovelnose and pallid sturgeon (Keenlyne 1993; H. Bollig, U.S. Fish and Wildlife Service, Gavins Point National Fish Hatchery, Yankton, South Dakota, personal communication) and green sturgeon (Van Eenennaam et al. 2001). However, only the

G.T.O. LeBreton et al. (eds.), Sturgeons and Paddlefish of North America, 277–311.
© 2004 *Kluwer Academic Publishers. Printed in the Netherlands.*

aquaculture of white sturgeon has become fully established since the mid-1990s, including closing of the reproductive cycle in captivity and spawning of F2 domestic broodstock, intensive rearing of juveniles and adults, and marketing of food fish and caviar. The material in this chapter is based primarily on the farming of white sturgeon, and modifications of the culture techniques presented are expected to be developed for other sturgeon species as their specific husbandry methodologies are researched.

13.2 White Sturgeon Culture

There are a few recent reviews of sturgeon culture in Europe and the Caspian Sea region (Ronyai and Varadi 1995; Bronzi et al. 1999; Hochleithner and Gessner 1999; Chebanov and Ballard 2001) but North American sturgeon culture is based primarily on the research and farming of white sturgeon (Doroshov 1985; Conte et al. 1988; Doroshov et al. 1997; Mims et al. 2002). The white sturgeon is considered to be one of the superior sturgeon species for aquaculture. This superiority is due to its high quality meat, its ready acceptance of artificial diets, its tolerance to high density and fast growth at water temperatures 20-23°C (Moberg and Doroshov 1992), and more recently because of its high quality caviar (Rodewald 2001; Stanley 2001).

13.3 Wild Broodstock

Sturgeon aquaculture begins with the capture and spawning of wild broodstock. Mature females and males captured on the spawning grounds are transported to hatcheries and induced to ovulate and spermiate, usually within a few days, but sometimes mature sturgeon are captured during winter and are held for several months prior to the spring spawning induction. In some cases, vitellogenic females or previously spawned males and females are held in captivity for a year or more, until they complete their next reproductive cycle (Mohler and Fletcher 1999; T. Patterson, College of Southern Idaho, Twin Falls, Idaho, personal communication). Wild-caught white sturgeon held in captivity at the College of Southern Idaho readily feed on live trout that are provided, but in general, it is very difficult to encourage wild sturgeon to resume feeding in captivity (DiLauro 1998).

13.3.1 CAPTURE AND HANDLING

Adult sturgeon are captured using a variety of methods, including gill nets, set lines, baited hooks, snagging, and trammel nets (Conte et al. 1988; Hochleithner and Gessner 1999). After capture, the fish are moved using stretchers made of canvas or some other similar material (Conte et al. 1988) and are transported in large (1000-2000L) insulated tanks. These tanks are supplied with supplemental aeration or oxygen to ensure that dissolved oxygen concentrations remain above 80% saturation and the insulation helps to maintain water temperatures similar to the temperature at the site of capture. Mims et al. (2002) report a safe loading rate for transportation of broodstock to be approximately 1kg of body weight per 4L of water. At the hatchery, the broodfish are maintained at spawning temperatures (14-16°C) using flow-through or recirculation systems.

13.3.2 SPAWNING INDUCTION AND EGG FERTILIZATION

Sturgeon will not spawn naturally under hatchery conditions due to the lack of environmental cues, therefore artificial spawning induction by injection of exogenous hormone is used. Male broodstock are usually procured in larger numbers than females and most can be readily induced to spermiate. However, spawning induction is only successful for female broodstock that have reached an advanced stage of gonadal maturity (Section 6.4).

Under hatchery conditions it is critical to control water temperature during the spawning season so that the female broodstock complete normal egg maturation. Optimal water temperature for spawning is species specific, but it generally ranges between 14-17°C for most sturgeon (Conte et al. 1988). Spawning at cooler water temperatures does not appear to be as detrimental as spawning at elevated temperatures and successful ovulations have been induced at 9-10°C. Hochleithner and Gessner (1999) report that sturgeon do not complete final maturation and will undergo ovarian atresia during continuous exposure to water temperatures below 6°C or above 18°C.

Ova and semen are obtained after injecting fish with common carp pituitary material (CCP) or with the mammalian gonadotropin-releasing hormone analog (GnRHa) [D-Ala6, Des-Gly10]-LH-RH Ethylamide (Lutes et al. 1987; Conte et al. 1988). These hormones are currently not approved for sturgeon and U.S. Food and Drug Administration (FDA) Investigational New Animal Drug (INAD) permits are required for their use. Wild broodstock captured on the spawning grounds usually require only a small amount of exogenous hormone to initiate ovulation, for example, a total dose of $10\mu g.kg^{-1}$ of GnRHa, reported by Goncharov et al. (1991). To reduce handling stress, the preferred injection site is in the musculature, anterior to the dorsal fin, given while the fish is in the water. Males receive a single injection and females are given two injections, 10% and 90% of the total dose approximately 12 hours apart.

Latency time for female white sturgeon after the second injection is about 20-21 hours, ranging from 18-28 hours, at water temperatures 15-16°C. After the first ovulated eggs are observed, approximately 1-2 hours are allowed to pass before egg collection begins. This delay is to ensure that ovulation has been completed. The egg collection techniques are discussed in section 6.5. Spermiation usually begins about 18-22 hours after injection when the fish are held at 15-16°C and continues for another 18-24 hours. Hochleithner and Gessner (1999) observed the best sperm collection time to be between 24-36 hours after hormonal injection and Williot et al. (2000) reported that 36 hours after hormonal stimulation provided the most motile spermatozoa. Semen collection techniques have been illustrated by Conte et al. (1988).

Fertilization is conducted within 5-10 minutes of egg collection, although fertilization success has been reported after 2-4 hours, if the eggs are held in coelomic fluid at spawning temperature (Dettlaff et al. 1993; Gisbert and Williot 2002). The semi-dry method of fertilization is used (Dettlaff et al. 1993), but care must be taken to remove any excess coelomic fluid prior to the addition of semen, because this fluid interferes with fertilization. Semen is usually diluted 1:200 with hatchery water and then without delay the diluted semen is added to the bowl of eggs. The exact dilution will depend on the quality of the individual males and may be adjusted accordingly. The eggs and semen are gently mixed for 2-3 minutes, or until a few eggs begin to stick to the sides of the bowl. At this time, the water and semen solution is poured off and a de-adhesion suspension is immediately added to the eggs.

Sturgeon eggs have an outer glycoprotein matrix (jelly) that hydrates upon contact with water at spawning (Cherr and Clark 1984, 1985). This jelly coat needs to be either covered with small particles or chemically removed, to prevent egg clumping during incubation (Conte et al. 1988). Fuller's earth (Sigma Chemical Company, St. Louis, Missouri) is an excellent substitute for the historically used river silt, and this material has very uniform small-sized particles (100-200 microns). Other materials used for de-adhesion include milk powder, talcum, starch (Hochleithner and Gessner 1999) and pottery clay. The amount of jelly coat hydration varies between females and between species, thus the length of time required for egg de-adhesion will vary from 30-60 minutes. The process is also temperature dependent such that cooler water temperatures require a longer period of de-adhesion. The de-adhesion suspension is changed every 10-15 minutes to maintain dissolved oxygen and water temperature at optimal levels. De-adhesion by silting can be automated by using an upwelling flow (Monaco and Doroshov 1983) or an aerated container (Dettlaff et al.1993). Methods for the chemical elimination of the jelly coat are also available (Kowtal et al. 1986; Bouchard and Aloisi 2002). After de-adhesion the eggs are treated with a surface disinfection of 100 ppm of iodine for 15 minutes (Yesaki at al. 2002) or the process of disinfection and de-adhesion can be combined (Bouchard and Aloisi 2002).

13.3.3 EGG INCUBATION AND LARVAL REARING

Most North American sturgeon hatcheries use adaptations of the MacDonald jar incubator or trout egg barrel for egg incubation (Conte et al. 1988). The quantity of eggs incubated is approximately 100-120 ml of eggs per litre of total jar volume, and a very gently upwelling water flow (approximately 2-3L.min^{-1} for the MacDonald jars) is used during the sensitive early life stages (cleavage through neurulation). The flow is regulated so that the top layer of eggs are gently moving but remain suspended in a horizontal plane, and are not pushed up into the water column. This water flow ensures that the eggs are continually exposed to oxygenated water, but the potential physical shock to the developing embryos from too much tumbling, is minimized. After neurulation, which occurs at 50-60 hours post-fertilization when incubated at 15-16°C, water flow can be safely increased to about 5-8L.min^{-1} to completely mix and roll the eggs and reduce the spread of fungus. Dead eggs will develop fungus and without proper water flow will spread to live eggs and form large fungal clumps. At first, individual eggs and couplets with fungus are more buoyant then the live eggs and can be easily siphoned out of the incubation jars, but any large clumps will tend to settle in the mid-region of the jar. Optimal embryonic development for white and lake sturgeon occurs at 14-17°C, while survival is reduced at temperatures above 18ºC (Wang et al. 1985). After hatch, yolk-sac larvae are more tolerant to warmer temperatures (Buddington et al. 1993) and white sturgeon larvae have been successfully acclimated to 18-22ºC over a 2-6 hour period. Egg incubation systems are constructed to allow for collection of the hatching larvae into individual or communal larval tanks. Most hatched larval sturgeon are pelagic, swimming up and out of the incubation jars and into the larval receiving tanks, without intervention by the hatchery personnel.

There are numerous systems for larval and juvenile culture and virtually all of them are site- and species- specific. In general, early larval culture tanks may be circular, square or rectangular, 1-2m in diameter with a water depth of 25-40cm, and with flow rates of 8-19L.min^{-1} (Conte at al. 1988). Recirculation water is not commonly used during larval culture because of the higher sensitivity of larvae to water quality parameters. Stocking densities of larvae are approximately 15-20 $fry.l^{-1}$ (Conte et al. 1988) or 2,000-5,000 $fry.m^{-2}$ (Hochleithner and Gessner 1999). However, some species, such as Atlantic sturgeon, perform best at lower densities, e.g., 7.4 $fish.l^{-1}$ (Mohler et al. 2000).

Because young sturgeon are typically benthic-oriented the area of the feeding surface is important in creating optimal conditions for feeding (Sbikin and Budayev 1991), and recent studies on stocking rates for juveniles are based on surface area of the tank bottom. Once sturgeon begin to feed they are transferred to larger tanks (3-6m diameter circular, 0.5-1.0m deep) and densities can range from $10kg.m^{-2}$ for sturgeon weighing 10g, to $20kg.m^{-2}$ for fish weighing 100g (Hochleithner and Gessner 1999). However, lower densities of less than one fish per litre or $3.7kg.m^{-2}$ was optimal for rearing lake sturgeon weighing 8-22g (Fajfer et al. 1999), and a density of $2.4kg.m^{-2}$ was recommended for successful culture of pallid and shovelnose sturgeon (H. Bollig, U.S. Fish and Wildlife Service, Gavins Point National Fish Hatchery, Yankton, South Dakota, personal communication). Jodun et al (2002) reared juvenile Atlantic sturgeon (mean weight 369g) at densities ranging from $6.5\text{-}16.3kg.m^{-2}$ and growth performance was found to be inversely proportional to density.

13.4 Growout Systems

White sturgeon are grown to market size (7-9kg) in approximately 3-4 years at water temperatures 18-22°C in a variety of systems, including circular fibreglass and concrete tanks, concrete and earthen raceways and water recirculation systems. A continual harvest system is used; as the faster growing "top-end" fish reach market size first, they are removed from the system.

Most white sturgeon rearing facilities use large circular tanks (7-15m diameter and 1.5-2.5m deep) or raceways (3-5m wide, 25-45m long and 1m deep), exposed to natural photoperiod and supplied with flow-through or partially re-circulated ground water, with temperatures ranging from 13-17°C (Idaho) and 18-22°C (California). One large re-circulation facility has produced about 400,000 kg of white sturgeon annually using large 13m x 13m x 3m concrete tanks and culture temperatures of 20-23°C (Michaels and Bunter 1996).

White sturgeon can be cultured at densities ranging from $40\text{-}80kg.m^{-3}$ depending on the system and the use of supplemental oxygen. Mims et al. (2002) reported large juveniles in meat production systems to be stocked at rates up to 27 $fish.m^{-2}$, and Hochleithner and Gessner (1999) reported that after reaching one kg in body weight, sturgeon could be reared at densities of $40\text{-}100kg.m^{-2}$, as long as water quality was maintained. An increased metabolic rate was demonstrated when densities of white sturgeon were increased from $6\text{-}22kg.m^{-2}$ (Ruer at al. 1987) and because dissolved oxygen is usually the first water quality parameter to limit tank carrying capacity; most farms use some type of oxygen injection system to maintain higher fish densities. The intense culture of white sturgeon is possible because they readily use the entire water column in growout tanks. Most of the other North American sturgeon species tend to be more benthic-oriented and at this time have not been cultured under commercially intensive conditions.

13.5 Feed and Feeding Methods

The rapid growth and high feed efficiency of white sturgeon was first reported in the late 1980s and was considered a vital characteristic for aquaculture (Hung 1991a). Information on sturgeon nutritional requirements, practical feeds, and feeding of larvae and juveniles has been reviewed by, Hung (1991a, b), Medale et al. (1995), and Hung and Deng (2002).

Aquaculturists have found that sturgeon larvae prefer a softer semi-moist feed compared to the harder dry diets. In addition, high growth rates (specific growth rates over 10%.day^{-1}) and survival (95%) has been achieved using semi-moist salmonid based diets (Hung and Deng 2002). Most species of sturgeon larvae initiate their external feeding 7-10 days after hatch, at 16-20°C (Wang et al. 1985; Chapman and Carr 1995), and white sturgeon larvae do not begin to feed until after the melanin plug is extruded (Buddington and Christofferson 1985; Buddington 1991; Gawlicka et al. 1995). However, the melanin plug extrusion cannot always be used as a criterion to determine the time of transition to active feeding, for example, in Siberian sturgeon the plug is eliminated after the first feeding, along with the first feces (Gisbert and Williot 1997). In addition, under commercial production, feeding may be initiated by individual fish over a 3-5 day period, depending on water temperature. Thus, with the difficulty in identifying exactly when sturgeon first begin to feed, most culturists believe it is crucial to offer small amounts of the diet several days before actual feeding, to sensitize the larvae to the feed and to ensure food is available for the very first feeding individuals.

Continuous feeding for 24 hours per day, using automatic feeders is considered the optimum feeding regime for white sturgeon (Cui et al. 1997) and some supplementation by hand, during onset and the first few days of exogenous feeding is common practice. In general, starter feeds are fed at rates of 25-30% of the total body weight per day and are decreased gradually to 3-4% during the next 30 days (Hung 1991a). However, the larval feeding period must be carefully monitored, and feeding rates adjusted based on food acceptance and amount of uneaten food left over (Conte et al. 1988).

Feeds and feeding protocols for sturgeon larvae are still being developed. Some species may accept semi-moist salmonid diets at the onset of feeding (Buddington and Doroshov 1984; Lutes et al. 1990; Hung 1991a). Whereas, some sturgeon have higher survival when these diets are supplemented with live feed (Van Eenennaam et al. 2001), and some species perform best when initially offered live food and then are weaned to a commercial diet after 1-6 weeks (Bardi et al. 1998; Mohler 2000; Mohler et al. 2000). Transition to a dry diet is usually done when the fish are being fed a #2 (0.8-1.4mm) size semi-moist diet, at approximately 4-8 weeks posthatch. Currently, white sturgeon culturists routinely obtain over 80% survival while transitioning larvae to exogenous feeding using semi-moist diets.

In a study on foraging behaviour, it was reported that live diets resulted in greater foraging activity in the upper water column compared to dry diets (Ross and Bennett 1997). A recent protocol for feeding larvae developed by investigators at the University of Florida and the Bozeman Fish Technology Centre, U.S. Fish and Wildlife Service, Bozeman, Montana is showing positive results and is gaining acceptance for sturgeon feeding (Bardi et al. 1998). This protocol begins with larvae initially being fed brine shrimp for one week, and then being weaned to an experimental sturgeon starter feed.

After a period of 3-6 weeks, the juveniles are transferred to a commercial diet of choice. Results are showing that larvae fed following this protocol have consistently high survival (>80%) and good growth rates (Mohler et al. 1996; Bardi et al. 1998). The success of this experimental starter feed is due to the balance of feed ingredients, their high nutritional quality, and excellent physical properties (primarily uniform size and texture). The uniform size of the diet, almost spherical in shape, is achieved through a Micro-extrusion Marumerization (MEM) manufacturing process (Barrows 2000).

Dry salmonid based feeds are typically used for growout through maturation and although the size of the mouth relative to the body of sturgeon is quite large, sturgeon prefer smaller size feed compared to other fish of similar body sizes (Hung 1991a). The largest pellet sizes used typically range from 6 to 11mm. There is no standard growout diet and aquaculturists have been testing diets containing variable amounts of protein (25-50%) and lipid (8-25%) trying to find an optimal diet, in terms of growth, feed efficiency and cost. From numerous studies it seems that feeds with 35-45% protein, and 12-16% lipid content are suitable (Ronyai and Varadi 1995). Hung et al. (1998) reported a 51% protein and 18% lipid diet had the best growth for white sturgeon fingerlings (72g), fed for eight weeks, and protein requirements for optimum growth of juvenile white sturgeon 145-300g has been reported to be 40% (Moore et al. 1988). A prototype feeding table for white sturgeon 50–1000 g, at water temperatures 10-26°C, has been developed by Cui and Hung (1995) and Table 13-1 provides further guidelines for fish over 1kg. The feed rates in Table 13-1 were based on feeding observations and feed conversions, at the Aquatic Facilities of the University of California, Davis campus. The adjustments of the feed rates, for the different water temperatures, were based on information available for salmonids. Automatic (12 or 24 hour) and demand feeders are most commonly used on commercial growout facilities, and usually 60-70% of the daily feed ration is provided in the late afternoon as sturgeon are more active at night. Feed conversion ratios of 1.0-1.4 are common for juveniles and 1.6-2.0 for adults (Mims et al. 2002).

The same salmonid diets used in growout facilities are fed to broodstock maintained for spawning or caviar production, and because females do not reach puberty until age 7-8 years, broodstock diets have been rarely studied. However, there is a keen interest in finding an optimal diet for females used in caviar production. This interest is based on the fact that most males are sold for meat after sexing and, during the past few years, a large variation in the amount of fatty adipose tissue that surrounds the mature oocytes has been found during caviar processing. This fatty tissue slows down the processing because it requires more rinsing and cleaning of eggs prior to salting and canning, compared to a lean ovary. In addition, the caviar yields are lower for females with "fatty" ovaries. Because of this "fatty-ovary" problem, the idea of phase feeding was investigated in a pilot study at the University of California. In the first phase, a grower feed (48% protein, 20% lipid) was used to promote rapid growth and maturation and the second phase provided a finishing diet (48%/10%) that was lower in lipid, theorizing the fish would use some of the lipid stored in the ovaries for continued growth. The control diet fed the entire experiment was 43% protein and 13% lipid. The study found a difference in finished caviar as a percent of ovary weight, 74% in the phase fed group versus 67% in the control. The lower percentage in the control group was most likely due to the higher fat levels observed in the ovaries and not due to fewer eggs (F. Barrows, Bozeman Fish Technology Centre, U.S. Fish and Wildlife Service,

Bozeman, Montana, personal communication). Based on these preliminary results, phase feeding appears promising and could improve yields and quality of caviar, but it requires further research.

Table 13-1. Guidelines for white sturgeon feeding rates, expressed as a percent of body weight per day for water temperatures 10 to 22 °C.

	Water Temperature (°C)					
Mean Body Weight (kg)	**10-12**	**12-14**	**14-16**	**16-18**	**18-20**	**20-22**
1.0 - 1.5	0.63	0.66	0.69	0.71	0.73	0.75
1.5 – 2.0	0.56	0.59	0.61	0.63	0.65	0.66
2.0 – 3.0	0.50	0.52	0.54	0.56	0.57	0.59
3.0 – 4.0	0.44	0.46	0.48	0.49	0.51	0.52
4.0 – 5.0	0.40	0.42	0.44	0.45	0.46	0.48
5.0 – 7.5	0.36	0.37	0.39	0.40	0.41	0.42
7.5 – 10.0	0.32	0.34	0.35	0.36	0.37	0.38
10.0–20.0	0.26	0.27	0.28	0.29	0.30	0.31
20.0–40.0	0.22	0.23	0.24	0.25	0.26	0.27

13.6 Water Quality

Water quality criteria for sturgeon hatcheries and for long term rearing of sturgeon juveniles have been reviewed (Conte et al. 1988; Hochleithner and Gessner 1999). In addition, Wedemeyer (1996) compiled a table of recommended water chemistry limits for cold and warm water fish in intensive culture, and Meade (1989) provided one of the first extensive lists of water quality standards for aquaculture. There are also a growing number of studies addressing specific criteria for sturgeon, including: water temperature (Wang et al. 1985, 1987; Lutes et al. 1990; Hung et al. 1993; Kelly and Arnold 1999; Webb et al. 1999, 2001), salinity (McEnroe and Cech 1985; Jenkins et al. 1993; Altinok et al. 1998; Krayushkina 1998; LeBreton and Beamish 1998; Altinok and Grizzle 2001a; Jarvis et al. 2001), temperature and hypoxia (Cech et al. 1984; Secor and Gunderson 1998), ammonia and nitrite toxicity (Fontenot et al. 1998), ammonia-nitrogen production rates (Thomas and Piedrahita 1998), oxygen consumption rates (Thomas and Piedrahita 1997; Gisbert et al. 2001), effect of high dissolved carbon dioxide (Crocker and Cech 1996), and predicting oxygen consumption, carbon dioxide and ammonia production in aquaculture systems (Thomas and Piedrahita 1996).

13.7 Domestic Broodstock Development

Sturgeon culturists have begun to develop "domesticated" broodstock because the availability of wild sturgeon is unpredictable, or in many cases, they are not obtainable due to their threatened or endangered status. Although puberty is accelerated in cultured sturgeon compared to their wild counterpart, the absence of external sexual dimorphism, the asynchronous onset of vitellogenesis in females, and the temperature-sensitive final gonadal maturation complicate any broodstock development program. The following techniques are used in white sturgeon aquaculture to monitor reproductive development and to achieve satisfactory spawning performance.

13.7.1 SEXING TECHNIQUES

Knowledge of the sturgeon's gender is important for broodstock management. However, it is difficult to determine gender because sturgeon lack any external sex characteristics, except for spawning broodfish, and DNA sex identification is currently not available (Van Eenennaam et al. 1999). Sexing can only be done after differentiation of the ovary or testis which occurs at 1-2 years of age in cultured fish. The current method of management for aquaculturists is to sell the excess males as food fish and to retain the females for caviar production and spawning induction. Although this fits well into the current production scheme of white sturgeon farms, because sexing sturgeon refers to market size fish (7-9kg) that have clearly differentiated gonads, it has yet to be attempted on a large scale with any other North American species.

To sex sturgeon the fish are removed from the culture tanks using mechanical and hand-held equipment, and are placed into an anaesthetic tank until they roll ventral side up and opercular movements slow down but do not cease (called a stage 4-5 level of anaesthesia; Summerfelt and Smith 1990). Tricaine methane sulfonate (3-aminobenzoic acid ethyl ester or MS-222) is a commonly used fish anesthetic, which is FDA approved in the U.S. (Varner 2000). For white sturgeon, a stage 4-5 level of anaesthesia is induced with MS-222 within 10-20 minutes at 50-100ppm. Anaesthesia baths are well aerated and fresh water is exchanged every couple hours to maintain water temperature, and MS-222 is added as needed, to maintain the desired concentration. The frequent water replacement is also done because each fish removes part of the oxygen and anesthetic from the water, and adds mucous, carbon dioxide and ammonia (Summerfelt and Smith 1990).

After capture and anaesthesia, the fish is placed ventral side up in a stretcher and under aseptic conditions, sexing is conducted using a minor abdominal incision following procedures modified from Doroshov et al. (1983). The site of the incision is approximately 3-4 ventral scutes anterior from the pelvic fin and 2.5-5.0cm from the ventral mid-line (Figure 13-1). The exact position varies among individuals and between different sturgeon species due to the variation in scute patterns and body conformation. The ideal incision site is lateral to the mid-line and above vascularized muscle. The muscle tissue should be about 1.0-1.5cm thick, which will enhance the healing process after suturing. This incision area should be directly above the gonad so that you can look down into the incision and clearly distinguish whether it is a male or female. An incision approximately 2.5-3.5cm long is made using a scalpel and if the existing natural or artificial light is not bright enough to clearly see the gonad, a flashlight, fibre-optic light or tripod mounted construction light can be used.

The female ovarian tissue is off-white to yellowish in colour but the distinguishing characteristic is the grainy ovigerous folds. Gonad colour will vary depending on whether the fish is from a wild or cultured stock and the husbandry conditions, such as diet ingredients. It is important to distinguish the ovarian tissue from the fatty adipose tissue, within which the developing ovarian folds are embedded, because the adipose tissue is often similar in colour to the ovarian tissue. Variable amounts of the fatty adipose tissue are also present in the male gonads but the distinct testicular tissue is smooth, turgid and whitish in colour.

Stages of sturgeon gametogenesis have been recently described with corresponding macro and micro photographs in Bruch et al. (2001). To close the incision, an Ethicon® Polydioxanone (PDS II) synthetic monofilament absorbable suture with a reverse cutting CP-1 needle is used to make a simple interrupted suture or a cross-suture (Knecht et al. 1987; Giddings 1997). The major advantages of this material are that it is a monofilament and therefore has less drag in the tissue, it has the least tissue reaction and it has the slowest loss of tensile strength (Knecht et al. 1987; Boothe 1998).

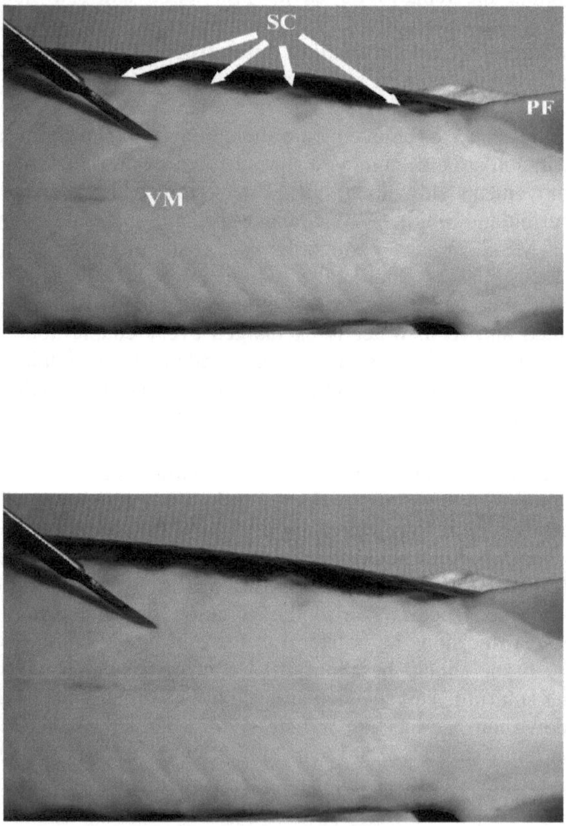

Fig. 13-1. Location of the abdominal incision on a white sturgeon, used to visually examine the gonads, during sex identification. The incision is made approximately 3 to 4 ventral scutes (SC) anterior from the pelvic fin (PF) and midway between the ventral mid-line (VM) and scutes.

Similar to largemouth bass incision healing (Gilliland 1994), white sturgeon have been found to heal faster and have only a slight tissue reaction to this material, compared to other sutures, when kept at water temperatures between 18-22°C. However, in colder water (12-15°C) the PDS dissolves very slowly, and may remain intact for over a year. Thus, at these cold-water sites, either chromic gut suture material is used or the PDS sutures are removed after 3-4 weeks. The use of surgical staples and tissue adhesives to close the incision, which was attempted with minimal success many years ago, is currently being re-examined.

Although this method of sexing is an invasive, labour-intensive method, it is the least expensive and is virtually 100% accurate with only incidental mortalities. Additionally, an experienced crew can sex 500-1000 fish per day. A crew usually consists of two people sexing and 3-6 people moving fish. This technique has been successfully used on white sturgeon farms but other more sensitive sturgeon species may not tolerate this procedure.

There are two other techniques that are being refined for sexing immature sturgeon. One is the analysis of blood plasma for sex steroid levels (Webb et al. 2002; Feist et al. 2004) but this technique currently has some limitations. The need to handle the fish twice is a concern to sturgeon culturists (once to collect the blood sample and tag the fish, and the second time to sort the sexes after steroid analysis is completed), and also the laboratory cost of determining the steroid levels in the plasma. The second technique is the use of ultrasonography. Interest in this technique is growing and success has been recently reported with *Acipenser stellatus* (Moghim et al. 2002) and shovelnose sturgeon (Colombo et al. 2004).

13.7.2 GROWTH AND MATURATION RATES

Cultured sturgeon have an accelerated growth rate and mature at a younger age, compared to their wild counterparts. This accelerated growth and maturation is due to husbandry conditions, including warmer water temperatures and high quality commercial fish feed. Gonadal sex differentiation in white sturgeon occurs at age 18 months and at a body weight of 1.1-2.3kg (Doroshov et al. 1997). Males mature typically at age 4 years and at a body weight of 12-18kg, and females at 7-8 years and body weight 35-45kg. Mature wild males and females are similar in body size but age of sexual maturation ranges from 10-12 years and 15-32 years, respectively (Doroshov et al. 1997).

13.7.3 TEMPERATURE REGIME FOR BROODFISH

Although constant warm water temperatures (18-22°C) accelerate sturgeon growth and maturation, these temperatures negatively affect late vitellogenesis and final maturation leading to follicular atresia and impaired ovulatory response to hormonal stimulation, in broodstock females (Webb et al. 1999; 2001). Similar problems (failed spermiation after hormone injection) have been observed in mature males. For females it appears that just a few days of elevated grow-out temperatures can arrest the final maturation process (i.e., germinal vesicle migration) and result in failed or poor ovulation. Elevated temperature effects on ovarian development include morphological (hypertrophy of granulose cells) and physiological (decrease in plasma estrogen, increase in the androgen/estrogen ratio) aspects of ovarian atresia (Linares-Casenave et al. 2002). To obtain successful spawning of sturgeon cultured in year-round warm water, a cold water source (9-12°C) should be provided to maturing broodfish during the late phase of vitellogenesis and spermatogenesis (Williot et al. 1991; Chebanov and Savelyeva 1999; Hochleithner and Gessner 1999; Webb et al. 1999; 2001). This vernalization (exposure to cold water) usually begins in November and lasts through the spawning season. In some cases the cool water is held constant (e.g., 12 ± 1°C) and on other sites the water temperature drops to a low of 9-10°C in January and slowly increases a degree or two every 3-4 weeks until optimal spawning temperature (14-16°C) is reached during March or April.

13.7.4 DETERMINING SPAWNING READINESS

Identification of spawning readiness in sturgeon broodstock is based on the diagnostic techniques described in Dettlaff et al. (1993). Females (and males) identified as mature broodstock should be kept in a separate tank and at low stocking densities. These tanks should have water flows maintaining dissolved oxygen at >80% saturation and temperatures ranging from 10-14°C. Although wild white sturgeon are spring breeders, domesticated females held in tanks with constant cold water have an extended spawning season, lasting through the summer and into early fall (Doroshov et al. 1997; Webb et al. 1999).

Whether a specific female is ready to spawn is determined by collecting a sample of oocytes to calculate their polarization index (PI), and to measuring their capacity to undergo germinal vesicle breakdown (GVBD) in the presence of a maturation-inducing steroid (Dettlaff et al. 1993; Mims et al. 2002). Oocytes are extracted by a catheter (Conte et al. 1988) and are kept in a sturgeon Ringer solution (Dettlaff et al. 1993) with penicillin ($0.03g.l^{-1}$) and streptomycin ($0.05g.l^{-1}$). The incision for inserting the catheter is made in the same area as described for sexing, although a much smaller (6-9mm) incision is made, and the incision is closed using a single suture. Females are returned to the broodstock tank until spawning or further sampling. Domestic white sturgeon females have been sampled in this manner 3-5 times, every 4-6 weeks, and have been successfully induced to ovulate, producing eggs with high fertility and hatch rates.

Oocytes to be used for PI determination are boiled for 5-6 minutes, chilled on wet ice for 20 minutes, stored in 10% buffered formalin overnight, and bisected along the animal-vegetal axis using a double-edged razor blade. The PI is calculated by measuring the distance of the germinal vesicle from the inner border of the oocyte chorion at the animal pole, and dividing this by the animal-vegetal oocyte diameter. These measurements, on the bisected oocytes, are usually made using some type of digital image analyses system and a dissecting microscope. However, there are numerous systems available to project the bisected oocyte image onto a computer or television screen and the measurements can be made, directly on the screen, using a simple ruler.

The PI alone provides a very good indication of female readiness to spawn, but it does not measure directly the capacity of the oocyte to mature. Thus, the catheterized oocytes are also analyzed for their capacity to undergo GVBD in the presence of a maturation-inducing steroid. To determine the capacity of the oocyte to undergo GVBD, the in vitro oocyte maturation assay, as described in Dettlaff et al. (1993) is conducted in sturgeon Ringer solution at 16°C for 16 hours. Fifteen oocytes are incubated with $5\mu g.ml^{-1}$ progesterone and 15 are incubated without progesterone to serve as a control. Each treatment typically has two replicates. After incubation, the oocytes are boiled, chilled on wet ice, stored in 10% buffered formalin, bisected and examined for GVBD.

Dettlaff et al. (1993) recommended an oocyte PI of 0.07 or less for selection of female broodstock for spawning, although domestic white sturgeon have been successfully induced to ovulate when the oocyte PI is less than 0.10 and there is 100% GVBD response in the egg maturation assay. Females that have oocytes with PIs greater than 0.10 and a GVBD response of less than 100% are typically re-sampled in 4-6 weeks, while held in water temperatures ranging from 10-14°C.

13.7.5 SPAWNING INDUCTION

Males and females of cultured white sturgeon have annual and biennial gonadal cycles, respectively (Doroshov et al. 1997). Females also appear to have a specific endogenous rhythm when cultured under similar conditions throughout their life, as many females have been spawned during the same month on consecutive reproductive cycles. When mature females are maintained at cold water sites, the spawning season can be extended because the low water temperature slows down the process of late vitellogenesis and final maturation. The earliest and latest successful spawning of domestic white sturgeon in California has been February 2 and November 7, respectively. But, in general, as spawning progresses into summer and fall, the effect of photoperiod starts to counteract the cold water exposure and over 50% of the females will start the process of atresia.

CCP and GnRHa have been used alone or in combination for spawning induction (Doroshov et al. 1997; Webb et al. 1999) and currently GnRHa at total doses of 10 and $20 \mu g.kg^{-1}$ for males and females, respectively, have been very successful in recent INAD field trials. When CCP was used to induce ovulation in domestic white sturgeon broodstock, good ovulation rates were obtained (>80%). However, fertility was marginal (average 65%) and mortality of the embryos prior to hatch was high (average hatch was only 8%). In comparison, females induced to spawn with GnRHa had similar ovulation rates as well as high fertility and hatch rates (82% and 64%, respectively) Thus, the CCP is an effective inducer of ovulation but may adversely affect egg quality, particularly when applied to domestic fish in a less advanced stage of oocyte development. Observations on GVBD, cortical reaction and egg cleavage suggest that the high dose of CCP may induce ovulation of eggs that did not fully complete cytoplasm maturation and did not establish activation competence (Nocillado et al. 1999). The effect of GnRHa, mediated by endogenous gonadotropin, appears to be more favourable with regards to egg quality.

Males selected for spawning should have firm white testes at the time of biopsy during the fall (Williot et al. 2000) or a good body conformation (full upper and lower abdomen) if they have not been biopsied. Generally, males that have reached full sexual maturity (age 4+) are not biopsied for repeated spawning because they mature annually. During spawning induction, 2-5 domestic males are injected per female. To ensure a supply of sperm both early and late in the spawning season, males should be held at water temperatures 5-6°C below the optimal spawning temperature. Manipulation of the water temperature to ensure a sperm supply can either be done by increasing the water temperature to spawning levels (15-16°C) after injection to stimulate spermiation and then decreased after semen collection, or by maintaining the water temperature at a constant 10-12°C, which will delay spermiation several hours. Interestingly, Williot et al. (2000) reported highest sperm motility at cooler water temperatures (10°C) for cultured Siberian sturgeon, compared to 15 and 17.5°C, which were considered optimal spawning temperatures. Semen is evaluated for quality based on criteria described in Conte et al. (1988) and Dettlaff et al. (1993). Besides having a high percent of motile sperm (> 75%), it is important that the sperm head trajectory has a distinct straight-line velocity (Toth et al. 1997; Billard et al. 2000; Cosson, et al. 2000). If activated sperm is moving very slowly or is shaking side-to-side while moving (or with no forward progression), it is considered poor quality and is not used. Techniques for short-term cold storage of semen have been established (Mims 1991; DiLauro et al. 1994) and are useful when spawning several females in a short time period.

Individual females selected for spawning are placed in separate circular spawning tanks (3-4m diameter), allowing the fish to swim freely. Similar to wild broodfish, induction of ovulation is accomplished by using two injections. Starting at 16 hours after the second injection, the presence of ovulated eggs in the spawning tank is checked every hour. Water temperature and oxygen levels must be maintained during spawning induction, as oxygen demand increases for a mature female during ovulation. A spray-bar with water flows of 40-80L.min^{-1} can provide the main source of dissolved oxygen; however, air stones may also be added to ensure sufficient levels of oxygen. The dissolved oxygen content at the outlet should be maintained at >80% saturation. As a precautionary measure all tanks require a secure cover to prevent the sturgeon from jumping out of the tanks and becoming injured.

Ovulated females are removed from the spawning tank by either tube-net and then transferred to a hooded stretcher, or by using only the hooded stretcher. An ice chest (75-95L), submersible water pump, and large diameter (2-3cm) tygon tubing, is used as a mini-recirculation system to anesthetize females to a stage 4-5 level of anaesthesia while in the stretcher. The gills are initially irrigated with water containing 50-75ppm MS-222 for about 10 minutes to anesthetize the fish, and then every 5 to 10 minutes the anaesthesia is replaced with fresh oxygenated water, to prevent deep anaesthesia and potential mortality.

Eggs are collected using one of four methods based on the hatchery objectives and the species: 1) euthanize the fish and remove all the eggs, 2) hand-stripping, 3) minimally invasive surgical technique (MIST), or 4) caesarian surgery. Euthanizing the fish and opening the abdominal cavity to collect all the ovulated eggs was done historically (Conte et al. 1988), but today this technique is rarely used. For small, delicate or rare species, or large species from which only a litre or two of eggs are needed, hand-stripping of ovulated eggs may be practical (Ireland et al. 2002). However, stripping only removes a portion of the eggs because of how the eggs are released from the fish. Sturgeon and paddlefish release their ovulated eggs into the body cavity, and the ova are transported into the funnels of the dual Müllerian ducts (Appendix 4 in Conte et al. 1988). The openings of the Müllerian ducts are located approximately in the mid-abdominal region and there is a muscular valve in the mid-part of the duct. Thus, stripping only removes eggs that are in the posterior region of the duct, that have passed the valve. Typically, 500-1000ml of eggs can be collected during each of the first few strips, but fertility and egg quality decrease over time during prolonged stripping. For the past ten years hand-stripping has been successfully used on the endangered stock of Kootenai River white sturgeon and the quantity of eggs collected from individual females has ranged from 16 to 100 thousand (J. Siple, Kootenai Tribe of Idaho, Bonners Ferry, Idaho, personal communication).

The development of the MIST overcomes some of the shortcomings of the hand stripping method. The MIST requires the insertion of a scalpel into the urogenital opening and cutting through the oviduct (Štěch et al. 1999). This cut permits the direct stripping of eggs from the body cavity through the vent, bypassing the Müllerian ducts and allowing for a shorter hand-stripping time. Depending on the quantity of coelomic fluid in the body cavity, about 55-80% of the eggs are easily and quickly collected (Mims et al. 2004).

The fourth method to remove sturgeon eggs is by caesarian surgery. This method is based on the procedure developed by Conte et al. (1988) with some modifications. After anesthetizing the female a 10cm incision is made in the abdomen using a scalpel and tissue forceps. The caesarian incision is located lateral to the mid-line, 4-6 ventral scutes anterior from the pelvic fin depending on the size and species of the female, because some smaller species may require a smaller incision closer to the vent. The incision region should have a muscle thickness of about 1.0-1.5cm for effective suturing. After the incision is made the eggs are removed using a plastic spoon. Caution is used when collecting eggs from the anterior portion of the body cavity where the liver, gall bladder, spleen and other vital organs are located, to avoid trauma to these organs. Egg collection (ranging from 60 to 80% of the total) is completed in 15 to 20 minutes with this method. The quantity of eggs collected from individual white sturgeon females has ranged from 60 to 300 thousand.

Once the eggs are collected using one of the above methods it is important that the eggs remain at spawning temperature, out of direct sunlight, with no addition of water prior to fertilization. About 1-2 litres of eggs are divided into separate stainless steel bowls, and each bowl of eggs is fertilized by an individual male to ensure that one predominant male (if the semen is pooled) does not fertilize the majority of the eggs from one female.

While the eggs are being fertilized and de-adhered, the female's caesarian incision is sutured closed, which takes about 20 minutes. The incision is closed by two sets of sutures for added strength, to ensure the peritoneum will be closed, and to help with apposition and rapid healing. The first is an internal suture used to bring the peritoneum and bottom half of the muscle together and the second is an external suture for the top part of the muscle and skin. The internal suture is made using single interrupted stitches with a PDS II #0 suture and a swaged-on CT-2 taper needle. The external stitches will use the same material but with a thicker (#1) suture and larger swaged-on CP-1 cutting needle, for cutting through the skin. The external sutures used are a special tension suture pattern called the "far-near-near-far" pattern (Knecht et al. 1987). The advantage of this suture is that it apposes the skin edges and provides a degree of tension. This tension is important for the larger sturgeon females because it assures that the suture is not torn when they become more active as they are healing. After suturing an intraperitoneal (IP) injection of antibiotic in the area of the incision is recommended, such as the Pfizer Liquamycin LA 200® which contains $200 mg.ml^{-1}$ oxytetracycline, at a dose of $5 mg.kg^{-1}$ body weight.

After egg collection, the females are placed into a clean recovery tank for at least four weeks. Earthen ponds and raceways are not used because they provide a source of potential pathogens and appear to prolong the healing process. Primary healing is complete in approximately one month, at which time the external sutures can be removed. During the first ten years of white sturgeon caesarian surgery, the most common cause of post-surgery mortality was improper suturing and prolonged anaesthesia. However, with the new technique of alternating between anesthetic water and fresh water and the improved internal and external suturing, the occurrence of post-surgery mortalities is rare. It should also be noted that many individual white sturgeon females have been spawned repeatedly (2-4 times) using these improved techniques, with no discernable complications during the subsequent reproductive cycle and caesarian surgery.

13.8 Commercial Production

In North America, aquaculture of white sturgeon exists in California, Oregon, Idaho and initial trials are occurring in British Columbia, Canada. In 2003, production was approximately 800t of sturgeon meat and 7t of caviar.

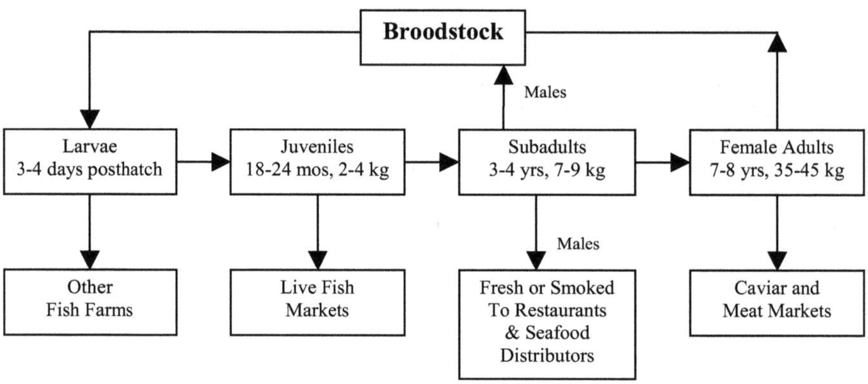

Fig. 13-2. The general production scheme of a commercial white sturgeon farm.

The general production scheme (Figure 13-2) begins by spawning domesticated broodstock and selling some yolk sac larvae to other fish farms. The remaining larvae are reared for about 18-24 months and then some are sold in the live fish markets at 2-4kg in body size. However, most of the sturgeon are cultured for another 1-2 years and are sexed as subadults (7-9kg). A select group of subadult males are then retained as future broodstock but the majority are sold as food fish. The females are reared until they reach sexual maturity, at which time they are used as either broodstock or sold for their caviar and meat. In this section we will give a brief overview of the aquaculture production milestones in North America.

13.8.1 HATCHLINGS

The first successful hatchery spawning of wild-caught white sturgeon took place in 1980 and the first sales of larvae and fingerlings occurred two years later. Production levels of larvae hatched from wild broodstock peaked during the early 1990s and the first spawning of domestic F1 female broodstock occurred in 1990. Subsequently, in five years time, the use of wild broodstock ceased (Figure 13-3). Production of larvae from domestic broodstock has remained relatively stable, although decreased production has occurred since year 2000 due to increased regulations for international shipments, decreased demand from other farms, and very high survival (80-95%) of larvae produced.

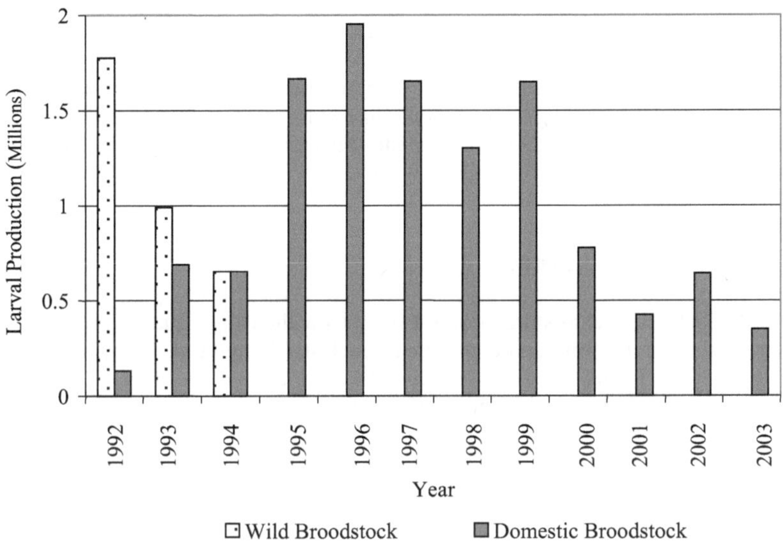

Fig. 13-3. California production of larvae hatched from wild and domestic white sturgeon.

13.8.2 FOODFISH

The first sales of food fish occurred in 1985 and, during the past several years, production has fluctuated around the 700-900t level. The primary product is a "bullet", which is an eviscerated fish with head, tail and fins removed, providing about a 65% yield from the whole live body weight. Another popular product is the skinless fillet, which yields about 40-45% and the proximate composition of the flesh is 77.3% moisture, 18.2% protein, 3.4% lipid and 1.2% ash (Price et al. 1989). However, the muscle of sturgeon can sometimes be tough in texture when the muscle is in rigor (Izquierdo-Pulido et al 1992). This tough texture can be minimized by allowing the sturgeon meat to go through rigor prior to cooking, which is accomplished by aging the meat for a couple days, prior to shipping the product. The relatively high price (US$4-6.$kg^{-1}$ whole weight) associated with most of the sturgeon products results in the majority of the production being sold to seafood distributors and the restaurant trade. A limited but increasing amount of product is being sold to the live fish markets at smaller fish sizes (2-4kg at 18-24 months of age), and some value-added product is being prepared by cold or hot smoking. Although sales of food fish (7-9kg) decline when facilities start retaining a larger percent of females for caviar production and sell only the males, the meat from females processed for caviar is sold during the caviar processing season (March-June).

13.8.3 CAVIAR

During the period of 1989-1997, the U.S. imported an average of 59t of sturgeon caviar per year (Hoover 1998), and imports were up to 74t with a declared value of US\$332.kg^{-1} during 2000 (Raymakers and Hoover 2002). The first processed caviar from California farmed white sturgeon was made in 1995 and production has been increasing each year (Figure 13-4). During 2003, production was approximately 7t and production goals in the near future are estimated to be 10-15t. At the time of writing, the retail price for domestic white sturgeon caviar was comparable to the traditional osetra grade caviar. Wholesale prices are approximately US\$300.kg^{-1} and retail prices are as high as US\$1000.kg^{-1}. Processed white sturgeon females were, in general, 7-8 years old, weighing 35-45kg, with a gonadosomatic index of 9-12% and the final yield of canned caviar was 6-9% of the body weight.

13.9 Paddlefish Culture

Paddlefish, similar to sturgeons, are sought for their high quality meat and caviar. Their meat is white, firm, and considered boneless (Wang et al. 1995), and their eggs are greyish-black and measure between 2.0 and 2.5mm in diameter (Ballard and Needham 1964; Shelton and Mims 1995). A female paddlefish weighing 12-18kg may contain 100 – 400 thousand eggs and could yield about 10% finished caviar, based on the whole fish body weight. Historically, the paddlefish fishery was never as large as the sturgeon fishery but the U.S. peak harvest was reported to be 1,105t during 1899 (Carlson and Bonislawsky 1981).

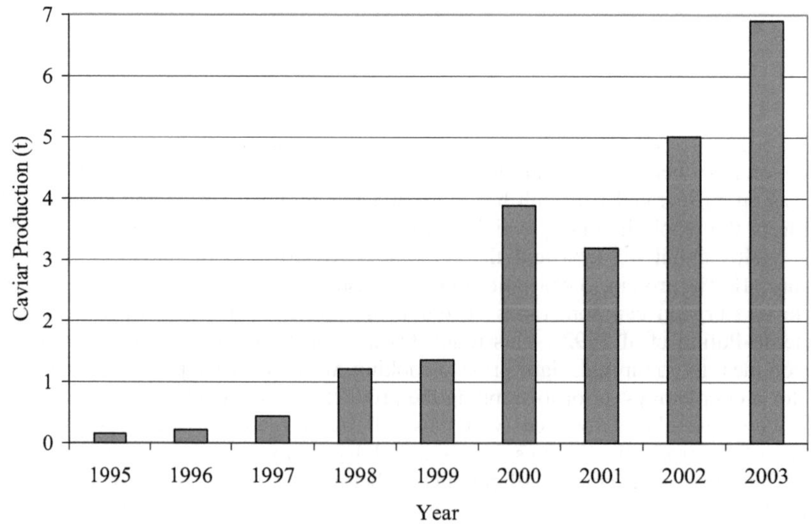

Fig. 13-4. California production of caviar (metric tons) from domestic white sturgeon.

Aquaculture of this species is still in the research and development phase but its potential is unquestionable. Two excellent reviews describing techniques for the production of paddlefish were recently published by Mims et al. (1999) and Mims (2001). Various aspects of the natural history and culture of paddlefish were also published by Dillard et al. (1986), and a bibliography of paddlefish is updated by Georgi and Dingerkus (2001). In addition, an informative video on the artificial propagation of paddlefish was produced by Mims and Durborow (1998), and is available through the U.S. Cooperative Extension Service, Kentucky State University. It is not the intent of this review to duplicate these authors' efforts but to highlight their recommendations and emphasize the unique aspects of paddlefish culture.

13.9.1 AQUACULTURE POTENTIAL

The paddlefish is a filter feeder, dependant primarily on zooplankton as its major food source (Eddy and Simmer 1929; Ballard and Needham 1964). The paddlefish can grow rapidly, up to 4.5kg per year (Mims et al. 1999) and this rapid growth is also exhibited under culture conditions. They can be raised extensively in water reservoirs or ponds, and intensively in raceways and tanks using commercial diet formulations (Webster et al. 1991). Paddlefish are easily propagated using artificial spawning techniques (Graham et al. 1986; Mims et al. 1997), including MacDonald jars to incubate embryos (Graham et al. 1986; Mims and Shelton 1998) and commercial diet formulations to feed larvae (Webster et al. 1991; Mims and Shelton 1998). These and other outstanding biological and marketing criteria support the culture of paddlefish as a potential aquaculture enterprise. Although the current culture of paddlefish is primarily in its developmental stage, the likelihood for a competent industry has been demonstrated, for the most part, in small-scale demonstration and commercial projects in Kentucky, Alabama, Missouri, and Oklahoma (Mims 2001).

13.9.2 BROODSTOCK DEVELOPMENT

Cultured stocks of paddlefish are still derived from broodfish collected primarily from the wild. However, there are a few domestic stocks that have been developed and these may become the brood source for other aquaculture farms (Kincaid et al. 1999). Paddlefish are native to the U.S. and are currently found in most large rivers and lakes within the Mississippi River basin. Paddlefish to be used for spawning induction are easily collected by snagging or using gill nets, with at least a 15cm bar mesh size. They are usually collected 1-2 months prior to spawning, depending upon water temperature. Water temperatures ranging from 12-16°C are preferred for the collection of wild fish because temperatures above 16°C will increase mortality during handling and transport. Broodfish are best transported in insulated fish hauling tanks containing chilled and well oxygenated water, and salt is often added (0.5-1.0%) to aid in osmoregulation, stress and disease control. Broodfish should be transported at low densities ($0.25kg.l^{-1}$) and care must be taken to reduce the possibility of injury, especially because of their elongated rostrum or 'paddle'.

13.9.3 SPAWNING INDUCTION

Paddlefish require 7-9 years to attain sexual maturity. Males mature earlier than females but females tend to live longer than males (Russell 1986). Paddlefish exhibit little external dimorphism between the sexes, but males are generally smaller than females and may be distinguished by the presence of minute tubercles on their heads and opercular flaps (Russell 1986).

Guidelines for selecting ripe paddlefish, inducing spawning, fertilization, and hatching are very similar to those developed and used in the husbandry of sturgeon (Doroshov et al. 1983; Dettlaff et al. 1993; this chapter sections 6.4, 6.5). Female broodstock are anesthetized with MS-222 (e.g., 80-100mg.l^{-1}) and samples of the gametes are collected with a catheter. Males are evaluated by the presence and motility of their sperm. Ripe females have slightly oval and distinctly pigmented oocytes, and the germinal vesicle is well polarized in the animal pole (Shelton and Mims 1995).

Ripe broodfish are induced to spawn using extracts of paddlefish or carp pituitaries at a dose of 2-4mg.kg^{-1} (Needham 1965) or synthetic forms of GnRHa are used at 50-100μg.kg^{-1} (Graham et al. 1986; Linhart et al. 2000). The latency period between hormone administration and spawning may vary between 12 and 24 hours at 16-18°C. Semen is collected by hand stripping or by suction. Semen collection by suction can be performed by using a small diameter (0.2cm) tube attached to a syringe. The sperm is evaluated for motility and paddlefish sperm may remain motile for up to 6 minutes after water activation (Linhart et al. 1995). Paddlefish semen can be stored undiluted, on ice or in a refrigerator, for a few days. The semen can also be stored for a short-term when diluted in saline water (0.9%) or buffered solutions (20μM TRIS-HCL; pH 8.5) containing glucose (100-150μM) and potassium chloride (5 μM) (Mims 1991; Linhart et al. 1995; Cosson and Linhart 1996). Although cryopreservation techniques appear promising, the best fertilization rate has been only about 16% using frozen and thawed semen (Brown and Mims 1999; Mims et al. 2000).

Female paddlefish, like sturgeons, have gymnovarian type ovaries. Eggs are ovulated into the abdominal cavity and are released outside the body through the Müllerian ducts located along the lateral walls of the body cavity. The same techniques used for removing eggs in sturgeon (see section 6.5) are used for paddlefish. The technique that is quickly gaining acceptance among hatchery managers is the MIST technique.

13.9.4 EGG FERTILIZATION AND INCUBATION

The eggs of paddlefish are fertilized using semen from more than three males, and only semen with greater than 75% motile sperm is recommended for egg fertilization (Cosson et al. 2000; Linhart et al. 2000). Depending on sperm density and to prevent polyspermy, one part semen is diluted with 100-200 parts water. As with sturgeon, paddlefish eggs become very sticky after coming in contact with water, therefore, clay is usually used for de-adhesion (Graham et al. 1986; Mims and Shelton 1999). The eggs are incubated in traditional MacDonald jars. Measured volumetrically, one litre of eggs contains about 50,000 embryos, and approximately 75-100 thousand embryos are placed in one incubating jar (Graham et al. 1986; Mims and Shelton 1999). Optimal temperatures for incubation of paddlefish embryos and culture of larvae, range from 16 to 20°C (Graham et al. 1986; Kroll et al. 1992; Shelton et al. 1997). Paddlefish embryos hatch in approximately 5-7 days, at 18°C (Graham et al. 1986), and another 5-6 days is required for yolk absorption and initiation of external feeding.

13.9.5 LARVAL AND JUVENILE REARING

Larval rearing of paddlefish may be carried out extensively in ponds or intensively in tanks (Webster et al. 1991). In ponds, the preferred foods for larvae are slow-moving zooplankton such as *Daphnia* (Mims et al. 1995b), but bird predation is a major concern in raising paddlefish in ponds (Graham et al. 1996). Larvae and juvenile paddlefish may also be weaned to readily accept commercial diets (Kroll et al. 1992; Mims et al. 1999). Paddlefish are best fed by hand, in combination with automatic feeders, at 5-10% of their body weight per day. Feed conversion efficiencies vary from 2-4kg of feed per kilogram of fish body weight (Mims et al. 1999; Mims 2001). Juvenile paddlefish are often stocked in fertilized ponds for final growout and a yield of approximately 100-200kg of paddlefish per hectare of surface water is expected. However, for maximum efficiency and productivity, reservoir ranching and polyculture with catfish in ponds are more practical systems for raising paddlefish for meat and caviar (Mims et al. 1999). Reservoir ranching systems for production of paddlefish involve releasing juvenile fish into constructed reservoirs. The paddlefish forage on natural foods, and are harvested later for meat or caviar (Mims et al. 1999). Ongoing research on ranching paddlefish in reservoirs (10-20 paddlefish per hectare) and polyculture with catfish (75 paddlefish and 10-12 thousand catfish per hectare) has led to estimates that yearly production yields may reach 55-170kg and 200-400kg per hectare of water, respectively (Mims et al. 1999; Mims 2001). Onders et al. (2001) reported that reservoir ranching is a viable method for producing market-size paddlefish within 20 months of stocking fingerlings (30-67cm total length), with an overall gross yield of 13.7kg per hectare.

13.9.6 WATER QUALITY REQUIREMENTS AND DISEASES

Little has been written on water quality requirements and diseases that afflict paddlefish in culture (Mims 2001). However, paddlefish appear to grow well under a wide range of water quality parameters (e.g., water temperatures 15-27°C; oxygen ranging from 30% to saturation; pH 6-9; total ammonia nitrogen less than $0.2mg.l^{-1}$), and few diseases have been isolated from them (e.g., Columnaris and Aeromonas sp. bacteria, and the parasite *Ichthyophthirius sp*). However, water quality criteria should follow strict physical, chemical, and biological criteria for warm water finfish, such as those established for best growth and reproduction of channel catfish.

13.9.7 COMMERCIAL PRODUCTION

The culture of paddlefish, particularly in reservoirs and polyculture systems, provides a potentially profitable alternative to other more traditional aquaculture systems, such as monoculture of catfish in ponds. Although several farms and reservoir ranching systems are in operation in North America, the current level of commercial production for paddlefish is unknown. Today, most natural stocks of paddlefish are strictly regulated, and the trade of their products restricted internationally by the United Nations Convention on International Trade in Endangered Species of Wild Fauna and Flora (CITES, Appendix II).

13.10 Culture of Other Sturgeon Species

A brief synopsis of the culture activities for the remaining North American sturgeon is presented, and although there is other information available, much of it is only accessible in the grey literature. Although the techniques being developed are modelled from those established for white sturgeon, there are some unique culture attributes to each species and more will undoubtedly be discovered in the future. Because a number of the species are listed as threatened or endangered, their potential for use as aquaculture candidates are limited. For some of these species, culture techniques are being established as part of their management plans for stock enhancement (Smith and Clugston 1997; St Pierre 1999; Smith et al. 2002) and to better understand aspects of their natural history and biology. Stock enhancement plans are also addressing the potential impacts of artificial selection and rearing on genetic variability, and the potential disease risk to the native population (LaPatra et al. 1999; Ireland et al. 2002). The great majority of these captive stocks are maintained at National Fish Hatcheries of the U.S. Fish and Wildlife Service, and a few state hatcheries at various locations across the U.S. Several stocks are also maintained at educational institutions, including: University of California at Davis, University of Florida, Ohio State University, and the Center for Great Lake Studies in Milwaukee. One of the main difficulties in the propagation of the rarer species is collecting mature females and males, and collecting them at the same time (Smith and Dingley 1984). Not only are the fish difficult to find and capture, but males usually migrate first to the spawning grounds and females follow a few weeks to months later (Carr et al. 1996).

13.10.1 SHORTNOSE STURGEON

Since the listing of the shortnose as endangered in 1973, fisheries data and surveys indicate that there are populations in 16 rivers along the eastern coast (Kynard 1997). After the first successful induced spawning and culture (Smith and Dingley 1984; Smith et al. 1986), little additional husbandry information has been published that is specific to this species, although for an endangered species, it has the most numerous captive stocks, which are now on their third captive bred generation. At present, the shortnose sturgeon is being developed as an aquaculture species in Canada, whereas in the U.S. it has been primarily used in a stocking enhancement program (Smith and Jenkins 1991; Smith et al. 1995; 2002).

13.10.2 ATLANTIC AND GULF STURGEON

During 1998, a moratorium on the Atlantic sturgeon harvest in U.S. waters was adopted by the Atlantic States Marine Fisheries Commission and, presently, most of the interest in advancing Atlantic sturgeon culture technology is motivated by management plans aimed at stock restoration (Smith and Clugston 1997; St. Pierre 1999). Recent studies on Atlantic sturgeon culture have included the effects of ration and temperature (Kelly and Arnold 1999), and the effects of feed types, feeding frequency and rearing density on growth and survival of juveniles (Mohler et al. 1996; Mohler et al. 2000; Jodun et al. 2002; Giberson and Litvak 2003). A culture manual for Atlantic sturgeon, which will be a culmination of the research conducted since 1991 (Mohler in press), is in the final stages of being published by the U.S. Fish and Wildlife Service. This manual will cover subjects such as broodstock handling and transport, spawning and egg incubation, and the rearing of larvae and juveniles.

The Gulf sturgeon is a subspecies of the Atlantic sturgeon, which occurs along the Gulf of Mexico from Florida Bay, Florida to the mouth of the Mississippi River and was listed as a federally threatened species in 1991. Although artificial spawning of the Gulf sturgeon has been accomplished (Parauka et al. 1991) and feeding trials with larvae initiated (Bardi et al. 1998) culture techniques are still in the research phase.

13.10.3 LAKE STURGEON

Artificial spawning and initial rearing trials of lake sturgeon began in the 1980s (Folz et al. 1983; Ćzeskleba et al. 1985). More recently, studies assessing the effect of formulated diets on larvae growth (DiLauro et al. 1998) and the effect of different densities on juvenile growth (Fajfer et al. 1999) have been conducted. Since the early 1980s the Wisconsin Department of Natural Resources has collected, hatched and reared lake sturgeon for research and population restoration (Fajfer et al. 1999). Even with the information available from research that could facilitate lake sturgeon aquaculture, no aquaculture production has been established. The absence of lake sturgeon culture is partly related to the prohibition of culture by private aquaculturists in some states, such as Wisconsin, and it is also related to observations that lake sturgeon are difficult to culture and commercial production may not be economically viable. However, the future of lake sturgeon aquaculture may be changing as described in the Wisconsin's Lake Sturgeon Management Plan (Anon. 2000). One objective of the management plan was to establish a cooperative for lake sturgeon propagation, between various state and federal government agencies, Native American tribes and the aquaculture industry. Together they plan to use established technical criteria to produce lake sturgeon for research and rehabilitation.

13.10.4 SHOVELNOSE AND PALLID STURGEON

The shovelnose sturgeon status is unique because in some states, it is a commercial species, in some a sport fish and in others it is fully protected (Keenlyne 1997). The pallid sturgeon, on the other hand, is protected throughout the U.S. since it was listed as endangered in 1990. Currently, there is no aquaculture of these two species and all culture related efforts have been conducted to study either the biology of the species or for stock enhancement purposes (Keenlyne 1993). Personnel at the Blind Pony State Hatchery in Missouri were first successful in spawning wild-caught broodstock during 1992 and their culture of pallid sturgeon continues (H. Bollig, U.S. Fish and Wildlife Service, Gavins Point National Fish Hatchery, Yankton, South Dakota, personal communication).

13.10.5 GREEN STURGEON

Until very recently there was virtually nothing known about the life history of this species, in fact historically, the flesh was considered poisonous (Kirsch and Fordice 1889). Although the flesh of wild-caught individuals is considered to be inferior to white sturgeon because of the high quantity of red muscle, it does support a minor commercial fishery on the Pacific coast and it is an important fishery to Native Americans (Van Eenennaam et al. 2001). The first information available on spawning induction and larval culture has found green sturgeon to have large eggs (4.0-4.7 mm), resulting in large, robust larvae (Van Eenennaam et al. 2001; Deng et al. 2002). Studies at the time of this writing have shown that this species has considerable growth potential in captivity, with individuals reaching a body weight of 2.5-3.0kg in one year at water temperatures of 18-19°C.

13.11 Challenges

The constraints to aquaculture expansion in general, have been reviewed by Carlberg (2001) and the potential ecological impacts of these activities have also been reviewed (Naylor et al. 2000; 2001; Black 2001). In addition to the general challenges faced by all aquaculturists, sturgeon farmers must also focus on the following unique challenges.

13.11.1 WATER REQUIREMENTS

Sturgeon require not only a warm water source for fast growth through initiation of maturation but also a cold water source during late vitellogenesis, final maturation and spawning, or caviar production. If there are no naturally occurring water sources of appropriate temperature then a substantial financial investment is required to install some type of recirculation system to heat and/or chill the existing water supply.

13.11.2 REGULATORY

The capture of wild broodstock for obtaining gametes is still necessary for all species of North American sturgeon, except the white sturgeon, but their supply is unpredictable and restricted by government regulations. These regulations are aimed at protecting native stocks from potential 'genetic contamination' and disease transmission, and have prohibited private aquaculture operations and commercialization of sturgeon in certain regions. However, these concerns are being addressed in current conservation-based aquaculture programs (LaPatra et al. 1999; St. Pierre 1999; Ireland et al. 2002).

As of April 1998, international trade in sturgeon must be in compliance with the provisions of CITES (Gnam 1998) and all sturgeon species are presently listed under Appendix II. These regulations apply to live fish, and all parts and derivatives of these species, including meat and caviar. The U.S. Fish and Wildlife Service administers CITES in the United States and all aquaculture facilities must apply for CITES export permits to ship sturgeon internationally. When shipping sturgeon products outside the U.S. the products must be brought through one of the designated ports, a U.S. Fish and Wildlife Service Declaration for the Importation or Exportation of Fish and Wildlife form must be submitted, and all shipments must be inspected and cleared (Gnam 1998). National trade of processed products is not regulated, however shipments of live sturgeon are under state-to-state regulations.

13.11.3 DISEASE CONTROL

In general, sturgeon appear to be more resistant to diseases that are frequently found in other fish species such as *Flavobacterium columnare* (Altinok and Grizzle 2001b). Nevertheless, the white sturgeon iridovirus (WSIV) and herpes virus type 2 (WSHV-2) are considered to be the major causes of juvenile mortality among cultured populations of white sturgeon (Hedrick et al. 1990; 1992; Watson et al. 1995; Georgiadis et al. 2000a). These viruses were first described among cultured white sturgeon in California, but it is now apparent that they are present among both cultured and wild stocks of white sturgeon in Oregon and Idaho (LaPatra et al. 1994). Both WSIV and WSHV-2 are assumed to be transmitted from infected adults to their progeny and recent epidemiological investigations with WSIV strongly support this mode of transmission, while tank to tank transmission is not a predominant way of spreading WSIV (Georgiadis et al. 2001).The occurrence of infections among white sturgeon juveniles after leaving the hatchery is associated with stressful events experienced during culture, including high density rearing, water temperature fluctuations, handling and transport (LaPatra et al. 1994; 1996; Georgiadis et al. 2000a; Georgiadis et al. 2001). Survivors of viral outbreaks can undergo a runting or slowed growth phase but rapid recovery is possible (Georgiadis et al. 2000b, 2000c). Concerns over possible spread of WSIV to other sturgeon populations in North America and Europe have prompted restrictions on exports of white sturgeon to other regions. To decrease the impact of these viruses on white sturgeon farms, modifications in husbandry practices have been made that have successfully decreased the frequency of outbreaks. In recent years, these diseases have not affected overall production levels and some of the key husbandry modifications have been to minimize handling and transport, and to maintain optimum water quality and densities.

13.11.4 GENETICS

Current husbandry techniques for domestic white sturgeon are very successful. In fact, the annual spawning of 4-5 females and a dozen males can provide enough larvae for the entire industry, due to the high fecundity of white sturgeon and excellent survival of larvae and juveniles. However, one year-class derived from just a few individuals could ultimately lead to high rates of inbreeding (Bentsen and Gjerde 1994; Bentsen and Olesen 2002), which could be detrimental to the species fitness. To prevent inbreeding, commercial producers of sturgeon will need to take into account genetic diversity when breeding their broodstock, and the results from ongoing genetic research could assist in this process (Rodzen 2001; Rodzen and May 2002; Rodzen et al. 2004).

The development of cryopreservation techniques for sturgeon (Tsvetkova et al. 1996; Dzuba et al. 1999; Jähnichen et al. 1999; Glogowski et al. 2002) and paddlefish spermatozoa (Brown and Mims 1999; Mims et al. 2000) will ultimately allow for the long-term preservation of valuable genetic resources. However, current techniques are inadequate at storing sufficient amounts of semen needed to fertilize large quantities of eggs under commercial aquaculture conditions. For example, cryopreservation of semen typically involves diluting the semen 1:1 with an extender-cryoprotectant and the diluted semen is then frozen in 0.25ml straws, but the ovulated eggs collected from one female white sturgeon requires about 60ml of semen. Therefore, 480 straws would be needed to fertilize the eggs from one female. However, protocols to freeze semen in 5ml straws are being tested. Although cryopreservation is currently not a practical technique for commercial sturgeon farms it can play a valuable role in preserving endangered sturgeon stocks throughout the world.

Monosex culture of female sturgeon would be very beneficial for commercial caviar production, and one technique that has been attempted to achieve this goal is gynogenesis. If female sturgeon are homogametic (XX), all the gynogenetic progeny will be female, but if the female sturgeon are heterogametic (ZW), gynogens will be male (ZZ), female (ZW), and super females (WW) (Van Eenennaam et al.1999). The first attempts at gynogenesis in sturgeon met with limited success (Romashov et al. 1963; Kowtal 1987), but more recently, success has been achieved in inducing gynogenesis in shovelnose (Mims and Shelton 1998) and white sturgeon (Van Eenennaam et al. 1996), and the white sturgeon results provided evidence of a ZW-female heterogametic genetic sex determination system (Van Eenennaam et al.1999). In the white sturgeon study, the gynogenetic progeny groups of four dams included a percentage of males ranging from 18 to 50% (Van Eenennaam et al.1999). In contrast, induced gynogenesis in paddlefish resulted in 100% females (Mims et al. 1997) and it is assumed that paddlefish have the XX-female sex determination system. Sex reversal of the gynogenetic female paddlefish has been accomplished (Mims et al. 1995a) and a breeding program using these neomales to produce all female offspring is in the final stages of development (Mims 2001).

13.11.5 ECONOMICS AND MARKETING

Bioeconomic analyses and production models of sturgeon aquaculture have revealed that meat and caviar production are profitable under a variety of different facility sizes, stocking and harvesting schedules, product mix and pricing regimes (Peykani 1993; Logan et al. 1995; Wade and Fadel 1997; Sanders et al. 2003). The first economic analysis and each subsequent study used different biological data, due to continued improvements in husbandry techniques resulting in increased growth rates, and as growth rates increased the rates of return improved. Other aspects of sturgeon aquaculture that could further improve the economic returns are earlier detection of sex, development of markets for smaller sized fish, and retention of more female fish for caviar production. Currently, meat sales occur at 7-9kg body weight, however the sale of smaller fish (2-4kg) to the live markets is being developed, and the percent of females retained for caviar production is increasing.

13.11.6 CAVIAR PRODUCTION

When white sturgeon females are eight years old, fifty percent or more may be mature and harvested for caviar. However, the definition of "true" caviar is the non-ovulated oocytes, prepared from the entire ovaries that have been removed from sacrificed individuals. Sturgeon caviar processing is very labour intensive because it is all prepared by hand. Sternin and Dore (1993) describe caviar processing in detail and because of the unique features of sturgeon oocytes (size, colour, taste) from individual females, ovaries from each fish must be processed and packed separately. There are some reports of non-lethal methods of making caviar, but these methods typically use ovulated oocytes from females that are hormonally induced to spawn (Kopylenco et al. 1999). This method is not FDA-approved in the U.S. due to the potential presence of residual hormones used to induce ovulation. In addition, the quality of the caviar made from ovulated oocytes is different from "true" caviar.

13.12 Summary

The aquaculture of white sturgeon is well established in North America. The culture techniques for other sturgeon are still undergoing research and development and will require further consideration and modification, depending on the species and production goals. Future research directions of sturgeon and paddlefish aquaculture will undoubtedly include: alternative sex identification techniques, breeding plans for selection of desirable traits, monosex production, post-harvest technology (i.e., for caviar), disease detection and control, and development of optimal feeds for growout, broodstock and caviar production.

The rapid decline in caviar supplies from wild populations and the increasing harvest restrictions, particularly in the Caspian Sea region (Stone 2002), will undoubtedly create a demand for other sources of caviar. Sturgeon and paddlefish aquaculture will almost certainly be the major source of caviar in the future. However, the continued expansion and sustainability of sturgeon farming for meat and caviar will ultimately depend on culture regulations, production costs, long-term demand, and price.

The development of aquaculture techniques will also be useful in management programs and enhancement of sturgeon and paddlefish populations. Knowledge gained from artificial rearing of these species can be used for the propagation of sturgeon in hatcheries. These hatchery reared individuals can then be used for enhancement or augmentation of wild stocks, and also provide fish for research and educational programs. Aquaculture by private industry and public hatcheries may be the only hope for the conservation of wild sturgeon, as the world's natural populations continue to decline.

Acknowledgements

We would like to thank Serge Doroshov for his inspiration, ideas, and pioneering research that formed the basis of this chapter. We are also grateful to Peter Struffenegger and Ken Beer for their reviews of the first draft. Additional thanks to Nancy Léonard, the editors and other reviewers for their helpful suggestions and Javier Linares for his assistance in preparing the final figures.

13.13 References

Altinok, I., Galli, S.M., and Chapman, F.A. 1998. Ionic and osmotic regulation capabilities of juvenile Gulf of Mexico sturgeon, *Acipenser oxyrinchus desotoi*. Comp. Biochem. Physiol., A. 120:609-616.

Altinok, I, and Grizzle, J.M. 2001a. Effects of brackish water on growth, feed conversion and energy absorption efficiency by juvenile euryhaline and freshwater stenohaline fishes. J. Fish Biol.59:1142-1152.

Altinok, I, and Grizzle, J.M. 2001b. Effects of low salinities on Flavobacterium columnare infection of euryhaline and freshwater stenohaline fish. J. Fish Dis. 24: 361-367.

Anonymous. 2000. Wisconsin's Lake Sturgeon Management Plan. Wisconsin Department of Natural Resources and the Bureau of Fisheries Management and Habitat Protection.

Ballard, W.W., and Needham, R.G. 1964. Normal embryonic stages of *Polyodon spathula* (Walbaum). J. Morphol. 114: 465-478.

Barannikova, I.A. 1987. Review of sturgeon farming in the Soviet Union. J. of Ichthyol. 27: 62-71.

Bardi, R.W. Jr., Chapman, F.A., and Barrows, F.T. 1998. Feeding trials with hatchery-produced Gulf of Mexico sturgeon larvae. Prog. Fish-Cult. 60:25-31.

Barrows, F.T. 2000. Larval feeds: two methods for production of one-size, microbound particles. Global Aquacult. Advocate 3: 61-63.

Bentsen, H.B., and Gjerde, B. 1994. Designs of fish breeding programs. In Proc. 4th World Congr. Genet. Appl. Livest. Prod., Edinburgh, Scotland. pp. 149-158.

Bentsen, H.B., and Olesen, I. 2002. Designing aquaculture mass selection programs to avoid high inbreeding rates. Aquaculture 204: 349-359.

Billard, R., Cosson, J., and Linhart, O. 2000. Changes in the flagellum morphology of intact and frozen/thawed Siberian sturgeon *Acipenser baeri* (Brandt) sperm during motility. Aquacult. Res. 31: 283-287.

Black, K.D. 2001. Environmental impacts of aquaculture. CRC Press LLC, Boca Raton, FL.

Boothe, H.W. 1998. Selecting suture materials for small animal surgery. Small Animal 20: 155-163.

Bouchard, H.J. III., and Aloisi, D.B. 2002. Investigations in concurrent disinfection and de-adhesion of lake sturgeon eggs. N. Am. J. Aquacult. 64:212-216.

Bronzi, P., Rosenthal, H., Arlati, G., and Williot, P. 1999. A brief overview on the status and prospects of sturgeon farming in Western and Central Europe. J. Appl. Ichthyol. 15: 224-227.

Brown, G.G., and Mims, S.D. 1999. Cryopreservation of paddlefish *Polyodon spathula* milt. J. World Aquacult. Soc. 30: 245-249.

Bruch, R.M., Dick, T.A., and Choudhury, A. 2001. A practical field guide for the identification of stages of lake sturgeon gonad development, with notes on lake sturgeon reproductive biology and management implications. Published by Sturgeon for Tomorrow, Fond du Lac, Wisconsin. Printed by Graphic Communications Center Inc., Appleton, Wisconsin.

Buckley, J., and Kynard, B. 1981. Spawning and rearing shortnose sturgeon from the Connecticut River. Prog. Fish-Cult. 43: 74-76.

Buddington, R.K. 1991. Ontogenic development of sturgeons: selected physiological examples. In Acipenser Actes du premier colloque international su l'esturgeon. Edited by P. Williot. CEMAGREF, Bordeaux, France. pp. 53-63.

Buddington, R.K., and Doroshov, S.I. 1984. Feeding trials with hatchery produced white sturgeon juveniles (*Acipenser transmontanus*). Aquaculture 36:237-243.

Buddington, R.K., and Christofferson, J.P. 1985. Digestive and feeding characteristics of the chondrosteans. Environ. Biol. Fish. 14: 31-41.

Buddington, R.K., Hazel, J.R., Doroshov, S.I., and Van Eenennaam, J.P. 1993. Ontogeny of the capacity for homeoviscous adaptation in white sturgeon (*Acipenser transmontanus*). J. Exp. Zool. 265: 18-28.

Carlberg, J.M. 2001. U.S. aquaculture: current status and future directions. Aquacult. Mag. 27:36-43.

Carlson, D.M, and Bonislawsky, P.S. 1981. The paddlefish (*Polyodon spathula*) fisheries of the Midwestern United States. Fisheries 6: 17-27.

Carr, S.H., Tatman, T., and Chapman, F.A. 1996. Observations on the natural history of the Gulf of Mexico sturgeon (*Acipenser oxyrinchus desotoi* Vladykov 1955) in the Suwannee River, southeastern United States. Ecol. Freshwater Fish 5:169-174.

Carter, E.N. 1904. Notes on sturgeon culture in Vermont. Trans. Am. Fish. Soc. 33: 60-75.

Cech, J.J. Jr., Mitchell, S.J., and Wragg, T.E. 1984. Comparative growth of juvenile white sturgeon and striped bass: effects of temperature and hypoxia. Estuaries 7: 12-18.

Chapman, F.A., and Carr, S.H. 1995. Implications of early life stages in the natural history of the Gulf of Mexico sturgeon, *Acipenser oxyrinchus desotoi*. Environ. Biol. Fish. 43: 407-413.

Chebanov, M.S., and Savelyeva, E.A. 1999. New strategies for brood stock management of sturgeon in the Sea of Azov basin in response to changes in patterns of spawning migration. J. Appl. Ichthyol. 15:183-190.

Chebanov, M., and Ballard, R. 2001. The culture of sturgeons in Russia: production of juveniles for stocking and meat for human consumption. Aquat. Living Resour. 14: 375-381.

Cherr, G.N., and Clark, W.H. Jr. 1984. Jelly release in the eggs of the white sturgeon, *Acipenser transmontanus*. An enzymatically mediated event. J. Exp. Zool., 230: 145-149.

Cherr, G.N., and Clark, W.H. Jr. 1985. Gamete interaction in the white sturgeon, *Acipenser transmontanus*: a morphological and physiological review. Environ. Biol. Fish. 14: 11-22.

Conte, F.S., Doroshov, S.I., Lutes, P.B., and Strange, E.M. 1988. Hatchery manual for the white sturgeon *Acipenser transmontanus* Richardson with application to other north American Acipenseridae. University of California, Division of Agriculture and Natural Resources, Oakland, California.

Colombo, R.E., Wills, P.S., and Garvey, J.E. 2004. Use of ultrasound imaging to determine sex of shovelnose sturgeon. N. Am. J. Fisheries Management 24: 322-326.

Cosson, J., and Linhart, O. 1996. Paddlefish (*Polyodon spathula*) spermatozoa: effects of potassium and pH on motility. Folia Zoologica 45: 361-370.

Cosson, J., Linhart, O., Mims, S.D., Shelton, W.L., and Rodina, M. 2000. Analysis of motility parameters from paddlefish and shovelnose sturgeon spermatozoa. J. Fish Biol. 56: 1348-1367.

Crocker, C.E., and Cech, J.J. Jr. 1996. The effects of hypercapnia on the growth of juvenile white sturgeon, *Acipenser transmontanus*. Aquaculture 147: 293-299.

Cui, Y., and Hung, S.S.O. 1995. A prototype feeding-growth table for white sturgeon. J. Appl. Aqua. 5: 25-34.

Cui, Y., Hung, S.S.O., Deng, D-F., and Yang, Y. 1997. Growth performance of juvenile white sturgeon as affected by feeding regime. Prog. Fish-Cult. 59: 31-35.

Ćzeskleba, D.G., AveLallemant, S., and Thuemler, T.F. 1985. Artificial spawning and rearing of lake sturgeon, *Acipenser fulvescens*, in Wild Rose State Fish Hatchery, Wisconsin, 1982-83. In North American Sturgeons: Biology and Aquaculture Potential. Edited by F.P. Binkowski and S.I. Doroshov, S.I. Dr. W. Junk Publishers, Dordrecht, pp 79-86.
Deng, X., Van Eenennaam, J.P., and Doroshov. S.I. 2002. Comparison of early life stages and growth of green and white sturgeon. Am. Fish. Soc. Symp. 28: 237-248.
Dettlaff, T.A., Ginsburg, A.S., and Schmalhausen, O.I. 1993. Sturgeon fishes: developmental biology and aquaculture. Springer-Verlag, New York.
DiLauro, M.N. 1998. Renewal of voluntary feeding by wild-caught Atlantic sturgeon juveniles in captivity. Prog. Fish-Cult. 60: 293-296.
DiLauro, M.N., Krise, W.F., Hendrix, M.A., and Baker, S.E. 1994. Short-term cold storage of Atlantic sturgeon sperm. Prog. Fish-Cult. 56: 143-144.
DiLauro, M.N., Krise, W.F., and Fynn-Aikins, K. 1998. Growth and survival of lake sturgeon fed formulated diets. Prog. Fish-Cult. 60:293-296.
Dillard, J.G., Graham, L.K., and Russell, T.R. (editors). 1986. The paddlefish: status, management, and propagation. North Central Division, Amer. Fish. Soc. Special Publ. 7.
Doroshov, S.I. 1985. Biology and culture of sturgeon Acipenseriformes. In Recent advances in aquaculture. Edited by J.F. Muir and R.J. Roberts. Westview Press, Boulder, Colorado. pp. 251-274.
Doroshov, S.I., Clark, W.H. Jr., Lutes, P.B., Swallow, R.L., Beer, K.E., McGuire, A.B., and Cochran, M.D. 1983. Artificial propagation of the white sturgeon, *Acipenser transmontanus*. Aquaculture 32: 93-104.
Doroshov, S.I., Moberg, G.P., and Van Eenennaam, J.P. 1997. Observations on the reproductive cycle of cultured white sturgeon, *Acipenser transmontanus*. Environ. Biol. Fish. 48: 265-278.
Dzuba, B.B., Kopeika, E.F., Cherepanov, V.V., and Drokin, S.I. 1999. Sturgeons sperm quality after 6 years of cryopreservation. J. Appl. Ichthyol. 15: 312.
Eddy, S., and Simer, P.H. 1929. Notes on the food of paddlefish and the plankton of its habit. Trans. Illinois St. Acad. Sci. 21: 59-68.
Fajfer, S., Meyers, L., Willman, G., Carpenter, T., and Hansen, M.J. 1999. Growth of juvenile lake sturgeon reared in tanks at three densities. N. Am. J. Aquacult. 61: 331-335.
Feist, G., Van Eenennaam, J.P., Doroshov, S.I., Schreck, C.B., Schneider, R.P., and Fitzpatrick, M.S. 2004.Early identification of sex in cultured white sturgeon, Acipenser transmontanus, using plasma steroid levels. Aquaculture 232:581-590.
Folz, D.J., Czeskleba, D.G., and Thuemler. 1983. Artificial spawning of lake sturgeon in Wisconsin. Prog. Fish-Cult. 45: 231-233.
Fontenot, Q.C., Isely, J.J., and Tomasso, J.R. 1998. Acute toxicity of ammonia and nitrite to shortnose sturgeon fingerlings. Prog. Fish-Cult. 60: 315-318.
Gawlicka, A., Teh, S.J., Hung, S.S.O., Hinton, D.E., and de la Noiie, J. 1995. Histological and histochemical changes in the digestive tract of white sturgeon larvae during ontogeny. Fish Physiol. Biochem. 14: 357-371.
Georgi, T.A., and Dingerkus, G. 2001. The bibliography of paddlefishes, family Polyodontidae. Doane College, Crete, Nebraska.
Georgiadis, M. P., Hedrick, R.P., Johnson, W.O., Yun, S., and Gardner, I.A. 2000a. Risk factors for white sturgeon iridovirus (WSIV) and white sturgeon herpesvirus-2 (WSHV-2) outbreaks in 3 commercial sturgeon farms. Am. J. Vet. Res. 61:1232-1240.
Georgiadis, M. P., Hedrick, R.P., Johnson, W.O., and Gardner, I.A. 2000b. Growth of white sturgeon (*Acipenser transmontanus*) following recovery from the stunted stage in a commercial farm in California, USA. Prev. Vet. Med. 43: 283-291.
Georgiadis, M. P., Hedrick, R. P., Johnson, W.O., and Gardner, I.A. 2000c. Mortality and recovery of runt white sturgeon (*Acipenser transmontanus*) in a commercial farm in California, USA. Prev. Vet. Med. 43:269-281.
Georgiadis, M.P., Hedrick, R.P., Carpenter, T.E., and Gardner, I.A. 2001. Factors influencing transmission, onset, and severity of outbreaks of white sturgeon iridovirus (WSIV) in a commercial hatchery. Aquaculture 194: 21-35.
Giberson, A.V., and Litvak, M.K. 2003. Effect of feeding frequency on growth, food conversion efficiency, and meal size of juvenile Atlantic sturgeon and shortnose sturgeon. N. Am. J. Aquacult. 65: 99-105.
Giddings, F.D. 1997. Surgical knots and suturing techniques. Giddings Studio Publishing, Fort Collins, Colorado.
Gilliland, E. 1994. Comparison of absorbable sutures used in largemouth bass liver biopsy surgery. Prog. Fish-Cult. 56: 60-61.
Gisbert, E., and Williot, P. 1997. Larval behaviour and effect of the timing of initial feeding on growth and survival of Siberian sturgeon (*Acipenser baeri*) larvae under small scale hatchery production. Aquaculture 156: 63-76.

Gisbert, E., and Williot, P. 2002. Influence of storage duration of ovulated eggs prior to fertilization on the early ontogenesis of sterlet (*Acipenser ruthenus*) and Siberian sturgeon (*Acipenser baeri*). Internat. Rev. Hydrobiol. 87: 605-612.
Gisbert, E., Cech, J.J. Jr., and Doroshov. S.I. 2001. Routine metabolism of larval green sturgeon (*Acipenser medirostris* Ayres). Fish Physio. Biochem. 25: 195-200.
Glogowski, J., Kolman, R., Szczepkowski, M., Horváth, Á., Urbányi, B., Sieczyński, P., Rzemieniecki, A., Domagala, J., Demianowicz, W., Kowalski, R., and Ciereszko, A. 2002. Fertilization rate of Siberian sturgeon (*Acipenser baeri* Brandt) milt cryopreserved with methanol. Aquaculture 211: 367-373.
Gnam, R. 1998. Implementation of 1997 CITES listing made effective April 1, 1998. In Proceedings of the symposium on the harvest, trade and conservation of north American paddlefish and sturgeon. Edited by D.F. Williamson, G.W. Benz and C.M. Craig. TRAFFIC North America/World Wildlife Fund, Washington, DC. pp. 192-214.
Goncharov, B.F., Igumnova, L.V., Polopan, I.S., and Savieleva, E.A. 1991. Induced oocyte maturation, ovulation and spermiation in sturgeons (Acipenseridae) using synthetic analogue of gonadotropin-releasing hormone. In Acipenser Actes du premier colloque international su l'esturgeon. Edited by P. Williot. CEMAGREF, Bordeaux, France. pp. 351-364.
Graham, L.K., Hamilton. E.J., Russell, T.R., and Hicks, C.E. 1986. The culture of paddlefish – a review of methods. In The paddlefish: status, management, and propagation. Edited by J.G. Dillard, L.K. Graham and T.R. Russell. Amer. Fish. Soc., North Central Division, Special Publ. 7, Bethesda, Maryland. pp. 78-94.
Harkness, W.J.K., and Dymond, J.R. 1961. The lake sturgeon: the history of its fishery and problems of conservation. Fish and Wildlife Branch, Ontario Department of Lands and Forests, Ontario.
Hedrick, R.P., Groff, J.M., McDowell, T., and Wingfield, W.H. 1990. An iridovirus infection of the integument of the white sturgeon *Acipenser transmontanus*. Dis. Aquat. Org. 8: 39-44.
Hedrick, R.P., McDowell, T.S., Groff, J.M., Yun, S., and Wingfield, W.H. 1992. Isolation and some properties of an Iridovirus-like agent from white sturgeon *Acipenser transmontanus*. Dis. Aquat. Org. 12: 75-81.
Hochleithner, M., and Gessner, J. 1999. The sturgeon and paddlefishes (Acipenseriformes) of the world. Biology and Aquaculture. Aquatech Publ., Kitzbuehel, Austria.
Hoover, C. 1998. Import and export of sturgeon and paddlefish in the United States. In Proceedings of the symposium on the harvest, trade and conservation of north American paddlefish and sturgeon. Edited by D.F. Williamson, G.W. Benz and C.M. Craig. TRAFFIC North America/World Wildlife Fund, Washington, DC. pp. 162-170.
Hung, S.S.O. 1991a. Nutrition and feeding of hatchery-produced juvenile white sturgeon (*Acipenser transmontanus*): an overview. In Acipenser Actes du premier colloque international su l'esturgeon. Edited by P. Williot. CEMAGREF, Bordeaux, France. pp. 65-77.
Hung, S.S.O. 1991b. Sturgeon, Acipenser spp. In Handbook of nutrient requirements of finfish. Edited by R.P. Wilson, R.P. CRC Press, Inc. Boca Raton, FL, USA. pp. 153-160.
Hung, S.S.O., and Deng, D.F. 2002. Sturgeon, Acipenser spp. In Nutrient requirements and feeding of finfish for aquaculture. Edited by C.D. Webster and C. Lim. CAB International Publ., Wallingford, UK. pp. 344-357.
Hung, S.S.O., Lutes, P.B., Shqueir, A.A., and Conte, F.S. 1993. Effect of feeding rate and water temperature on growth of juvenile white sturgeon (*Acipenser transmontanus*). Aquaculture 115: 297-303.
Hung, S.S.O., Conte, F.S., and Frost, F. 1998. Growth of white sturgeon fed seven commercial feeds. Aquaculture 28: 551-557.
Ireland, S.C., Anders, P.J., and Siple, J.T. 2002. Conservation aquaculture: an adaptive approach to prevent extinction of an endangered white sturgeon population. In Biology, management, and protection of north American sturgeon. Edited by W. Van Winkle, P.J. Anders, D.H. Secor and D.A. Dixon. American Fisheries Society, Symposium 28, Bethesda, Maryland. pp. 211-222.
Izquierdo-Pulido, M.L., Hatae, K., and Haard, N.F. 1992. Nucleotide catabolism and changes in texture indices during ice storage of cultured sturgeon, *Acipenser transmontanus*. J. Food Biochem. 16: 173-192.
Jähnichen, H., Warnecke, D., Trölsch, E., Kohlmann, K., Bergler, H., and Pluta, H.J. 1999. Motility and fertilizing capability of cryopreserved *Acipenser* ruthenus L. sperm. J. Appl. Ichthyol. 15: 204-206.
Jarvis, P.L., Ballantyne, J.S., and Hogans, W.E. 2001. The influence of salinity on the growth of juvenile shortnose sturgeon. N. Am. J. Aquacult. 63:272-276.
Jenkins, W.E., Smith, T.I.J., Heyward, L.D., and Knott, D.M. 1993. Tolerance of shortnose sturgeon, *Acipenser brevirostrum*, juveniles to different salinity and dissolved oxygen concentrations. Proc. Annu. Conf. Southeast Assoc. Fish and Wildl. Agencies 47: 476-484.
Jodun, W.A., Millard, M.J., and Mohler, J. 2002. The effect of rearing density on growth, survival, and feed conversion of juvenile Atlantic sturgeon. N. Am. J. Aquacult. 64: 10-15.

Keenlyne, K.D. 1993. Recent north American studies on pallid sturgeon, *Scaphirhynchus albus* (Forbes and Richardson). In Proceedings of the 2nd International Symposium on Sturgeons. Edited by A.D. Gershanovich and T.I.J. Smith. VNIRO Publishing, Moscow. pp. 225-234.

Keenlyne, K.D. 1997. Life history and status of the shovelnose sturgeon, *Scaphirhynchus platorynchus*. Environ. Biol. Fish. 48: 291-298.

Kelly, J.L., and Arnold, D.E. 1999. Effects of ration and temperature on growth of age-0 Atlantic sturgeon. N. Am. J. Aquacult. 61: 51-57.

Kincaid, H.L., Mengel, L., Gray, M., and Brimm, S. 1999. National Fish Strain Register-Paddlefish and Sturgeon. Fish and Wildlife Service Resource Publ. NFSR-PS. Also available through the internet at: http://159.189.37.201/newstu/paddle.htm

Kirsch, P.H., and Fordice, M.W. 1889. A review of the American species of sturgeons (Acipenseridae). Proc. Acad. Natural Sci. Philadelphia 2: 245-257.

Knecht, C.D., Allen, A.R., Williams, D.J., and Johnson, J.H. 1987. Fundamental techniques in veterinary surgery. 3rd edition. Saunders, Philadelphia.

Kopylenco, L, Gromova, V., Ponamarev, M., and Kopylenco, I. 1999. A method for canning ovulated sturgeon eggs. J. Appl. Ichthyol. 15: 329.

Kowtal, G.V. 1987. Preliminary experiments in induction of polyploidy, gynogenesis, and androgenesis in the white sturgeon, *Acipenser transmontanus* Richardson. In Proceedings of the World Symposium on Selection, Hybridization, and Genetic Engineering in Aquaculture. Edited by K. Tiews. Heenemann Verlagsgesellschaft, Berlin, Vol. II, pp. 317-324.

Kowtal, G.V., Clark, W.H., Jr., and Cherr, G.N. 1986. Elimination of adhesiveness from the white sturgeon, *Acipenser transmontanus*: chemical treatment of fertilized eggs. Aquaculture 55: 139-143.

Krayushkina, L.S. 1998. Characteristics of osmotic and ionic regulation in marine diadromous sturgeon *Acipenser brevirostrum* and *A. oxyrinchus* (Acipenseridae). J. Ichthyol. 38:660-668.

Kroll, K.J., Van Eenennaam, J.P., Doroshov, S.I., Hamilton, J.E., and Russell, T.R. 1992. Effect of water temperature and formulated diets on growth and survival of larval paddlefish. Trans. Amer. Fish. Soc. 121: 538-543.

Kynard, B. 1997. Life history, latitudinal patterns, and status of the shortnose sturgeon, *Acipenser brevirostrum*. Environ. Biol. Fish. 48: 319-334.

LaPatra, S.E., Groff, J.M, Jones, G.R., Munn, B., Patterson, T.L., Holt, R.A., Hauck, A.K., and Hedrick, R.P. 1994. Occurrence of white sturgeon iridovirus infections among cultured white sturgeon in the Pacific Northwest. Aquaculture 126: 201-210.

LaPatra, S.E., Groff, J.M., Patterson, T.L, Shewmaker, W.D., Casten, M., Siple, J., and Hauck, A.K.1996. Preliminary evidence of sturgeon density and other stressors on manifestation of white sturgeon iridovirus disease. J. Appl. Aquacult. 6: 51-58.

LaPatra, S.E., Ireland, S.C., Groff, J.M., Clemens, K.M., and Siple, J.T. 1999. Adaptive disease management strategies for the endangered population of Kootenai River white sturgeon. Fisheries 24: 49-56.

LeBreton, G.T.O., and Beamish, F.W.H. 1998. The influence of salinity on ionic concentrations and osmolarity of blood serum in lake sturgeon, *Acipenser fulvescens*. Environ. Biol. Fish. 52:477-482.

Linares-Casenave, J., Van Eenennaam, J.P., and Doroshov, S.I. 2002. Ultrastructural and histological observations on temperature-induced follicular ovarian atresia in the white sturgeon, *Acipenser transmontanus*. J. Appl. Ichthyol.18:382-390.

Linhart, O., Mims, S.D., and Shelton, W.L. 1995. Motility of spermatozoa from shovelnose sturgeon and paddlefish. J. Fish Biol. 45: 902-909.

Linhart, O., Mims, S.D., Gomelsky, B., Hiott, A.E., Shelton, W.L., Cosson, J., Rodina, M., and Gela, D. 2000. Spermiation of paddlefish (*Polyodon spathula*) Acipeneriformes stimulated with injection of LHRH analogue and carp pituitary powder. Aquat. Living Resour. 13:1-6.

Logan, S.H., Johnston, W.E., and Doroshov, S.I. 1995. Economics of joint production of sturgeon (*Acipenser transmontanus* Richardson) and roe for caviar. Aquaculture 130: 299-316.

Lutes, P.B., Doroshov, S.I., Chapman, F., Harrah, J., Fitzgerald, R., and Fitzpatrick, M. 1987. Morpho-physiological predictors of ovulatory success in white sturgeon, *Acipenser transmontanus* Richardson. Aquaculture 66: 43-52.

Lutes, P.B., Hung, S.S.O., and Conte, F.S. 1990. Survival, growth, and body composition of white sturgeon larvae fed purified and commercial diets at 14.7 and 18.4°C. Prog. Fish-Cult. 52: 192-196.

McEnroe, M., and Cech, J.J. Jr. 1985. Osmoregulation in juvenile and adult white sturgeon *Acipenser transmontanus*. Environ. Biol. Fish. 14: 23-30.

Meade, J.W. 1989. Aquaculture management. Van Nostrand Reinhold, New York, New York.

Medale, F., Corraze, F., and Kaushik, S.J. 1995. Nutrition of farmed Siberian sturgeon. In Proceedings of the third international symposium on sturgeons. Edited by A. D. Gershanovick and T.I.J. Smith. VNIRO Publ., Moscow, Russia. pp. 289-298.

Meehan, W.E. 1909. Experiments in sturgeon culture. Trans. Am. Fish. Soc. 39: 85-91.

Michaels, J.T., and Bunter, A. 1996. White sturgeon culture at Sierra Aquafarms, Inc. In: Culture and Management of Sturgeon and Paddlefish. Edited by S. Doroshov, F. Binkowski, T. Thuemler and D. MacKinlay. Physiology Section of the American Fisheries Society. pp. 113-116.

Mims, S.D. 1991. Evaluation of activator solutions, motility duration, and short-term storage of paddlefish spermatozoa. J. World Aqua. Soc. 22: 224-229.

Mims, S.D. 2001. Aquaculture of paddlefish in the United States. Aquat. Living Resour. 14: 391-398.

Mims, S.D., and Durborow, R.M. 1998. Propagation and culture of the American paddlefish *Polyodon spathula*. Cooperative Extension Service, Land Grant Program, Kentucky State University, Frankfort, KY. U.S.A. Video (VHS) no. 1000.

Mims, S.D., and Shelton, W.L. 1998. Induced meiotic gynogenesis in shovelnose sturgeon. Aquacult. Int. 6: 323-329.

Mims, S.D., and Shelton, W.L. 1999. Monosex culture of paddlefish and shovelnose sturgeon. In Proceedings of the symposium on the harvest, trade and conservation of north American paddlefish and sturgeon. Edited by D.F. Williamson, G.W. Benz and C.M. Craig. TRAFFIC North America/World Wildlife Fund, Washington, DC. pp. 67-76.

Mims, S.D., Shelton, W.L., and Clark, J.A. 1995a. Steroid-induced sex reversal of paddlefish. In Proceedings of the 5th International Symposium of Reproductive Physiology of Fish. Edited by F. Goetz and P. Thomas. University of Texas, Austin, Texas. pp. 129.

Mims, S.D., Clark, J.A., Williams, J.C., and Lovshin, L.L. 1995b. Food selection by larval paddlefish *Polyodon spathula* supplied with rice bran to promote production of live foods, with prepared diets, or with their combination in earthen ponds. J. World Aquacult. Soc. 26: 438-446.

Mims, S.D., Shelton, W.L., Linhart, O., and Wang, C. 1997. Induced meiotic gynogenesis of paddlefish, *Polyodon spathula*. J. World Aquacult. Soc. 28: 334-343.

Mims, S.D., Shelton, W.L., Wynne, F.S., and Onders, R.J. 1999. Production of paddlefish. Southern Regional Aquaculture Center Publ. No. 437.

Mims, S.D., Tsvetkova, L.I., Brown, G.G., and Gomelsky, B.I. 2000. Cryopreservation of sperm of sturgeon and paddlefish. In Cryopreservation in Aquatic species. Edited by T.R. Tiersch and P.M. Mazik. World Aquaculture Society, Baton Rouge, LA, USA. pp. 121-129.

Mims, S.D., Lazur, A., Shelton, W.L., Gomelsky, B., and Chapman, F. 2002. Species profile. Production of Sturgeon. Southern Regional Aquaculture Center Publ. No. 7200.

Mims, S.D., Onders, R. J., Gomelsky, B., and Shelton, W.L.. 2004. Effectiveness of the minimally invasive surgical technique (MIST) for removal of ovulated eggs from first-time and second-time MIST-spawned paddlefish. N. Amer. J. Aquacult. 66: 70-72.

Moberg, G.P., and Doroshov, S.I. 1992. Reproduction in cultured white sturgeon, *Acipenser transmontanus*. NOAA Technical Report NMFS 106: 99-104.

Moghim, M., Vajhi, A.R., Veshkini, A., and Masoudifard, M. 2002. Determination of sex and maturity in *Acipenser stellatus* by using ultrasonography. J. Appl. Ichthyol. 18:325-328.

Mohler, J.W. 2000. Early culture of the American Atlantic sturgeon *Acipenser oxyrinchus oxyrinchus* Mitchill, 1815 and preliminary stocking trials. Bol. Inst. Esp. Oceanogr. 16: 203-208.

Mohler, J.W. In press. Culture manual for the Atlantic sturgeon *Acipenser oxyrinchus oxyrinchus*. U.S. Fish and Wildlife Service, Hadley, MA.

Mohler, J.W. and Fletcher, J.W. 1999. Induced spermiation in wild Atlantic sturgeons held captive up to six years. N. Am. J. Aquacult. 61: 70-73.

Mohler, J.W., Fynn-Aikins, K., and Barrows, R. 1996. Feeding trials with juvenile Atlantic sturgeons propagated from wild broodstock. Prog. Fish-Cult. 58:173-177.

Mohler, J.W., King, M.K., and Farrell, P.R. 2000. Growth and survival of first-feeding and fingerling Atlantic sturgeon under culture conditions. N. Am. J. Aquacult. 62: 174-183.

Monaco, G., and Doroshov, S.I. 1983. Mechanical de-adhesion and incubation of white sturgeon eggs (*Acipenser transmontanus* Richardson). Aquaculture 35: 117-123.

Moore, B.J., Hung, S.S.O., and Medrano, J.F. 1988. Protein requirement of hatchery-produced juvenile white sturgeon (*Acipenser transmontanus*). Aquaculture 71: 235-245.

Naylor, R.L., Goldburg, R.J., Primavera, J.H., Kautsky, N., Beveridge, M.C.M., Clay, J., Folke, C., Lubcheno, J., Mooney, H., and Troell, M. 2000. Effects of aquaculture on world fish supplies. Nature 405:1017-1024.

Naylor, R.L., Williams, S.L., and Strong, D.R. 2001. Aquaculture-a gateway for exotic species. Science 294:1655-1656.

Needham, R.G. 1965. Spawning of paddlefish induced by means of pituitary material. Prog. Fish-Cult. 27: 13-19.

Nocillado, J.N., Van Eenennaam, J.P., and Doroshov, S.I. 1999. Final ovarian maturation and egg quality in hormonally-induced white sturgeon (*Acipenser transmontanus*). J.Appl. Ichthyol. 15: 319.
Onders, R.J., Mims, S.D., Wang, C., and Pearson, W.D. 2001. Reservoir ranching of paddlefish. N. Amer. J. Aquacult. 63: 179-190.
Parauka, F.M., Troxel, W.J., Chapman, F.A., and McBay, L.G. 1991. Hormone-induced ovulation and artificial spawning of Gulf of Mexico sturgeon (*Acipenser oxyrinchus desotoi*). Prog. Fish-Cult. 53: 113-117.
Peykani, G.R. 1993. Economic feasibility analyses of sturgeon aquaculture in the United States: an application of dynamic linear programming. PhD dissertation, University of Florida, Gainesville, Florida.
Post, H. 1890. The sturgeon, some experiments in hatching. Trans. Am. Fish. Soc. 19: 36-40.
Price, R.J., Hung, S.S.O., Conte, F.S., and Strange, E.M. 1989. Processing yields and proximate composition of cultured white sturgeon (*Acipenser transmontanus*). J. Food Sci. 54:216-217.
Prince, E.E. 1905. The Canadian sturgeon and caviar industries. Can. Sess. Pap. No. 22. Spec. Append. Rep., liii-lxx. Government Printing Bureau, Ottawa.
Pruter, A.T. 1972. Review of commercial fisheries in the Columbia River and in contiguous ocean waters. In The Columbia River estuary and adjacent ocean waters: bioenvironmental studies. Edited by A.T. Pruter and D.L. Alverson. University of Washington Press, Seattle. pp. 81-120.
Purkett, C.A., Jr. 1963. Artificial propagation of paddlefish. Prog. Fish Cult. 25(1): 31-33.
Raymakers, C., and Hoover, C. 2002. Acipenseriformes: CITES implementation from Range States to consumer countries. J. Appl. Ichthyol. 18: 629-638.
Rodewald, J. 2001. Do good while eating well. Gourmet 61: 152.
Rodzen, J.A. 2001. Preservation of genetic variation within aquaculture stocks of white sturgeon (*Acipenser transmontanus*). Ph.D. Dissertation. University of California, Davis, California, USA.
Rodzen, J.A., and May, B. 2002. Inheritance of microsatellite loci in the white sturgeon (*Acipenser transmontanus*). Genome 45: 1064-1076.
Rodzen, J.A., Famula, T., and May, B.2004. Estimation of parentage and relatedness in the polyploidy white sturgeon (Acipenser transmontanus) using a dominant marker approach for duplicated microsatellite loci. Aquaculture 232: 165-182.
Romashov, D.D., Nikolyukin, N. I., Belyaeva, V.N., and Timofeeva, N.A. 1963. Possibilities of producing diploid radiation-induced gynogenesis in sturgeons. Radiobiology 3: 145-154. (Translated from Radiobiologiya 3: 104-109).
Ronyai, A., and Varadi, L. 1995. The sturgeons. In Production of aquatic animals. Edited by C.E. Nash and A.J. Novotny. Elsevier Science B.V., Amsterdam, The Netherlands. pp. 95-108.
Ross, R.M., and Bennett, R.M. 1997. Comparative behaviour and dietary effects in early life phases of American sturgeons. Fish. Manage. Ecol. 4: 17-30.
Ruer, P.M., Cech, J.J. Jr., and Doroshov, S.I. 1987. Routine metabolism of the white sturgeon, *Acipenser transmontanus*: effect of population density and hypoxia. Aquaculture 62: 45-52.
Russell, T.R. 1986. Biology and life history of the paddlefish- a review. In The paddlefish: status, management, and propagation. Edited by J.G. Dillard, L.K. Graham, and T.R. Russell. Amer. Fish. Soc., North Central Division, Special Publ. 7, Bethesda, Maryland. pp. 2-21.
Ryder, J.A. 1890. The sturgeon and sturgeon industries of the eastern coast of the United States, with an account of experiments bearing upon sturgeon culture. U.S. Fish. Comm. Bull. 8:231-238.
Sanders, B.J., Fadel, J.G., and Wade, E.M. 2003. Economic optimization modeling of white sturgeon (*Acipenser transmontanus*) caviar and meat production under different management conditions. Aquaculture 217: 409-430.
Sbikin, Y.N., and Budayev, S.V. 1991. Some aspects of the development of feeding relationships in groups of young sturgeons (Acipenseridae) during artificial rearing. J. Ichthyol. 31: 23-30.
Secor, D.H., and Gunderson, T.E. 1998. Effects of hypoxia and temperature on survival, growth, and respiration of juvenile Atlantic sturgeon, *Acipenser oxyrinchus*. Fish. Bull. 96: 603-613.
Shelton, W.L., and Mims, S.D. 1995. Oocyte staging in paddlefish, *Polyodon spathula*. Trans. Kentucky Acad. Sci. 56: 22-27.
Shelton, W.L., Mims, S.D., Clark, J.A., Hiott, A.E., and Wang, C. 1997. A temperature-dependent index of mitotic interval (t0) for chromosome manipulation in paddlefish and shovelnose sturgeon. Prog. Fish-Cult. 59: 229-234.
Smith, T.I.J., and Dingley, E.K. 1984. Review of biology and culture of Atlantic (*Acipenser oxyrinchus*) and shortnose sturgeon (*A. brevirostrum*). J. World Maricult. Soc. 14: 210-218.
Smith, T.I.J., and Jenkins, W.E. 1991. Development of a shortnose sturgeon, *Acipenser brevirostrum*, stock enhancement program in north America.. In Acipenser Actes du premier colloque international su l'esturgeon. Edited by P. Williot. CEMAGREF, Bordeaux, France. pp. 329-336.
Smith, T.I.J., and Clugston, J.P. 1997. Status and management of Atlantic sturgeon, *Acipenser oxyrinchus*, in North America. Environ. Biol. Fish. 48: 335-346.

Smith, T.I.J., Dingley, E.K., and Marchette, D.E. 1980. Induced spawning and culture of Atlantic sturgeon. Prog. Fish-Cult. 42: 147-151.

Smith, T.I.J., Dingley, E.K., and Marchette, D.E. 1981. Culture trials with Atlantic sturgeon, *Acipenser oxyrinchus*, in the USA. J. World Maricult. Soc.12: 78-87.

Smith, T.I.J., Dingley, E.K., Lindsey, R.D., Van Sant, S.B., Smiley, R.A., and Stokes, A.D. 1986. Spawning and culture of shortnose sturgeon, *Acipenser brevirostrum*. J. World Maricult. Soc. 16: 104-113.

Smith, T.I.J., Heyward, L.D., Jenkins, W.E., and Collins, M.R. 1995. Culture and stock enhancement of shortnose sturgeon, *Acipenser brevirostrum*, in the southern United States. In Proceedings of the 2nd International Symposium on Sturgeons. Edited by A.D. Gershanovich and T.I.J. Smith. VNIRO Publishing, Moscow. pp. 204-214.

Smith, T.I.J., Collins, M.C., Post, W.C., and McCord, J.W. 2002. Stock enhancement of shortnose sturgeon: a case study. In Biology, management, and protection of north American sturgeon. Edited by W. Van Winkle, P.J. Anders, D.H. Secor, and D.A. Dixon. American Fisheries Society, Symposium 28, Bethesda, Maryland. pp. 31-44.

St. Pierre, R.A. 1999. Restoration of Atlantic sturgeon in the northeastern USA with special emphasis on culture and restocking. J. Appl. Ichthyol. 15: 180-182.

Stanley, L. 2001. Good eggs. Wine Spectator. 26: 126-131.

Štěch, L., Linhart, O., Shelton, W.L., and Mims, S.D. 1999. Minimally invasive surgical removal of ovulated eggs from paddlefish. Aqua. Int. 7: 129-133.

Sternin, V., and Dore, I. 1993. Caviar – the resource book. Cultura, Moscow, Russia.

Stone, L. 1900. The spawning habits of the lake sturgeon (*Acipenser rubicundus*). Trans. Am. Fish. Soc. 29: 118-128.

Stone, R. 2002. Caspian ecology teeters on the brink. Science 295:430-433.

Summerfelt, R.C., and Smith, L.S. 1990. Anesthesia, surgery, and related techniques. In Methods for fish biology. Edited by C.B. Schreck and P.B. Moyle. American Fisheries Society, Bethesda, Maryland. pp. 213-272.

Thomas, S.L., and Piedrahita, R.H. 1996. Measurements of metabolic rates of sturgeon and striped bass under commercial culture conditions. Aquacultural engineering society proceedings II: Successes and failures in commercial recirculating aquaculture. 2: 481-493.

Thomas, S.L., and Piedrahita, R.H. 1997. Oxygen consumption rates of white sturgeon under commercial conditions. Aquacult. Eng. 16: 227-237.

Thomas, S.L., and Piedrahita, R.H. 1998. Apparent ammonia-nitrogen production rates of white sturgeon (*Acipenser transmontanus*) in commercial aquaculture systems. Aquacult. Eng. 17: 45-55.

Toth, G.P., Ciereszko, A., Christ, S.A., and Dabrowski, K. 1997. Objective analysis of sperm motility in the lake sturgeon, *Acipenser fulvescens*: activation and inhibition conditions. Aquaculture 154: 337-348.

Tsvetkova, L.I., Cosson, J., Linhart, O., and Billard, R. 1996. Motility and fertilizing capacity of fresh and frozen-thawed spermatozoa in sturgeons *Acipenser baeri* and *A. ruthenus*. J. Appl. Ichthyol. 12: 107-112.

Van Eenennaam, A.L., Van Eenennaam, J.P., Medrano, J.F., and Doroshov, S.I. 1996. Rapid verification of meiotic gynogenesis and polyploidy in white sturgeon (*Acipenser transmontanus* Richardson). Aquaculture 147: 177-189.

Van Eenennaam, A.L., Van Eenennaam, J.P., Medrano, J.F., and Doroshov, S.I. 1999. Evidence of female heterogametic genetic sex determination in white sturgeon. J. Heredity 90: 231-233.

Van Eenennaam, J.P., Webb, M.A.H., Deng, X., Doroshov, S.I., Mayfield, R.B., Cech, J.J. Jr., Hillemeier, D.C., and Willson, T.E. 2001. Artificial spawning and larval rearing of Klamath River green sturgeon. Trans. Am. Fish. Soc. 130: 159-165.

Varner, P.W. 2000. Anesthetics. In Encyclopedia of Aquaculture. Edited by R.R. Stickney. John Wiley & Sons, Inc. New York, New York. pp. 33-38.

Wade, E.M., and Fadel, J.G. 1997. Optimization of caviar and meat production from white sturgeon (*Acipenser transmontanus*). Agricult. Syst. 54: 1-21.

Wang, C., Mills, S.D., and Xiong, Y.L. 1995. Consumer acceptability of paddlefish, a potential aquaculture species. Meat Focus Int. 4: 8-9.

Wang, Y.L., Binkowski, F.P., and Doroshov, S.I. 1985. Effect of temperature on early development of white and lake sturgeon, *Acipenser transmontanus* and *Acipenser fulvescens*. In North Americans sturgeons: Biology and aquaculture potential. Edited by F.P. Binkowski and S.I. Doroshov. Dr. W. Junk Publishers, Dordrecht, The Netherlands. pp. 43-50.

Wang, Y.L., Buddington, R.K., and Doroshov, S.I. 1987. Influence of temperature on yolk utilization by white sturgeon, *Acipenser transmontanus*. J. Fish Biol. 30: 263-271.

Watson, L.R., Yun, S.C., Groff, J.M., and Hedrick, R.P. 1995. Characteristics and pathogenicity of a novel herpes virus isolated from adult and sub adult white sturgeon *Acipenser transmontanus*. Dis. Aquat. Org. 22: 199-210.

Webb, M.A.H., Van Eenennaam, J.P., Doroshov, S.I., and Moberg, G.P. 1999. Preliminary observations on the effects of holding temperature on reproductive performance of female white sturgeon, *Acipenser transmontanus* Richardson. Aquaculture 176: 315-329.

Webb, M.A.H., Van Eenennaam, J.P., Feist, G., Linares-Casenave, J., Fitzpatrick, M., Schreck, C.B., and Doroshov, S.I. 2001. Effects of thermal regime on ovarian maturation and plasma sex steroids in farmed white sturgeon. Aquaculture 201: 137-151.

Webb, M.A.H., Feist, G.W., Foster, E.P., Shreck, C.B., and Fitzpatrick, M.S. 2002. Potential classification of sex and stage of gonadal maturity of wild white sturgeon using blood plasma indicators. Trans. Am. Fish. Soc. 131: 132-142.

Webster, C.S., Mims, S.D., Tidwell, J.H., and Yancey, D.H. 1991. Comparison of live food organisms and prepared diets as first food for paddlefish, *Polyodon spathula* (Walbaum), fry. Aquacult. Fish. Manage. 22: 155-163.

Wedemeyer, G.A. 1996. Physiology of fish in intensive culture systems. Chapman and Hall, New York, New York.

Williot, P.R., Brun, T., Rouault, T., and Rooryck, O. 1991. Management of female spawners on the Siberian sturgeon, *Acipenser baeri* Brandt: first results. In Acipenser Actes du premier colloque international su l'esturgeon. Edited by P. Williot. CEMAGREF, Bordeaux, France. pp. 365-380.

Williot, P., Kopeika, E.F., and Goncharov. 2000. Influence of testis state, temperature and delay in semen collection on spermatozoa motility in the cultured Siberian sturgeon (*Acipenser baeri* Brandt). Aquaculture 189: 53-61.

Yesaki, T.Y., Ek, R., Siple, J., Van Eenennaam, J.P., and Doroshov, S.I. 2002. The effects of iodophor disinfection and transportation on the survival to hatch of fertilized white sturgeon (*Acipenser transmontanus*) eggs. J. Appl. Ichthyol. 18: 639-641.

CHAPTER 14

CONCLUSION: ENSURING THE FUTURE OF STURGEON

KEVIN G. BUTTERWORTH[1] AND NANCY J. LÉONARD[2]

[1]The Centre for Aquaculture and the Environment, University of British Columbia, 4160 Marine Drive, West Vancouver, British Columbia, V7V1N6 Canada

[2]Great Lakes Fishery Commission & Michigan State University, Michigan, U.S.A.

Prior to the European settlement of North America, sturgeon were an important resource for the First Nations. These fish were harvested by any means possible and utilized as a source of oil, roe, isinglass and meat. Unfortunately these unique species were decimated due to overexploitation by European settlers. Subsequent habitat degradation continued to decrease stocks. However, the First Nations have retained the sturgeon as a cultural icon, and have been involved in ongoing programs to manage the recovery of some of these fisheries.

Sturgeon and paddlefish have a slow growth rate, which, coupled with long periods between spawning and their relative late maturation, contributes to their vulnerability to the impact of anthropogenic processes on their habitat. Hence, it is imperative that research is undertaken to reveal as much as possible about sturgeon life history, biology, and behaviour to facilitate the implementation of a well informed management program directed at conserving the remaining stocks.

An examination of the available literature reveals that sturgeon and paddlefish have a highly adaptable feeding strategy. The barbels and abundance of electroreceptors allow these fish to feed irrespective of the turbidity of the water. In addition, their specialized jaw mechanisms allow them to opportunistically utilize temporally available sources of food in their respective benthic or pelagic habitats. These adaptations have historically contributed to the proliferation of stocks. Analysis has revealed that "much of the morphological variation documented among contemporary stocks was likely established thousands of years ago" (Chapter 5). Hence, Acipenseriformes are very successful at occupying their highly specialized environmental niches.

The morphological variations in Acipenseriformes have fascinated researchers, not least because they exhibit anatomical features of both selachians and teleosts. A closer examination of the metabolism of sturgeon and paddlefish reveals that aspects of their biochemistry resemble those of the teleosts, however, key hormones and the endocrinology of these fish resemble that of the chondrichthyans. In addition, the Acipenseriformes exhibit a lower rate of gene evolution, and express pathways that have been lost by teleosts, hence providing insights into vertebrate evolution.

The impact of overexploitation on the genetic diversity of sturgeon and paddlefish is difficult to assess; combined with the possibly detrimental effects of restoration

G.T.O. LeBreton et al. (eds.), Sturgeons and Paddlefish of North America, 312–314.
© 2004 *Kluwer Academic Publishers. Printed in the Netherlands.*

efforts on genetic diversity, this necessitates a cautious conservational approach, using management strategies that include a genetic component.

Currently there are some discrepancies in the jurisdictional management of remaining sturgeon stocks. The need for close coordination between agencies, both nationally and internationally, will increase as the demand for sturgeon products rise due to globalisation. Currently, international cooperation is occurring through CITES (www.cites.org; Wijnstekers 2001; Raymakers and Hoover 2002). Nonetheless, management and enforcement cooperation is needed on the North American level, or, more specifically, among jurisdictions sharing responsibility for common stocks. This cooperation will allow for more comprehensive research and assessment projects. Furthermore, these efforts will result in effective and efficient surveillance and enforcement of management programs, and consequently enhance the sustainability of our populations.

Expanding human populations have resulted in an increase in the demand for hydroelectric power and freshwater resources, leading to an ever increasing pressure to create reservoirs of fresh water that can be used to generate power. Unfortunately, the construction of dams reduces spring flooding, and severely restricts the ability of sturgeon and paddlefish to access traditional spawning grounds. To address this problem, ongoing telemetry research is increasing our knowledge of migration patterns as well as habitat preference. Future conservation efforts need to focus on the construction of fishways to facilitate migration to spawning grounds. In addition, the establishment of artificial spawning channels and minimal stream flows would also contribute to relieving the negative pressure on the survival of Acipenseriformes.

Proactively, knowledge gained in the aquaculture of sturgeon can be used in hatcheries for the artificial rearing of sturgeon bred for eventual release into the wild. These "hatchery" sturgeon can enhance remaining stocks. However, as restrictions on the wild harvest of sturgeon are tightened, there will be an ever increasing demand for sturgeon products. Thus there is the potential for the cultivation of sturgeon to not only occupy this niche, but to ultimately develop into the leading source of sturgeon on the market.

Integral to any aquaculture venture is a thorough grounding in the life history of that species. Research into the embryogenesis of Acipenseriformes can be used as a solid foundation upon which decisions about brood stock efficacy can be based. Unfortunately, current knowledge on the effect of environmental parameters on the early development of sturgeon is inadequate. However, knowledge gained developing effective aquaculture techniques can be transferred to conservation programs where the restoration of optimal habitat conditions for sturgeon early life stages is of paramount importance. In addition, the relative ease of availability of sturgeon life stages from aquaculture sources will facilitate research into the evolutionary importance of Acipenseriformes.

In compiling the historical and current knowledge of sturgeon biology and management in one volume, we aimed to produce a reference that serves as a guide to the restoration of these species. Additionally, our authors depicted within their chapters the research needs regarding the biology of these individuals, their habitat requirements, and management protocols to ensure the sustainability of these unique species.

The future of Acipenseriformes in North America depends on a coordinated and concerted research effort by scientists and policy makers in all jurisdictions. With the growing public interest in protecting these species, and the implementation of an ecosystemic multijurisdictional management program, it appears hopeful that we will be successful in conserving and enhancing our North American Acipenseriformes. We trust that this book will be a useful reference and stimulate interest amongst individuals who will ultimately further contribute to our knowledge and management of these highly valuable and unique species.

References

Hoover, C. and C. Raymakers. 2000. Acipenseriformes: CITES implementation from Range States to consumer countries, . J. Appl. Icthyol. 18(4-6): 629-638

Wijnstekers, W. 2001. The evolution of CITES: a reference to the convention on international trade in endangered species of wild fauna and flora, 6th edition. CITES Secretariat, Switzerland. pp.492.

APPENDIX

IDENTIFICATION KEY TO STURGEONS AND PADDLEFISH OF NORTH AMERICA

MARTIN HOCHLEITHNER[1] AND PAUL VECSEI[2]
[1]AquaTech, Unterbrunnweg 3, Kitzbuehel, Austria 6370,
E-mail: aquaculture@nextra.at
[2]Warnell School of Forest Resources, University of Georgia, Athens, Georgia 30602, USA. E-mail: pjv3402@owl.forestry.uga.edu

Author's note

The dichotomous key is a tool used by ichthyologists, fisheries biologists, and students for identifying unknown specimens to the level of species, genus or family. Constructed from a series of couplets, each consists of two statements describing characteristics of a particular organism or group of organisms. The reader is offered a choice between the two statements that best fits the unidentified specimen in question (Timme 1991)

The statements contained in the couplets initially cover broad characteristics and become narrower or more focused as more choices become available. As the reader progresses through the couplets, the key choices are ordered from broad to narrow characteristics until only a single choice remains. If the correct or relevant couplet is applicable and the unknown specimen is included in the key, positive identification is possible.

Sturgeon researchers require familiarity with the terminology used to describe sturgeon morphology. A good key should be well designed, and maintain simple but appropriate terminology. This key may be lacking in this respect; several of the rules pertaining to dichotomous keys have been broken during the creation of this key. In reality, no single sturgeon key exists that works: every key concerning Acipenser, whether worldwide or regional, has some major flaw.

Our priority was to construct a key that worked, and it proved very difficult. We had to disregard the couplets of equivalent features, and structured the key so that the reader should consider couplets only until they were superseded by a new couplet. In constructing the key, we used the following criteria whenever possible:

- Morphological and meristic characters that are constant.
- Characteristics that are generally visible on preserved as well as live specimens.
- Inclusion of osteological characters (bony plates) whenever possible. They are consistent and easy to observe or count.

G.T.O. LeBreton et al. (eds.), Sturgeons and Paddlefish of North America, 315–318.
© 2004 *Kluwer Academic Publishers. Printed in the Netherlands.*

Skull cartilaginous, but covered by a large number of bony plates. Axial skeleton cartilaginous, vertebrae lacking centra, notochord persistent, extending into upper lobe of caudal fin. Pelvic fins abdominal without true spines. V-shaped fulcrae cover the axis of each unpaired fin: ... .(Order: Acipenseriformes)
Two Families in North America: 1

1
Body without rows of bony scutes,
two small barbels (1 pair) on the underside of the rostrum anterior to the mouth,
snout long and flattened (more than 70 % of the head length): ... (Family: Polyodontidae)
One Species in North America: Polyodon spathula
Body with five rows of bony scutes,
four barbels (2 pairs) arranged in a single transverse row anterior to the mouth,
snout short (less than 70 % of the head length): ... (Family: Acipenseridae)
Eight Species in North America: 2

2
Snout conical with rounded edges,
spiracle and pseudobranchiae present,
caudal peduncle short, partly covered with bony scutes,
lower lip interrupted medially and with two lobes,
barbels not or slightly fimbriated: ... (Genus: Acipenser) 3
Snout spatulate with sharp edges,
spiracle and pseudobranchiae absent,
caudal peduncle long, fully covered with bony scutes,
lower lip continous and with four lobes,
barbels highly fimbriated: ... (Genus: Scaphirhynchus) 7

3
Plates present along both sides of the anal-fin base,
3 or more post-dorsal and post-anal plates (some in pairs),
mouth opening usually less than 60 % of interorbital width: ... *Acipenser oxyrinchus*
No plates at the sides of the anal-fin base,
3 or fewer post-dorsal and post-anal plates,
mouth opening usually more than 60 % of interorbital width: ... 4

4
An olive-green lateral band along lateroventral surface (between lateral and ventral scutes),
barbels closer to mouth and slightly fimbriated,
36 or fewer lateral scutes: ... *Acipenser medirostris*
No olive-green lateral band between scute rows,
barbels not fimbriated: ... 5

5
Post-dorsal and post-anal plates absent,
5 or more pre-anal plates,
36 or more lateral scutes,
42 or more dorsal fin rays,
barbels closer to the tip of the snout: ... *Acipenser transmontanus*
Post-dorsal and post-anal plates present,
4 or less pre-anal plates: ... 6

6
Fewer than 23 anal fin rays,
lateral scutes are lighter than the body background,
base of barbels closer to the tip of the snout,
42 or less dorsal fin rays: ... *Acipenser brevirostrum*
More than 23 anal fin rays,
lateral scutes almost same color as the body background: ... *Acipenser fulvescens*

7
Dorsal fin with more than 38 rays and anal fin with more than 24 fin rays,
caudal fin with more than 65 fin rays,
outer-barbel length more than 1.5 times of inner barbel length,
base of outer-barbels well behind inner-barbels: ... *Scaphirhynchus albus*
Dorsal fin with less than 38 rays and anal fin with less than 24 fin rays,
caudal fin with less than 65 fin rays,
outer-barbel length less than 1.5 times of inner-barbel length,
base of barbels nearly in one row: ... 8

8
Belly covered with large exposed bony plates,
spines on the tip of the snout and between the small eyes: ... *Scaphirhynchus platorhynchus*
Belly covered with small embedded bony plates,
no spines on the tip of the snout and between the large eyes: ... *Scaphirhynchus suttkusi*

Meristic Characters in Sturgeons and Paddlefish of North America

Species	DF	AF	DS	LS	VS	GR	poD	poA	prA
Acipenser brevirostrum	30-42	19-22	7-13	21-35	6-11	22-29	1-3	1-2	2-3
Acipenser fulvescens	35-45	25-30	9-17	29-42	7-12	25-40	1-2	1-2	1-2
Acipenser medirostris	29-44	19-32	7-12	20-36	5-12	15-26	1-2	1-2	1-4
Acipenser oxyrinchus	38-46	23-30	7-16	24-35	6-14	15-27	3-9	3-9	2-6
Acipenser transmontanus	42-52	25-32	11-14	36-48	9-12	23-36	0	0	5-9
Scaphirhynchus albus	37-43	23-28	14-18	40-48	9-13	?	8-9	7-8	4-5
Scaphirhynchus platorhynchus	29-39	18-25	13-19	38-50	9-14	?	8-9	8-9	4-6
Scaphirhynchus suttkusi	29-34	18-22	15-21	40-49	11-14	?	7-9	7-8	4-6
Polyodon spathula	?	?	0	0	0	?	0	0	0

Abbreviations: DF = Dorsal fin rays, AF = Anal fin rays, DS = Dorsal scutes, LS = Lateral scutes, VS = Ventral scutes, GR = Gill rakers, poD = Plates between dorsal- and caudal fin, poA = Plates between anal- and caudal fin, prA = Plates between anus and anal fin;

References

Hochleithner M., Gessner J., 2001: The Sturgeons and Paddlefishes (Acipenseriformes) of the World: Biology and Aquaculture. AquaTech Publications, Kitzbuehel/Austria. 207 pages.

Mayden R.L., Kuhajda B.R., 1996: Systematics, Taxonomy, and Conservation Status of the Endangered Alabama Sturgeon, Scaphirhynchus suttkusi Williams and Clemmer (Actinopterygii, Acipenseridae). Copeia, 2: 241-273.

North J.A., Farr R.A., Vecsei P., 2002: A comparison of meristic and morphometric characters of green sturgeon *Acipenser medirostris*. J. Appl. Ichthyol., 18(4-6): 234-239.

Scott W.B., Crossman E.J., 1973: Freshwater Fishes of Canada. Fisheries Research Board of Canada. 966 pages.

Timme, S.L. 1991. How to construct and use a dichotomous key. Pages 101-110, *in* Tested studies for laboratory teaching. Volume 12. (C. A. Goldman, Editor). Proceedings of the 12th Workshop/Conference of the Association for Biology Laboratory Education, 218 pp

Vecsei P., Charette R., Trukshin I., Maliepaard T., Hochleithner M., Lafleur Y., 2001: CITES Identification Guide - Sturgeons and Paddlefish. Environment Canada. 180 pages.

Glossary

Acid-base characteristics: acid (H^+) and base (OH^-) equivalents, often expressed as pH (negative $logarithm_{10}$ of the hydrogen ion activity, in solutions.)

Amphihaline: fish whose life cycle include both a freshwater and marine component.

Anadromous: fish ascending rivers from the sea to breed.

Anaerobic metabolism: metabolism in the absence of free oxygen.

Animal pole/hemisphere: the end/half of the oocyte or early embryo containing relatively more cytoplasm and less yolk. Frequently the location of the mature germinal vesicle, and thus fertilization. The animal-vegetal axis is established maternally during oogenesis.

Archenteron: the primitive gut cavity that forms within the embryo as a result of gastrulation.

Arterial: P_{50}: half-saturation value of blood, under arterial CO_2 and pH conditions.

Atresia: the resorption of the ovarian follicles by degenerative processes during late vitellogenesis, or before the female undergoes final maturation and spawning.

Blastomeres: individual cells within an early (pre-gastrula) embryo.

Blastula: a multicellular embryo that has mostly or entirely completed cleavage, but in which the large-scale movements of gastrulation have not yet begun.

Blood O_2 affinity: binding affinity of the blood for oxygen, expressed as P_{50}.

Breeding interval: time between consecutive breedings of individual.

Chloride cells: mitochondria-rich cells (e.g., in gill tissue) that move chloride (and sodium) against their concentration gradients.

Chorion: a tough protective extracellular layer surrounding the oocyte and developing embryo.

Cortical reaction: the process in which the cortical alveoli fuse to the plasma membrane of the oocyte, and release their contents into the space between the plasma membrane and egg chorion. The cortical vesicles contain enzymes, structural proteins and other glycoproteins. The cortical reaction renders separation of the chorion from the plasma membrane, imparts tonicity and strengthens the chorion ('water hardening'), and prevents fertilization by multiple sperm ('block to polyspermy').

Cortisol: glucocorticoid hormone secreted by the interrenal gland and involved in stress response

Dimorphism: the appearance of two distinct forms such as size, structure, colour, etc. among animals or plants of the same species. For example, the sex of an individual can be determined by its external appearance; said to exhibit sexual dimorphism.

Ectoderm: the outermost embryonic germ layer, which gives rise to the central nervous system, epidermis, neural crest, sensory placodes, etc.

Ectothermic: cold-blooded, with no regulation of body temperature.

Electrolytes: charged molecules (ions) that act as conductors.

Endoderm: the innermost embryonic germ layer, which gives rise to the digestive system lining, respiratory system, certain endocrine organs, etc.

Epiboly: expansion of the surface area of a tissue or cell layer; may occur via cell division, shape change, or rearrangement.

Epinephrine: catecholamine hormone (sympathoadrenal system) involved in "fight or flight" response

Fecundity: number of eggs laid by a fish.

Final maturation: the final process of maturation in the development of an egg (oogenesis). This process is associated with both morphological and physiological changes such as the resumption of meiosis (from the dictyate oocyte to the metaphase II), yolk polarization, nucleus migration and breakdown (GVBD), and expulsion of the first polar body. Final maturation may last from hours to a few days and is believed to be the period in which the egg acquires much of its competence to fuse with the sperm during fertilization. Final maturation can be induced in vitro with progesterone.

Gastrulation: the series of cell and tissue movements that establish the three embryonic germ layers and generate the archenteron or primitive gut cavity.

Germinal vesicle breakdown (GVBD): the breakdown and disintegration of the nucleus (GV) during final maturation. Occurs during the period of final egg maturation and signals the resumption of meiosis.

Glucogenic amino acids: amino acids serving as a substrate for gluconeogenesis

Gluconeogenic processes: metabolic pathway for the formation of glucose or glycogen from noncarbohydrate sources.

Gymnovarian: the condition of the female gonads in which the ovary is open, and the eggs are released into the body cavity. The eggs gain access to the outside of the body through the Müllerian or paramesonephric ducts. Found in ancient fishes such as sturgeon.

Gynogenesis: gynogenesis is a ploidy manipulation technique used in fish species with female homogamety (XY{male}:XX{female}), resulting in all female progeny. To induce gynogenesis the male genetic component is inactivated, usually through the use of ionizing or UV-irradiation of the sperm, and diploidy is restored through second polar body retention by treating eggs with either a temperature shock, hydrostatic pressure or chemical treatments, subsequent to the entry of the inactivated sperm. Gynogenetic offspring are inbred (inheritance is entirely maternal) but this problem can be overcome by following gynogenesis with sex-reversal to produce XX males (neomales) that can then be crossed with normal females.

Hypercapnia: increased carbon dioxide partial pressures and contents.

Hyperosmoregulatory capability: able to regulate body water and solutes in dilute solutions (e.g., fresh water).

Hypertonic: having a higher total dissolved solute concentration (e.g., compared with that of the environment or, conversely, in an organism).

Hypometabolism: below-normal metabolic rate.

Hypoxia: low oxygen partial pressure and content.

Ionic regulation: regulation of specific ions (e.g., Na^+, Cl^-, Ca^{2+}, K^+) in body fluids when the fish is in either more dilute or more concentrated solutions.

Lactate: a metabolic end-product of the (anaerobic) glycolysis pathway (e.g., as result of hypoxia or of white muscle contractions, as in struggling or in burst swimming).

Long-day breeder: fish with seasonal reproduction during a long day length.

Mesoderm: the middle embryonic germ layer, which gives rise to tissues of the musculoskeletal system, excretory organs, reproductive ducts and gonads, circulatory system, etc.

Müllerian ducts: these paired ducts open internally into the coelomic cavity and externally at the urogenital opening. These funnel-shaped openings, positioned along the lateral walls of the body cavity, serve as passage ways for ovulated eggs to be discharged. In the female the duct has a one-way valve near the upper third of its length, and serves to control the discharge of eggs from the coelomic cavity during spawning. The ducts are also present in males, but their function is unknown. Because of the Müllerian duct structure, the location in the coelomic cavity, and the presence of the valve, eggs cannot be completely hand-stripped from the body cavity, as is done with salmonids.

Neomales: genetic females that are phenotypic and functional males, produced by an exposure to androgens, usually via an implant or by an application of androgens to the feed. This steroid treatment is only effective during the labile period before sexual differentiation.

Neural crest: a highly migratory and multipotent population of embryonic ectoderm cells that contributes to a wide range of derivatives including visceral skeleton, pigment cells, and ganglia.

Norepinephrine: (see epinephrine).

Normal table (or staging table): a description of successive steps during normal embryonic and larval development of a species; usually based on morphological characteristics. Note that discrete "stages" are necessarily somewhat arbitrary divisions of the continuous process of development.

Normocapnia: carbon dioxide partial pressures and contents that approximate atmospheric levels.

Normoxia: oxygen partial pressures and contents that approximate atmospheric levels.

O_2 injection: the infusion, under pressure, of pure oxygen into a column of water, practiced in fish culture.

Osmolality: total solutes in mOsm kg^{-1}.

Osmoregulation: regulation of body water and solutes.

Ovulation: discharge of matured eggs from the ovary into the ovarian or abdominal cavity.

Partial pressure of gas: pressure exerted by a particular gas (O_2, CO_2, N_2), in proportion to the total gas (e.g., barometric) pressure.

Photoperiod: a recurring cycle of light and dark periods.

Photophobic: animal avoiding light or sensitive to light.

Photosynthesis: formation of carbohydrates in the chlorophyll-containing plants exposed to light.

Placodes: local thickenings of ectoderm, usually via cell columnarization in response to inductive signals from underlying mesoderm. Specific placodes give rise to the ectodermal components of major sensory structures such as the eyes and olfactory organs.

Plasma glucose: monosaccharide in blood that is a source of energy and maintained at a relatively stable levels.

Phospholipid: lipid component of cell membranes containing a phosphoric acid residue.

Potamodromous: fish spending their entire lives in freshwater.

Proteolysis: the hydrolysis of proteins or peptides resulting in formation of soluble products.

Purified diets: diets that are constituted from purified ingredients, such as casein for protein, corn oil for lipid, corn starch for carbohydrate, cellulose for fibre, and vitamins and minerals premixes from chemically known compounds.

Radio-telemetry: the practice of locating or tracking organisms that carry tags that emit radio-wave signals, using a radio receiver and directional antenna.

Repletion ration: a satiation ration.

Resting-routine O_2 consumption: a measure of metabolic (measured by oxygen consumption) rate of a routinely active animal.

Salinity, ppt: concentration of salts in water expressed in "parts-per-thousand", i.e. gram per litre.

Specific Growth Rate (SGR): a parameter commonly used in nutritional energetics to express the effect of a ration on growth, in other words the growth rate relative to feed intake of the fish; it may reflect the nutritional value or quality of the diet. Also referred as the average daily percentage gain: $SGR = (\ln(Wt) - \ln(Wt1)/t2 - t1) \times 100$.

TL : "total length" of fish, measured from tip of snout to tip of caudal fin (upper lobe in sturgeon).

Ultrasonic telemetry: the practice of locating or tracking organisms that carry tags that emit characteristic, high-frequency sounds, using a directional hydrophone.

Vegetal pole/hemisphere: the end/half of the oocyte or early embryo containing relatively more yolk and less cytoplasm. The animal-vegetal axis is established maternally during oogenesis.

Vitellogenesis: production of egg yolk involving the hepatic synthesis and secretion of the yolk precursor protein (vitellogenin) in response to circulating estrogen. The blood transports vitellogenin to the ovary, whereby the selective uptake of vitellogenin occurs into growing oocytes via receptor-mediated endocytosis. Vitellogenin is then enzymatically cleaved into yolk proteins and is deposited as platelet yolk into the oocytes. Vitellogenic growth of the oocytes takes about 16-18 months in white sturgeon and oocytes increase in diameter from 0.6 to 3.5 mm.

MIX
Papier aus verantwortungsvollen Quellen
Paper from responsible sources
FSC® C105338
FSC
www.fsc.org

If you have any concerns about our products,
you can contact us on
ProductSafety@springernature.com

In case Publisher is established outside the EU,
the EU authorized representative is:
Springer Nature Customer Service Center GmbH
Europaplatz 3, 69115 Heidelberg, Germany

Printed by Libri Plureos GmbH
in Hamburg, Germany